Lecture Notes in Computer Science 4728

Commenced Publication in 1973
Founding and Former Series Editors:
Gerhard Goos, Juris Hartmanis, and Jan van Leeuwen

Symeon Bozapalidis   George Rahonis (Eds.)

Algebraic Informatics

Second International Conference, CAI 2007
Thessaloniki, Greece, May 21-25, 2007
Revised Selected and Invited Papers

 Springer

Volume Editors

Symeon Bozapalidis
George Rahonis
Aristotle University of Thessaloniki
54124 Thessaloniki, Greece
E-mail: {bozapali, grahonis}@math.auth.gr

Library of Congress Control Number: Applied for

CR Subject Classification (1998): F.3.1-2, F.4, D.2.1, I.1

LNCS Sublibrary: SL 1 – Theoretical Computer Science and General Issues

ISSN   0302-9743
ISBN-10   3-540-75413-X Springer Berlin Heidelberg New York
ISBN-13   978-3-540-75413-8 Springer Berlin Heidelberg New York

Springer is a part of Springer Science+Business Media

springer.com

© Springer-Verlag Berlin Heidelberg 2007
Printed in Germany

Typesetting: Camera-ready by author, data conversion by Scientific Publishing Services, Chennai, India
Printed on acid-free paper   SPIN: 12168668   06/3180   5 4 3 2 1 0

Preface

CAI 2007 was the 2nd International Conference on Algebraic Informatics. It was intended to cover the topics of algebraic semantics on graphs and trees, formal power series, syntactic objects, algebraic picture processing, infinite computation, acceptors and transducers for strings, trees, graphs, arrays, etc., and decision problems.

CAI 2007 was held during May 21–25, 2007 in Thessaloniki, Greece hosted by the Department of Mathematics of Aristotle University of Thessaloniki. The opening lecture was given by Jean Berstel and the other eight invited lectures by Jürgen Albert, Frank Drewes, Dora Giammarresi, Jozef Gruska, Jarkko Kari, Oliver Matz, Ulrike Prange (on behalf of Hartmut Ehrig), and Guo-Qiang Zhang. This volume contains eighth papers of the nine invited lectures and the accepted papers. We received 29 submissions, the contributors being from 14 countries. The Program Committee selected ten papers.

We are grateful to the members of the Program Committee for the evaluation of the submissions and the numerous referees who assisted in this work. We should like to thank all the contributors of CAI 2007 and especially the honorary guest and the invited speakers who kindly accepted our invitation to present their important work. Special thanks are due to Alfred Hofmann, the Editorial Director of LNCS, who gave us the opportunity to publish the proceedings of our conference in the LNCS series, as well as to Anna Kramer from Springer for the excellent cooperation. We are also grateful to the members of the Organizing Committee and a group of students who helped us with several organizing jobs. Last but not least we want to express our gratitude to Arto Salomaa for his constant interest in CAI and his support in Springer.

July 2007

Symeon Bozapalidis

Organization

CAI 2007 was organized by the Department of Mathematics, Aristotle University of Thessaloniki.

Program Committee

Symeon Bozapalidis, Thessaloniki (Chair)
Manfred Droste, Leipzig
Zoltan Ésik, Szeged/Tarragona
Werner Kuich, Vienna
Antonio Restivo, Palermo
Paul Spirakis, Patras
Heiko Vogler, Dresden

Organizing Committee

Symeon Bozapalidis (Chair)
Archontia Grammatikopoulou
Antonios Kalampakas
Costas Lolas
George Rahonis (Co-chair)

Referees

M. Bartha	D. Kuske	P. Pournara
S. Bozapalidis	C. Lutz	I. Petre
R. Diaconescu	A. Maletti	G. Rahonis
F. Drewes	C. Mathissen	A. Restivo
M. Droste	O. Matz	L. Rosaz
Z. Ésik	I. Maeurer	A. Sifaleras
P. Fiser	I. Meinecke	P. Spirakis
Z. Gazdag	D. Mitsche	Gh. Stefanescu
R. Gentilini	T. Mossakowski	S. Vagvolgyi
A. Grammatikopoulou	F. Z. Nardelli	H. Vogler
A. Kalampakas	K. Ogata	C. Umans
D. Kirsten	A. Pluhar	
W. Kuich	A. Papistas	

Table of Contents

On Generalizations of Weighted Finite Automata and Graphics Applications

Jürgen Albert and German Tischler

Department of Computer Science, University of Würzburg
Am Hubland, D-97074 Würzburg, Germany
{albert,tischler}@informatik.uni-wuerzburg.de

Abstract. Already computations of ordinary finite automata can be interpreted as discrete grayscale or colour images. Input words are treated as addresses of pixel-components in a very natural way. In this well understood context already meaningful operations on images like zooming or self-similarity can be formally introduced. We will turn then to finite automata with states and transitions labeled by real numbers as weights. These Weighted Finite Automata (WFA), as introduced by Culik II, Karhumäki and Kari, have turned out to be powerful tools for image- and video-compression. The recursive inference-algorithm for WFA can exploit self-similarities within single pictures, between colour components and also in sequences of pictures. We will generalize WFA further to Parametric WFA by allowing different interpretations of the computed real vectors. These vector-components can be chosen as grayscale or colour intensities or e.g. as 3D-coordinates. Applications will be provided including well-known fractal sets and 3D polynomial spline-patches with textures.

1 Introduction

In standard textbooks on formal languages and automata theory the most common examples for finite automata deal with the analysis or transformation of strings, which appear as sequences of input symbols. Real world applications with finite automata are found e.g. in UNIX tools like grep, lex and many others. If it comes to more numerically motivated applications one can find e.g. counting modulo(k) for some given constant k, or the well-known finite machine over the input alphabet $\{(0,0),(0,1),(1,0),(1,1)\}$ adding two arbitrary long binary numbers from right to left. But the pumping-lemma also makes clear that the numerical capabilities of finite state devices are limited e.g. there is no finite machine computing correctly the product of two arbitrary binary numbers. But, as we will see in our example later on, it does not take drastic generalizations to achieve this by some simple weighted automaton.

Before introducing those Weighted Finite Automata we will relate finite acceptors with concepts from computer graphics (cf. [22]) like pixel-addressing, zooming, multi-resolution-properties, lossy compression etc. This should provide a clearer separation of the generalization steps to WFA and Parametric

S. Bozapalidis and G. Rahonis (Eds.): CAI 2007, LNCS 4728, pp. 1–22, 2007.
© Springer-Verlag Berlin Heidelberg 2007

WFA (PWFA), which inherit much of their descriptive power already from the finite acceptors.

For the following we will only assume some basic knowledge about finite automata and elementary mathematics.

2   Finite Acceptors and Raster Images

We will start here with a minimalistic yet powerful approach, where the input-alphabet is always just $\Sigma = \{0,1\}$, and the rasterized images consist only of either black or white picture elements. In our very first step we will even restrict ourselves to "1-dimensional images" embedded into the unit-interval.

2.1   Inputstrings as Addresses

For some given natural number $r \geq 0$ consider all strings in Σ^r, i.e. all binary strings of length r. In the 1-dimensional case we can associate an input word with a half-open interval:

$x = b_1 \, b_2 \, \ldots \, b_r, \; b_i \in \{0,1\}$

$H(x) = [0. \, b_1 \, b_2 \, \ldots \, b_r, \; 0. \, b_1 \, b_2 \, \ldots \, b_r + 2^{-r})$

of length 2^{-r} within the unit interval $[0,1)$. This way the string 1011 stands for the interval $[\frac{11}{16}, \frac{12}{16})$. Increasing the length r of the strings by 1 therefore doubles the number of half-open sub-intervals of the unit interval.

Given any finite automaton A over Σ, some $r \geq 0$ and $x \in \Sigma^r$, we can assign the colour black to $H(x)$, iff x is accepted by A, $x \in L(A)$; otherwise the colour white is assigned to $H(x)$.

More formally, we assume the following representation for $A = (Q, \Sigma, M, I, F)$:

1. Q is a set of n states,
2. $\Sigma = \{0,1\}$ is the binary input alphabet
3. $M = (M_0, M_1), M_i \in \{0,1\}^{n \times n}$ are the transition matrices for the input-symbols 0, 1 resp. Here $M_i[s,t] = 1$ iff there is a transition from state s to state t labeled by input symbol i
4. $I \in \{0,1\}^{n \times 1}$ is the initial vector. This is a row-vector (and a unit-vector), where the component for the start-state $I[s_0] = 1$, all others are 0
5. $F \in \{0,1\}^{1 \times n}$ is the final column-vector, where a component $F[t] = 1$ iff the corresponding state t is a final state.

It should be obvious, that this notation is equivalent to the common definition of finite automata – i.e. finite state acceptors – if we declare acceptance for A as follows:

For $x = b_1 \, b_2 \, \ldots \, b_r \in \{0,1\}^r$ we have $x \in L(A)$ iff the function $f_A : \Sigma^* \to \mathbb{N}$ defined by

$$f_A(x) = I \times M_{b_1} \times M_{b_2} \cdots \times M_{b_r} \times F$$

yields some value $f_A(x) > 0$.

Fig. 1. Graph for A and input sequences of length 4

2.2 Image Generation by Finite Acceptors

Let us demonstrate the usefulness of this notation by the following finite automaton $A = (Q, \Sigma, M, I, F)$, where

1. $Q = \{q_1, q_2, q_3\}$,
2. $\Sigma = \{0, 1\}$,
3. the initial vector $I = (1, 0, 0)$,
4. the final vector $F^T = (1, 0, 0)$,
5. and for $M = (M_0, M_1)$ we have the transition matrices:

$$M_0 = \begin{pmatrix} 0 & 1 & 0 \\ 1 & 0 & 0 \\ 1 & 0 & 0 \end{pmatrix}, M_1 = \begin{pmatrix} 0 & 0 & 1 \\ 1 & 0 & 0 \\ 0 & 0 & 0 \end{pmatrix}.$$

The language accepted by this automaton is $L(A) = \{00, 01, 10\}^*$, as can be seen easily from the transition-graph for A (cf. Fig. 1).

For the input string 1011 our function f_A yields $f_A(1011) = I \times M_1 \times M_0 \times M_1 \times M_1 \times F = 0$. By our convention, since 1011 is not accepted by A, the colour white is assigned to the interval $[\frac{11}{16}, \frac{12}{16})$. On the other hand the string 1000 is accepted, $f_A(1000) = I \times M_1 \times M_0 \times M_0 \times M_0 \times F = 1$, and $[\frac{8}{16}, \frac{9}{16})$ is painted black. (cf. Fig. 1).

If we look at the pattern of intervals for all words it is easy to see that for all odd lengths r of the input nothing is accepted and for $r = 0, 2, 4, 6, \ldots$ we can describe this informally as dividing all black intervals of stage r into four equal half-open parts and painting the last one white to arrive at stage $r + 2$.

This sounds of course very familiar if compared to the construction of the well-known Cantor set C, frequently called "Cantor dust" as well. Starting with the closed unit interval $[0, 1]$ successively remove the middle thirds (as open intervals):

$$G_0 = [0, 1]$$

$$G_1 = [0, \frac{1}{3}] \cup [\frac{2}{3}, 1]$$

$$G_2 = [0, \frac{1}{9}] \cup [\frac{2}{9}, \frac{1}{3}] \cup [\frac{2}{3}, \frac{7}{9}] \cup [\frac{8}{9}, 1]$$

$$\ldots$$

Then the Cantor set C is defined as

$$C = \bigcap_{n=1}^{\infty} G_n,$$

where C is compact, has Lebesgue-measure 0 and frequently serves as a fundamental example of a fractal. So, it is no surprise that our example automaton A also shows fractal patterns. This will become even more apparent, if we change our interpretation of input words for A from 1-dimensional to 2-dimensional addresses.

2.3   Bi-level Images in 2D

The hierarchical form of addressing introduced above is generalized easily from the unit interval to the unit square $[0, 1) \times [0, 1)$, and further to any d-dimensional hypercube $[0, 1)^d$, $d \geq 1$. The so-called Morton- or Z-order can achieve this desired hierarchical addressing in a very natural and intuitive way. Fig. 2 shows the numbering sequences for the unit square and address lengths of 2, 4 and 8.

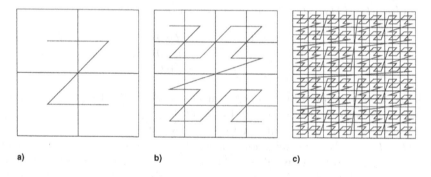

a)   b)   c)

Fig. 2. Morton-Order (Z-Order)

In the common raster-scan-order pixels are arranged in a rectangular matrix and visited row-wise starting in the upper left corner. Thus, the Z-order at least matches with the raster-scan for the very first and last pixel. Depending on the applications it might be more favorable that the origin of the image is placed in the lower left corner e.g. to display graphs of functions or relations in the common way. Thus, the Morton-order might also become an N-order.

For our example of the finite automaton A above, it should be noted that we do not have to change anything within the definition or computation of the results to apply the 2D-interpretation to the addresses. Now the sequence 1011 leads into the white square $[\frac{3}{4}, 1) \times [\frac{1}{4}, \frac{1}{2})$ and analogously 1000 into the black square $[\frac{1}{2}, \frac{3}{4}) \times [0, \frac{1}{4})$. The corresponding pictures are given in Fig. 3 for the resolutions of $2^2 \times 2^2$ and $2^8 \times 2^8$ pixels. Thus, our example automaton generates the well-known Sierpinski-triangle.

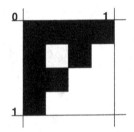 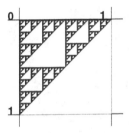

Fig. 3. Images generated by A of resolution 4×4 and 256×256

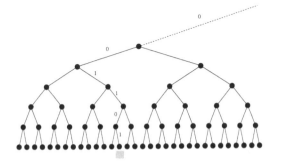 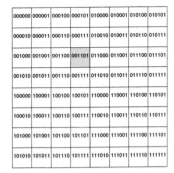

Fig. 4. Hierarchical addressing by bintrees

2.4 Bintrees for Addressing

In the 2D-interpretation an input-sequence of length $2r$ addresses a square of size $2^{-r} \times 2^{-r}$, and a sequence of length $2r + 1$ then a rectangle of size $2^{-r-1} \times 2^{-r}$, as shown in Fig. 4 for $r = 6$.

The hierarchical and fractal nature of the Morton-order is apparent in representing the addresses in complete binary trees (or bintrees) of depth $2r$.

It is verified easily, that the hierarchical addressing works with the bit-wise interleaving of the x- and y-coordinates. Again, if we assume the origin of the image at the upper left corner, then any input address for a square of size $2^{-r} \times 2^{-r}$ like $w = y_1\, x_1\, y_2\, x_2 \ldots y_r\, x_r$ is found at the coordinates $x = x_1\, x_2 \ldots x_r$, $y = y_1\, y_2 \ldots y_r$.

Since each node of the bintree represents the image given by its leaves, zooming into this sub-image quite naturally can be associated with removing the common prefix-string u from all the pixel-addresses uv of that sub-image.

Returning to our example for the Sierpinski triangle we can convince ourselves also, that the concept of multi-resolution makes sense for images defined by finite acceptors. We can use the image of size $2^r \times 2^r$ pixels, scale it by one half in each dimension and place a copy into each of the three first sub-quadrants of the image with $2^{r+1} \times 2^{r+1}$ pixels. These steps of scaling and copying are used as an intuitive description for a basic form of fractal image generation (see [39]);

Fig. 5. Dithering and Inverting

similar approaches are found in [31], [32]; a theoretical discussion of the concepts can be found in [37].

The reader is invited to generate as an exercise variations of the Sierpinski triangle automaton for the diagonal line in the unit square and the upper left black triangle.

Some more detailed remarks are in order here: Though quadtrees, octtrees, etc. are found more frequently in literature for hierarchical addressing of higher-dimensional data ([3], [14]), bintrees have measurable advantages for image- and video-compression. The Morton-order is superior in general to the raster-scan addressing, since it can exploit spatial redundancies much better. The Hilbert-order, where only direct neighbour pixels are visited during the traversal, can be the preferred traversal-order, if spatial redundancies without directional bias are present in the raw data. It should be noted, that the Hilbert-order is self-similar and hierarchical too and can be mapped to the Morton-order by a simple finite state transduction. Many other variations of the traversal-order exist, like the Hilbert-Peano-order, the triangular or the circular order, which can serve for special applications.

Another meaningful operation for the graphical interpretation of regular languages is of course the complementation $\Sigma^* - L(A)$, which yields for any fixed resolution just the inverted pixel values.

The example displayed in Fig. 5 also shows the effects of dithering, which is frequently used in the printing processes to create the illusion of a greater colour-depth or – as in our case now – higher number of available grayness values.

2.5 Bit-Planes for Grayscale and Colour-Images

Whereas in the multi-resolution hierarchy for black and white images we can possibly see the same patterns in different sizes, we consider now short stacks of images of the same resolution. In many image formats which are frequently called "raw formats" each pixel-value with its components is stored in a fixed

Fig. 6. Most significant, second significant and least significant bit-plane for 512×512 grayscale image lena

number of bits. These are often 8 bits for gray and 24 for colour images or even 32 bits if transparency values are specified for the pixel positions. The red, green and blue component is usually coded in 8 bits then. Each of the bit-positions thus defines a separate bit-plane and if the bit sequences are intensity values coded in binary the positions range from the "Most Significant Bit", MSB, to the "Least Significant Bit", LSB.

Therefore, the whole image is representable as some stack of bilevel images (see Fig. 6). And since for any given image of finite resolution there is only a finite amount of bitplanes and bits to be coded, it is obvious, that finite state acceptors are sufficient in principle for the representation of common digital images. We can, for example, interpret the complete bintrees discussed above as transition graphs for those finite state acceptors. Starting with such a bintree for a bilevel image one can reduce the number of states by the classical state minimization algorithm, which will produce some directed acyclic graph (DAG). Remember that all accepted words are addresses of same length. Instead of starting out with complete bintrees it is usually better to begin with so-called region bintrees, where an inner node becomes a final node, if all the leaves in that subtree are final. For this region bintree again the state minimization algorithm can be applied. These approaches can be viewed as rough sketches for the WFA inference algorithm, where pictures consist of real-valued pixels and transitions in the finite automaton carry real-valued weights.

Several compression algorithms for the whole stack of bit-planes have been developed in the past for lossless and lossy image reproduction. For cartoon-like images with only a few different colours these can simply be variations of runlength-encoding or in the general case also sophisticated predictive methods, where considering several neighbouring bit-planes for encoding bitvalues can be employed as in the JBIG-standard ([27]). Lossy compression methods can take advantage of the fact that the bit-planes of the LSB or near to it mostly carry noise and can be neglected in the coding-process.

A totally different approach can be taken by following the contours of shapes in the style of turtle-graphics as picture defining languages. Even then many self-similarities occur, which can be exploited by WFA-variants, see [7], [33].

3   Weighted Finite Automata

Although definitions of finite automata, where transitions are labeled by real numbers can be found in the classical textbooks on formal language theory like [1], this was mainly for the purposes of describing probabilistic behavior of finite state acceptors or for the study of formal power series ([2], [28]). The generation of digitized images from finite automata appears later in [3] and then in [5], [6], [9], where a recursive WFA-inference algorithm with remarkable compression-results had been presented. Since then several improvements have led to competitive WFA-codecs with performance-figures in general superior to the JPEG image compression standard ([35]) and depending on image characteristics on a par with advanced wavelet codecs like embedded zerotree wavelet coding ([8]) or the renowned JPEG2000 standard ([36]). Especially, Daubechie wavelets, which are used in several successful image compression schemes, are also generated by WFA with a small number of states as shown in [24]. Due to the simple mathematical structure of the WFA approximation, the image reconstruction can also be done faster than in wavelet based codecs which have to rely on a fast inverse wavelet transform. This makes the WFA-approach a good choice for low bit-rate image and video coding ([10], [11], [12]).

In the following we will present the standard Weighted Finite Automata as an extension of our previous vector-/matrix-notation of finite state acceptors. We refer the interested reader to [5], [34], [13] and [23] for a more rigorous mathematical treatment and especially for fundamental results about the families of real-valued functions that can be generated by WFA.

We define $A = (Q, \Sigma, W, I, F)$ to be a (standard) n state, k label Weighted Finite Automaton (WFA), if for some $n, k \in \mathbb{N}$

1. Q is a set of n states, in general numbered from 1 to n,
2. $\Sigma = \{0, 1, ..., , k-1\}$ is an input alphabet with k labels
3. $W = (W_0, W_1, \ldots W_{k-1}), W_i \in \mathbb{R}^{n \times n}$ are the weight matrices for transitions and the input-symbols $0, 1, \ldots k-1$ resp. Here $W_i[s, t]$ has a non-zero value iff there is a transition from state s to state t labeled by input symbol i with weight $W_i[s, t]$
4. $I \in \mathbb{R}^{n \times 1}$ is the initial distribution.
5. $F \in \mathbb{R}^{1 \times n}$ is the final distribution.

Each matrix A_i corresponds to one label i. In addition, two n-dimensional vectors, I and F, called the initial and final distribution vectors, are given. The initial distribution vector is a row-vector (i.e., has size $1 \times n$) and the final distribution vector is a column vector (of size $n \times 1$).

Formally, WFA A assigns a real number to each word over the label alphabet $\Sigma = \{0, 1, \ldots, k-1\}$. The value associated to word $w = i_1 i_2 \ldots i_l$ is

$$f_A(i_1 i_2 \ldots i_l) = I \times W_{i_1} \times W_{i_2} \times \cdots \times W_{i_l} \times F.$$

To simplify notation let us denote by A_w the corresponding product $A_{i_1} \times A_{i_2} \times \cdots \times A_{i_l}$ where $w = i_1 i_2 \ldots i_l$. Now we can write:

$$f_A(w) = I \times A_w \times F.$$

For our first introductory WFA example we refer to the well-known fact, that there does not exist a finite automaton (or finite machine) correctly multiplying pairs of arbitrary long binary numbers.

We consider the following WFA A with four states,
input alphabet $\Sigma = \{0, 1\}$,
initial distribution $I = (0.0, 0.0, 0.0, 1.0)$,
final distribution $F^T = (1.0, 0.5, 0.5, 0.25)$
and the weight matrices

$$W_0 = \begin{pmatrix} 1.0\ 0.0\ 0.0\ 0.0 \\ 0.0\ 0.0\ 0.5\ 0.0 \\ 0.0\ 1.0\ 0.0\ 0.0 \\ 0.0\ 0.0\ 0.0\ 0.5 \end{pmatrix}, \quad W_1 = \begin{pmatrix} 1.0\ 0.0\ 0.0\ 0.0 \\ 0.5\ 0.0\ 0.5\ 0.0 \\ 0.0\ 1.0\ 0.0\ 0.0 \\ 0.0\ 0.5\ 0.0\ 0.5 \end{pmatrix}$$

The transition graph in Fig. 7 reveals some simple underlying (weakly) monotonous structure of the transition directions, with state number 4 depicted on top as the unique start-state according to the initial distribution I.

For any natural number $r \geq 0$ and the binary values $x = 0.\ x_1\ x_2\ ...\ x_r$, $y = 0.\ y_1\ y_2\ ...\ y_r$ with $x_i, y_i \in \{0, 1\}$ $f_A(w) = x \times y$ is computed for the input sequence $w = x_1\ y_1\ x_2\ y_2\ ...\ x_r\ y_r$ with its given resolution.

It should be mentioned that we here actually compute the average values for the half-open x- and y-intervals of length 2^{-r}; thus, in fact we only approximate the function value for the infinite inputs:

$x = 0.\ x_1\ x_2\ ...\ x_r 000...$,

$y = 0.\ y_1\ y_2\ ...\ y_r 000...$.

This choice for our final distribution vector $F^T = (1.0, 0.5, 0.5, 0.25)$ nicely supports the multi-resolution property of our WFA, which is good for displaying the function-graph. Choosing $F^T = (1.0, 0.0, 0.0, 0.0)$ instead would yield the appropriate interpretation for the finite product of $x = 0.\ x_1\ x_2\ ...\ x_r$ and $y = 0.\ y_1\ y_2\ ...\ y_r$.

In Fig. 7 please note that here the origin of the square is in the lower left corner and hence the interleaving of the x- and y-coordinates in w starts with x. For better visibility of $f_A(\cdot)$ the contour lines in the second picture the grayness values are spread cyclically over the range $[0, 1]$.

3.1 WFA and Polynomials

The transition graph in Fig. 7 deserves some closer inspection w.r.t. to the functions represented by the remaining states. Which functions $f_i(x, y)$ will be generated, if we exchange the initial distribution $I = (0.0, 0.0, 0.0, 1.0)$ by $I_1 = (1.0, 0.0, 0.0, 0.0)$, $I_2 = (0.0, 1.0, 0.0, 0.0)$ or $I_3 = (0.0, 0.0, 1.0, 0.0)$ resp.?

The answer for I_1 trivially is the constant function $f_1(x, y) = 1$ and given the interleaving of x and y in w it is not hard to see that we have $f_2(x, y) = x$ and $f_3(x, y) = y$, the linear slopes in x- and y-direction.

This observation has been made in the early paper by Culik and Karhumäki (see [13]) and generalized to the family of all real-valued polynomials. The

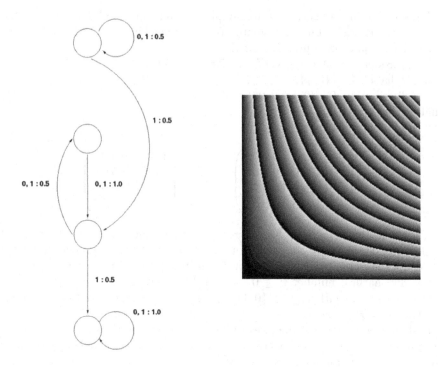

Fig. 7. $f_A(w) = x \times y$, WFA and function-values with cyclically shifted grayness contour lines

corresponding transition graphs for polynomials over a single variable $x \in [0, 1)$ are so-called line-automata. For any polynomial $p(x)$ of degree n at most $n + 1$ states are needed in a WFA to compute $p(x)$. More precisely spoken, for any $x_0 \in [0, 1)$ $p(x_0)$ can be approximated with arbitrary precision.

Let $x_1 = 0.b_1b_2\ldots b_m \in \{0, 1\}^m$ and $x_2 = 0.b_2b_3\ldots b_m \in \{0, 1\}^{m-1}$. Intuitively spoken, x_1 is x_2 shifted to the right with $\frac{1}{2}b_1$ added. Then

$$x_1{}^n = \left(\tfrac{1}{2}(b_1 + x_2)\right)^n \qquad\qquad (1)$$
$$= \tfrac{1}{2^n} \sum_{i=0}^{n} \binom{n}{i} b_1^{n-i} x_2^i$$

which means

$$x_1{}^n = \begin{cases} \frac{1}{2^n} x_2^n & \text{for } b_1 = 0 \\ \frac{1}{2^n} x_2^n + \frac{1}{2^n} \sum_{i=0}^{n-1} \binom{n}{i} x_2^i & \text{for } b_1 = 1 \end{cases} \qquad (2)$$

Since in the binomial formula only powers of x of the same or of lower degrees are needed for the computation, this implies the afore-mentioned structure of the line-automata. From numerical mathematics it is well-known that polynomials are very well suited for interpolations and approximations of functions which are "reasonably smooth". So this is a good start-point for applications of WFA in lossy image compression.

Also in some other respect WFA-generated functions and polynomials are intimately connected. It was shown in [13] and [23] that the polynomials are the only smooth functions generated by WFA, where "smooth" now means that all derivations exist everywhere in the domain. This rules out – somehow surprisingly – also the square-root-function for which was shown in [26], that it can be matched by WFA-generated functions extremely well.

We will return to the topics of the square-root and the applications of higher-dimensional polynomials in our next chapter on the more general concepts of Parametric WFA.

3.2 Image- and Video-Compression with WFA

The above sections have shown that Weighted Finite Automata and even the classical finite acceptors are interesting mechanisms for the generation of images, be it bilevel, grayscale or colour.

Now we will turn to the question of WFA-inference and image-compression, i.e. given some image, effectively find a small WFA A whose function $f_A(\cdot)$ approximates the image well. For simplicity we assume that the given image is grayscale of some resolution of $2^k \times 2^k$ pixels. This is no severe restriction since any colour image can be treated then as three grayscale images for the red, green and blue component. Furthermore, other resolution sizes can e.g. be embedded in some picture size $2^r \times 2^r$ for some suitable $r \geq 0$, when the unused pixels are left black. We will give a brief sketch of the inference algorithm invented by Culik and Kari in 1993 and refer to the seminal papers [5], [6], [25] and [9] for the implementation details and the remarkable test-results.

One can imagine that the inference algorithm reduces the bintree for the grayscale image wherever a sub-image belonging to an inner node of the tree can be approximated well. The algorithm is initialized with a small set of basis functions (e.g. the constant and the linear functions) and the entire image, correspondingly the root node of the bintree.

Two different approximation methods are tested recursively for every sub-node. First the block is approximated with a linear combination of the images which are available up to now.

In the alternative method the current sub-image is subdivided into halves, which means visiting the two sub-nodes, where the coder tries to find good approximations for the new smaller sub-images. After the recursion has returned from the subtrees both alternatives are compared and the better one is used.

The local decision, whether to subdivide further or to approximate by a linear combination, is evaluated by some cost-function C; C takes into account the errors produced by the current approximations versus the storage costs (counted in the number of bits to be used). Therefore, C uses a global parameter q to control the efficiency of the encoding process. Depending on the user-settings large values of q will produce better approximations while small values will produce high compression ratios.

In Fig. 8 the result of the inference algorithm is depicted as the set of image regions for which the cost-function decided that a linear combination was more

Fig. 8. Image regions of the bintree coded as linear combinations in a WFA

Fig. 9. Capturing motion in bintree decompositions

favorable than subdividing the current image. In this case the setting of the parameter q was aimed at a high compression rate. For low error rates it would be quite common to have several hundred or even thousands of states in the resulting WFA.

The next figure (Fig. 9) explains that these encoding principles can be carried over to the compression of sequences of pictures, i.e. video-clips. The "temporal redundancies" are modelled by so-called macro-blocks and motion-vectors (as in the MPEG-standards). One tries to find parts of the picture, which have been moved to nearby locations as a whole block. The translation of such a block is not much different from a linear combination in the approximation process. Thus, an encoding of the WFA inference algorithm produced quite satisfying results, especially for higher compression rates ([11]), [12]).

4   Parametric Weighted Finite Automata

As could be seen in the sections above the addressing of pixel-positions by input-sequences for (weighted) finite automata is a very powerful and flexible way to generate pictures and even video-clips. We will generalize now the way input words can represent pixel positions. Informally, instead of using the fixed binary (or k-ary) address representation we allow that pixel positions are also computed by some WFA ([15]).

We call these automata Parametric Weighted Finite Automata, or PWFA for short, because the input string acts as a parameter binding the functions for different dimensions together. Instead of computing single real values to input strings, in Parametric Weighted Finite Automata we can get points of higher dimensional real spaces $\mathbb{R}^d$. To do that, we have to change only one item in the definition for WFA, namely our initial distribution vector which becomes an initial distribution matrix of size $n \times d$ for some $d > 0$.

The quintuple $A = (Q, \Sigma, W, I, F)$ is an (n state, k label, d-dimensional) Parametric Weighted Finite Automaton (PWFA), if for some $n, k, d \in \mathbb{N}$

1. Q is a set of n states, in general numbered from 1 to n,
2. $\Sigma = \{0, 1, ..., , k-1\}$ is an input alphabet with k labels,
3. $W = (W_0, W_1, \ldots W_{k-1}), W_i \in \mathbb{R}^{n \times n}$ are the weight matrices for transitions and the input-symbols $0, 1, \ldots k-1$ resp. $W_i[s, t]$ has a non-zero value iff there is a transition from state s to state t labeled by input symbol i with weight $W_i[s, t]$,
4. $I \in \mathbb{R}^{n \times d}$ is the initial distribution matrix,
5. $F \in \mathbb{R}^{1 \times n}$ is the final distribution.

The transition diagram for A is the same as for WFAs with the exception that inside every node d initial distribution values and one final distribution value are inserted.

Now each input string $w = i_1 i_2 \ldots i_l \in \Sigma^*$ leads to a point in the d-dimensional space $\mathbb{R}^d$. The formula to compute $f_A(w)$ looks exactly as in the case of WFA, though we now use an initial distribution matrix I of size $n \times d$:

$$f_A(w) = I \times A_w \times F.$$

For the given PWFA A over the alphabet Σ let $S_n(A)$ denote the set of points computed by A on inputs of length n, and let $S_{\geq n}(A)$ be the set of points computed on inputs of length at least n:

$$S_n(A) = \{f_A(w) \mid w \in \Sigma^n\}$$

$$S_{\geq n}(A) = \bigcup_{i=n}^{\infty} S_i(A)$$

Now there are several options how topologically closed sets can be associated to PWFA. We will use as definition that the set $S(A)$ computed by a PWFA A is

$$S(A) = \bigcap_{n=0}^{\infty} \overline{S_{\geq n}}$$

where $\overline{S_{\geq n}}$ is the topological closure of $S_{\geq n}$. In other words, a point $\overline{x} \in \mathbb{R}^d$ ("a real-valued vector") is in $S(A)$ if either
(i) there exist infinitely many words w such that $f_A(w) = \overline{x}$, or
(ii) there exist points $f_A(w) \neq \overline{x}$ arbitrarily close to $\overline{x}$.

At first glance this might look overly complicated, compared to the simple cumulative approach, where one could define the set $\underline{S}(A)$ of points associated to a PWFA A as

$$\underline{S}(A) = \bigcup_{i=0}^{\infty} S_i(A)$$

But it turns out quickly that the second option is weaker in the sense, that any resulting set of points $\underline{S}(A)$ can be generated as $S(B)$ by some PWFA B under the first option. Technically spoken, all one has to do is to introduce a new label, which is used for "waiting loops" in each automaton state. The chosen first option also has the flavour of an attractor set in the context of fractals, which is useful as well.

Multidimensional sets given by a Parametric WFA can be interpreted as relations or images in many different ways. If $d = 2$ it is natural to interpret points (x, y) as points of the Euclidean plane, so $S(A)$ becomes a bilevel image. In case $d = 3$ we might have a set of points (x, y, z) of a 3D-object, a description of pixel locations x, y and intensities i of a 2D image, or a description of a moving 2D bilevel object where the third dimension is interpreted as the time coordinate. Case $d = 4$ could be a description of a 3D grayscale object, or a 2D grayscale video etc. In all cases, decoding consists of computation of d-dimensional points, followed by their interpretation. This new degree of freedom by separating generation of real values from their interpretation as coordinates or (colour-)intensities yields a high descriptional power and can lead to extremely compact representations. PWFA-applications include up to now e.g. representations of multidimensional wavelets ([16]), shapes of figures ([18]) and 3D-animations ([19]).

4.1 PWFA over a Unary Alphabet

Since WFA over a unary alphabet do not define useful functions, it may not be clear, what can be expected from PWFA over a single label 0.

Consider the following one-label, two-states PWFA C and its transition graph in Fig. 10.

The corresponding weight matrix A_0 defines a rotation of the plane $\mathbb{R}^2$ by the angle $\alpha = \cos^{-1}(0.8)$. The quotient of α and π is irrational, so iterating the rotation always finds unvisited points of the circle.

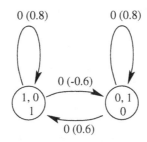

Fig. 10. Rotation by $\cos^{-1}(0.8)$

Fig. 11. (a) First 50 points of the circle, (b) the full circle $S(A)$

When the rotation is applied to the final distribution point $(1, 0)$ over and over again the unit circle gets drawn. Therefore,

$$S(A) = \{(x, y) \mid x^2 + y^2 = 1\} = \{(\cos(t), \sin(t)) \mid t \in \mathbb{R}\}.$$

Please note that in this case both options to define the set of points for the PWFA C coincide:

$$S(C) = \bigcap_{n=0}^{\infty} \overline{S_{\geq n}}$$

$$\underline{S}(C) = \bigcup_{i=0}^{\infty} S_i(C)$$

This holds true since each point in $S_i(C)$ is also an accumulation point in $\overline{S_{\geq n}}$, which is due to the irrationality of the quotient α/π and the true rotation $(0.8^2 + 0.6^2 = 1)$.

The two options would differ obviously, if we replace each occurrence of 0.8 in the weight matrix above by say 0.79. $\underline{S}(C)$ would yield a spiral towards the origin, but $S(C)$ would just consist of the origin, since there is no other accumulation point or any other point visited infinitely often.

The unary alphabet case for PWFA produces a genuine subfamily of the PWFA generated set. It has been characterized by decidability results and closure properties. Furthermore, the number of labels in Σ do not span a hierarchy, two labels actually suffice. On the other hand, the number of states gives rise to an infinite hierarchy, as can be concluded from the facts about the set of polynomials of true degree $m \geq 0$.

4.2   Simulation of Iterated Function Systems

Consider any Iterated Function System (IFS) with k contractive affine maps of $\mathbb{R}^2$. A PWFA simulating the IFS needs k labels, one label corresponding to each affine transform. Two states are needed to represent the x- and y-coordinates. In the more general case of a d-dimensional IFS we would use here d states. And in addition, one state is included to represent the constant function.

For example, the well-known dragon is generated by the following 3-state, 2-label PWFA shown in Figure 12:

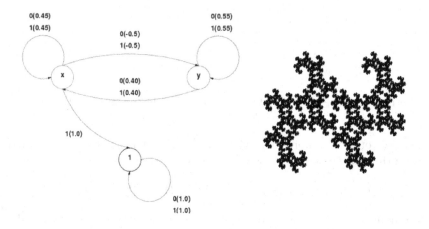

Fig. 12. Simulating IFS for dragon

PWFA are strictly more powerful generators than IFS and one can show that their family coincides with the more general Mutual Recursive Function Systems (MRFS) [4].

4.3   Curves and Segments with Parametric Polynomial Representation

If each of the d functions computed by a PWFA is a polynomial, we can produce a very compact automaton for the corresponding polynomial curve in $\mathbb{R}^d$.

For example, consider the 4-state, 2-label, 2-dimensional PWFA P as given in Fig. 13. It has initial distributions $(1, -1, 0, 0)$ and $(0, 1, -1, 0)$, if the states are numbered from left to right.

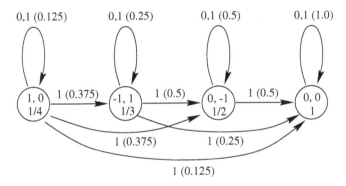

Fig. 13. Parametric WFA for $\{(t^3 - t^2, t^2 - t) \mid 0 \le t \le 1\}$

In the standard binary representation of a WFA the four states – from left to right – would compute the functions $f(t) = t^3, t^2, t$ and 1 over the interval $[0, 1)$, respectively. Let us interpret now the two dimensions of the Parametric WFA P as the x- and y-coordinates of points. Then the given PWFA computes the points of the curve segment as shown in the first image in Fig. 14. More precisely, the pixels computed by the PWFA approach this set of points as the lengths of the input words increase.

The second image is also generated by a PWFA, which essentially holds two copies of the PWFA P, three extra labels plus two helper states. Then for two points on the curve of P the line between those points is gradually filled with black pixels in some random (fractal) fashion. For the displayed picture the computation was stopped intentionally to leave some pixels of the interior untouched.

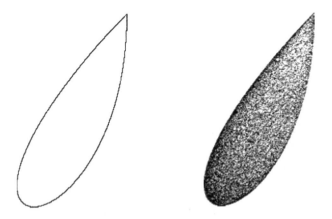

Fig. 14. Polynomial curve, filling the interior

We had mentioned already that any polynomial $p(x)$ of degree m can be computed by a standard one-dimensional WFA with $m + 1$ states, as shown in [13]. Therefore, any d-dimensional curve

$$\{(p_1(t), p_2(t), \ldots, p_d(t)) \mid 0 \le t \le 1\}$$

with parametric representation using the polynomials $p_1(t), p_2(t), \ldots, p_d(t)$ is computable by a PWFA. Furthermore, if the highest degree of the polynomials $p_1(t), p_2(t), \ldots, p_d(t)$ is m, the PWFA will only need $m + 1$ states again.

4.4   Spline Curves and 3D-Patches

For many practical purposes it is essential, that a set of single points (say in the plane $\mathbb{R}^2$) can be represented or approximated by smooth curves. Frequently the sets of points are segmented into small groups and cubic polynomials are employed for the local approximation. This piecewise approach normally requires some smoothness conditions at the segment-borders, i.e. function values and derivations up to a given degree have to match at the borders.

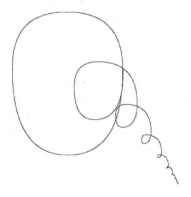

Fig. 15. Smooth curves from parabola chunks

The example presented in Fig. 15 uses a chunk of a quadratic parabola and some affine transformation to put the pieces together in a smooth way. This can be done with a PWFA with just 9 states and 2 labels.

In a more general set-up figures in the plane (like drawings of animals, font letters, ...) can be approximated by following their contours with polynomial splines. This has been studied in [17] and [18] for the cases of Catmull-Rom-splines, B-splines and Bezier-curves ([21], [20], [40]).

The next step to represent polynomial surfaces in 3D is then close at hand. Working with the popular Bezier-spline surfaces requires grids of control points. Those grids can be derived from the 2D cases in a canonical fashion as in Fig. 16. For an eighth of a sphere-surface, actually an approximation thereof, this grid is shown here. To combine eight of these (rotated) patches for a complete sphere is easy then.

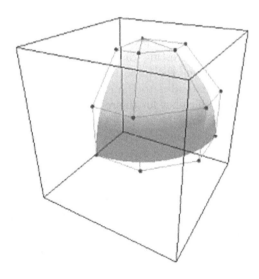

Fig. 16. Bezier patch for sphere approximation

If we like to have such 3D-patches covered by some texture, we can employ the WFA inference algorithm to generate an approximation of the given texture-image from a weighted finite automaton which we include in our PWFA. There we still have to solve the problem to assign a pixel-value at a given position to some voxel of that patch, but this is not too hard within the mechanics of the PWFA. Fig. 17 shows the result of compressing the map by a WFA and applying the corresponding WFA to the complete sphere consisting of 8 Bezier-patches.

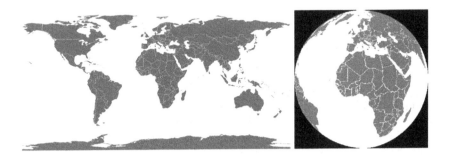

Fig. 17. Building and texturing the Bezier sphere

5 Conclusions and Open Problems

For finite acceptors and their extensions as Weighted Finite Automata and Parametric WFA several relations to graphics applications have been demonstrated here. Whereas WFA are nowadays pretty well understood and have efficient inference algorithms, this cannot yet be stated for PWFA. Up to now PWFA have been studied with respect to inclusion properties and decidability questions and a small number of hopefully interesting "hand-made" examples have been provided. For practical applications the important question is whether the WFA inference algorithm can be extended to PWFA or at least to some interesting sub-families of PWFA. The efficiency for the representation of 3D-spline-patches could be such an interesting topic for a PWFA-sub-family.

For some of the published examples of PWFA it seemed essential, that irrational weights can be employed. In a strict sense it is arguable here, whether the attribute "finite" is indeed justified for those Parametric Finite Automata, since we do not generate the irrational number by some kind of finite state device. There are results on language families and decidability questions for Integer Weighted Finite Automata by Halava and Harju ([29], [30]), but PWFA with rational weights should still be studied in detail.

And we look forward to numerous applications of PWFA in augmented reality, e.g. animated 3D-objects.

References

1. Eilenberg, S.: Automata, Languages and Machines, vol. A. Academic Press, New York (1974)
2. Salomaa, A., Soittola, M.: Automata-Theoretic Aspects of Formal Power Series. Springer, Heidelberg (1978)
3. Berstel, J., Nait Abdullah, A.: Quadtrees generated by finite automata. In: AFCET 61-62, pp. 167–175 (1989)
4. Culik, K., Dube, S.: L-systems and mutually recursive function systems. Acta Informatica 30, 279–302 (1993)
5. Culik, K., Kari, J.: Image compression using weighted finite automata. Computers and Graphics 17(3), 305–313 (1993)
6. Culik II, K., Kari, J.: Image-data compression using edge-optimizing algorithm for wfa inference. Journal of Information Processing and Management 30, 829–838 (1994)
7. Culik II, K., Valenta, V.: Finite automata based compression of bi-level and simple color images. Computers and Graphics 21, 61–68 (1997)
8. Shapiro, J.M.: Embedded image coding using zerotrees of wavelet coefficients. IEEE Transactions on Signal Processing 41(12), 3445–3462 (1993)
9. Kari, J., Fränti, P.: Arithmetic coding of weighted finite automata. Theoretical Informatics and Applications 28(3-4), 343–360 (1994)
10. Hafner, U.: Refining image compression with weighted finite automata. In: Storer, J.A., Cohn, M. (eds.) Proc. Data Compression Conference, pp. 359–368 (1996)
11. Hafner, U.: Image and video coding with weighted finite automata. In: Proc. of the IEEE International Conference on Image Processing, pp. 326–329. IEEE Computer Society Press, Los Alamitos (1997)
12. Hafner, U., Albert, J., Frank, S., Unger, M.: Weighted finite automata for video compression. IEEE Journal on Selected Areas in Communications 16(1), 108–119 (1998)

13. Culik, K., Karhumäki, J.: Finite automata computing real functions. SIAM J. Comput. 23(4), 789–814 (1994)
14. Fisher, Y.: Fractal image compression with quadtrees. In: Fisher, Y. (ed.) Fractal Image Compression, pp. 55–77. Springer, Heidelberg (1995)
15. Albert, J., Kari, J.: Parametric Weighted Finite Automata and Iterated Function Systems. In: Proc. Fractals in Engineering, Delft (1999)
16. Tischler, G., Albert, J., Kari, J.: Parametric Weighted Finite Automata and Multidimensional Dyadic Wavelets. In: Proc. Fractals in Engineering, Tours, France (2005)
17. Tischler, G.: Properties and Applications of Parametric Weighted Finite Automata. JALC 10(2/3), 347–365 (2005)
18. Tischler, G.: Parametric Weighted Finite Automata for Figure Drawing. In: Domaratzki, M., Okhotin, A., Salomaa, K., Yu, S. (eds.) CIAA 2004. LNCS, vol. 3317, pp. 259–268. Springer, Heidelberg (2005)
19. Tischler, G.: Refinement of Near Random Access Video Coding with Weighted Finite Automata. In: Ibarra, O.H., Yen, H.-C. (eds.) CIAA 2006. LNCS, vol. 4094, pp. 46–57. Springer, Heidelberg (2006)
20. De Boor, C.: A practical guide to splines. Springer, Heidelberg (1978)
21. Catmull, E., Rom, R.: A class of local interpolating splines, Computer Aided Geometric Design, pp. 317–326. Academic Press, London (1974)
22. Pavlidis, T.: Algorithms for Graphics and Image Processing. Computer Science Press (1982)
23. Droste, M., Kari, J., Steinby, P.: Observations on the Smoothness Properties of Real Functions Computed by Weighted Finite Automata. Fundamenta Informaticae 73(1,2), 99–106 (2006)
24. Culik II, K., Dube, S.: Implementing Daubechies Wavelet Transform with Weighted Finite Automata. Acta Informatica 34(5), 347–366 (1997)
25. Culik II, K., Kari, J.J.: Inference Algorithms for WFA and Image Compression. In: Fisher, Y. (ed.) Fractal Image Compression: Theory and Application, Springer, Heidelberg (1995)
26. Derencourt, D., Karhumäki, J., Latteux, M., Terlutte, A.: On Computational Power of Weighted Finite Automata. In: Havel, I.M., Koubek, V. (eds.) Mathematical Foundations of Computer Science 1992. LNCS, vol. 629, pp. 236–245. Springer, Heidelberg (1992)
27. ITU-T: Recommendation T.82 - Coded representation of picture and audio information - Progressive bi-level image compression (1993)
28. Berstel, J., Reutenauer, C.: Rational Series and Their Languages. Springer, Heidelberg (1988)
29. Halava, V., Harju, T.: Undecidability in Integer Weighted Finite Automata. Fundamenta Informaticae 38(1-2), 189–200 (1999)
30. Halava, V., Harju, T.: Languages Accepted by Integer Weighted Finite Automata. In: Karhumäki, J., Maurer, H.A., Paun, G., Rozenberg, G. (eds.) Jewels are Forever, pp. 123–134. Springer, Heidelberg (1999)
31. Culik II, K., Dube, S.: Affine Automata: A Technique to Generate Complex Images. In: Rovan, B. (ed.) Mathematical Foundations of Computer Science 1990. LNCS, vol. 452, pp. 224–231. Springer, Heidelberg (1990)
32. Shallit, J., Stolfi, J.: Two methods for generating fractals. Computers and Graphics 13(2), 185–191 (1989)
33. Culik II, K., Valenta, V., Kari, J.: Compression of Silhouette-like Images based on WFA. Journal of Universal Computer Science 3(10), 1100–1113 (1997)

34. Culik II, K., Kari, J.: Computational Fractal Geometry with WFA. Acta Informatica 34(2), 151–166 (1997)
35. Wallace, G.K.: The JPEG still picture compression standard. Communications of the ACM 34(4), 30–44 (1991)
36. International Organization for Standardization: ISO 15444: Information Technology — JPEG 2000 Image Coding System (2002)
37. Hutchinson, J.E.: Fractals and self-similarity. Indiana University Mathematics Journal 30(5), 713–747 (1981)
38. Watt, A.: Computer Graphics. Addison-Wesley, Reading (2000)
39. Barnsley, M.: Fractals everywhere, 2nd edn. Academic Press, London (1993)
40. Farin, G.: Curves and surfaces for computer aided geometric design, 2nd edn. Academic Press, London (1990)

Sturmian and Episturmian Words
(A Survey of Some Recent Results)

Jean Berstel

Institut Gaspard Monge, Université Paris-Est, Marne-la-Vallée, France

Abstract. This survey paper contains a description of some recent results concerning Sturmian and episturmian words, with particular emphasis on central words. We list fourteen characterizations of central words. We give the characterizations of Sturmian and episturmian words by lexicographic ordering, we show how the Burrows-Wheeler transform behaves on Sturmian words. We mention results on balanced episturmian words. We give a description of the compact suffix automaton of central Sturmian words.

1 Introduction

Sturmian words are combinatorial objects that are quite remarkable by the number of different characterizations they have, formulated in terms coming from different mathematical frameworks.

Sturmian words have a geometric description as digitized straight lines. Computer representation of lines has been an active subject of research, although early theory of Sturmian words remained unnoticed in the patter recognition community. The paper by [1] is a review of recognition of straight lines with respect to interaction with other disciplines. The natural generalization would be here to digitized planes, and as counter part to Sturmian bisequences.

Sturmian words have an arithmetic description, as rotations on the torus, a combinatorial description, as aperiodic words that are balanced, a description from the point of view of dynamical systems, as aperiodic words of minimal factor (subword) complexity, and so on. Many of these descriptions are known since the years 1940 and the fundamental paper [2], and a new widely disseminated research on these words has been started about thirty years ago.

In all these cases, the description given is a characterization, that is the condition stated fully describes the set of Sturmian words. Other, less known characterizations of this kind have been given. For instance, Sturmian words are characterized by the number of their return words, or by their palindromic complexity, that is the number of palindromic factors they have.

Theoretical computer scientists have contributed the point of view of effective computation. These have been studied and developed for the class of characteristic Sturmian words, where amazing computational descriptions have been provided. The special class of characteristic Sturmian words has itself some characterizations of several kinds.

S. Bozapalidis and G. Rahonis (Eds.): CAI 2007, LNCS 4728, pp. 23–47, 2007.

The richness of the theory of Sturmian words, as the meeting point of tools from different mathematical descriptions, and as extremal point of various families of infinite words, has of course led to tentatives of generalizations to other situations, especially with the objective to capture the essence of what makes the Sturmian words so special.

One of the limitation of Sturmian words is that they are over a binary alphabet. Among the extensions to larger alphabet, the so called episturmian words have appeared to be best suited family by the number of properties of Sturmian words they share.

Another extension is to two dimensions, that is to what are discretized or digital planes. This is quite interesting from the applications to pattern recognition, and is an ongoing research topic.

Another generalization is to trees. This is just at its beginnings (see [3]). Another extension is obtained when the reversal operator is replaced by an arbitrary involutory automorphism of the free monoid, see [4].

2   Sturmian and Episturmian Words

Before starting, we give some notational conventions. Given a nonempty word w, we denote by w^- the word without its last letter. If w has at least two letters, then we write $w^=$ instead of w^{--}. Thus, for instance $abaab^= = aba$.

Given a finite or infinite word w, the set of letters that occur in w is denoted by $\mathrm{Alph}(w)$. If w is infinite, $\mathrm{Ult}(w)$ denotes the set of letters that occur infinitely many often in w.

Finally, we denote by $w(k)$ the letter at position k $(k \geq 0)$ in the word w.

2.1   Complexity

Let w be an infinite word on some alphabet A. We denote by $F(w)$ the set of (finite) factors of w, and by $F_n(w) = F(w) \cap A^n$ the set of factors of length n of w. The *complexity function* c_w of w is defined by

$$c_w(n) = \mathrm{Card}(F_n(w))$$

This complexity is also called *subword* or *factor* or *block complexity*. The (right) *degree* $\deg_w(x)$ of a finite word x in w is the number of letters a such that xa is a factor of w:

$$\deg_w(x) = \mathrm{Card}\{a \in A \mid xa \in F(w)\}$$

Similarly, the left degree of w is the number of $a \in A$ with $ax \in F(w)$. Clearly, $\deg_w(x) \geq 1$ for each factor x of w. Also, $\deg_w(xy) \leq \deg_w(y)$ for all x, y. Clearly

$$c_w(n+1) = \sum_{x \in F_n(w)} \deg(x).$$

A factor x is right special (left special) if its degree (left degree) is strictly greater than 1. Any suffix of a right special factor is again right special. Observe that

$$c_w(n+1) - c_w(n) = \sum_{x \in S_n(w)} \deg(x) - 1. \tag{1}$$

where $S_n(w)$ is the set of right special factors of length n. An infinite word w is *episturmian* if the set $F(w)$ is closed under reversal, and if, for every $n \geq 1$ there exists at most one right special factor of length n. It is *aperiodic episturmian* if it is episturmian and aperiodic, that is not eventually periodic. This is equivalent to require that there is exactly one right special factor of each length. The word w is *strict episturmian* if w is aperiodic episturmian and if all its right special factors have the same degree. If this degree is k, then it follows from (1) that for $n \geq 1$

$$c_w(n) = kn + 1.$$

Below, we will give a more detailed description the Tribonacci word which is a strict episturmian word. The theory of episturmian words and morphisms has been developed in three basic papers [5,6,7] by Justin and Pirillo, the first with Droubay, see also [8].

Recall that an infinite word w is *recurrent* if each factor of w occurs infinitely many often in w, and it is *uniformly recurrent* if each factor occurs infinitely many often with bounded gaps between consecutive occurrences. In other terms, w is uniformly recurrent if, for every n, there exists N such that each factor of w of length N contains all factors of w of length n, in symbols $F_n(w) = F_n(u)$ for all $u \in F_N(w)$. Any episturmian word is uniformly recurrent.

Strict episturmian words are also called *Arnoux-Rauzy* words or *AR-words*. They were introduced and studied in [9], mainly in the case of three letters. Strict episturmian words over two letters are exactly the *Sturmian* words. These are aperiodic words of minimal block complexity in view of the well-known.

Theorem 1. [2,10] *An infinite word w is eventually periodic if and only if there exists an integer $n \geq 1$ such that $c_w(n) \leq n$.*

2.2 Other Complexity Functions

Several other measures of complexity of infinite words have been defined and compared to the block complexity.

The *palindrome complexity function* p_w of an infinite word w associates to each integer $n \geq 0$ the number of distinct palindromes of length n in w.

A general exposition of palindrome complexity together with new results is given in [11]. In particular, it is shown in this paper that if w is an aperiodic infinite word, then

$$p_w(n) < \frac{16}{n} c_w\left(n + \left\lfloor \frac{n}{4} \right\rfloor\right).$$

Thus in particular if $c_w(n) = O(n)$ then w has bounded palindromic complexity. This holds for Sturmian and episturmian words, for automatic words, and for words that are fixed points of primitive morphisms. For uniformly recurrent words, there is a more precise formula given in [12]. They prove that, provided the set of factors $F(w)$ is closed under reversal,

$$p_w(n) + p_w(n+1) \leq 2 + c_w(n+1) - c_w(n).$$

This is sharp for Sturmian words: these are characterized by the fact that $p_w(n) = 1$ if n is even and $p_w(n) = 2$ if n is odd [13], and also for AR-words over $r > 2$ letters: these words have palindrome complexity $p_w(n) = 1$ if n is even and $p_w(n) = r$ if n is odd [14].

Another complexity function is arithmetical complexity introduced in [15]. Given an infinite word $w = w(0)w(1)\cdots$, the *arithmetical complexity* function a_w associates to $n \geq 0$ the number of distinct words of the

$$w(k)w(k+d)w(k+2d)\cdots w(k+(n-1)d)$$

for $k \geq 0$, $d \geq 1$. The arithmetical complexity of a Sturmian word depends only on its slope (see below), since two Sturmian words have the same set of factors if and only if they have same slope. So, it is convenient to write a_α instead of a_w for a Sturmian word of slope α.

Theorem 2. [16] *For any Sturmian word of slope α, one has*

$$a_\alpha(n) \leq h(n)$$

where

$$h(n) = 2 + \binom{n+1}{3} + 2\sum_{i=1}^{n-1}(n-i)\phi(i).$$

Here ϕ is Euler's totient function. In fact, the authors give the exact expression for the arithmetical complexity of Sturmian words for $1/3 < \alpha < 1/2$ (note that exchanging the two letters in a Sturmian words replaces the slope α by $1 - \alpha$ without changing the complexity, so the result holds also for $1/2 < \alpha < 2/3$). Denote by (r_k) the decreasing sequence of rational numbers given by $r_k = k/(3k-1)$, for $k \geq 2$. Thus $r_2 = 2/3$, $r_3 = 3/8$.

Theorem 3. [16] *For any irrational α with $1/3 < \alpha < 1/2$, one has*

$$a_\alpha(n) = \begin{cases} h(n) - 8 & \text{if } n \text{ is odd} \\ h(n) - 9 & \text{otherwise} \end{cases}$$

for $n \geq 3k$, where k is such that $r_{k-1} > \alpha > r_k$.

For other results concerning arithmetical complexity, see [17].

A more general measure is the *maximal pattern complexity*. A window τ of size k is a sequence $0 = \tau_0 < \tau_1 < \cdots < \tau_{k-1}$ of integers. The τ-pattern at position n in w is the word

$$w(n+\tau_0)w(n+\tau_1)\cdots w(n+\tau_{k-1}).$$

Denote by $F_\tau(w)$ the set of τ-pattern occurring in w. The τ-complexity of w is the number $c_w(\tau) = \operatorname{Card} F_\tau(w)$, and the maximal pattern complexity is

$$c_w^*(k) = \sup_{|\tau|=k} c_w(\tau),$$

where $|\tau|$ denotes the size of τ. There is an analogue of Theorem 1 for the maximal pattern complexity:

Theorem 4. [18] *An infinite word w is eventually periodic if and only if $c_w^*(k) < 2k$ for some $k \geq 1$.*

Words with maximal pattern complexity $2k$ have been called pattern Sturmian words and are studied in [18]. Sturmian words are special cases of pattern Sturmian words. Generalizations are given in [19,20].

There is a variation of block complexity considered by [21,22]. Instead of counting the number of factors of given length in an infinite word, they count the number of factors of this length that occur infinitely many often in the word. If the word is uniformly recurrent, the complexities are the same. For skew words, as defined later, they are different.

2.3   Palindromic Closure

The *right palindromic closure* of a word w is the shortest palindrome which has w as a prefix. It is denoted by $w^{(+)}$. For instance, the right palindromic closure of 01011 is 0010110100. It is easy to prove that

$$w^{(+)} = uv\tilde{u},$$

where v is the longest palindrome suffix of w. In the example, the longest palindrome suffix of $w = 001011$ is 11, and therefore $w^{(+)} = 0010\,11\,0100$. The notion was introduced and used by de Luca [23,24] for the analysis of finite Sturmian words.

Given a finite word d, the *right iterated palindrome* produced by d is the word $P(d)$ defined as follows. $P(\varepsilon) = \varepsilon$ and for a word d and a letter a,

$$P(da) = (P(d)a)^{(+)}. \tag{2}$$

For example, for the word $abbaab$ one gets successively

d	$P(d)$
a	a
ab	aba
abb	$ababa$
$abba$	$ababaababa$
$abbaa$	$ababaababaababa$
$abbaab$	$ababaababaabababababaababaababa$

The word d is the *directive word* of $P(d)$. A *right iterated palindrome* is a right iterated palindrome w produced by some word d. If d is over at most two letters, then the word w is *binary*.

If d is an infinite word, the *right iterated palindrome* produced by d is the infinite word which has as prefixes all right iterated palindromes produced by the finite prefixes of d. This makes sense because $P(x)$ is a prefix of $P(xy)$ for all words x, y.

If a does not occur in d, then (2) gives simply $P(da) = P(d)aP(d)$. There is another way to compute (2) when the letter a occurs in d. Let pa be the longest prefix of d ending with the letter a, and define the word s by $P(pa) = P(p)s$. Then $P(da) = P(d)s$. In our example, for $db = abbaab$, one has $p = ab$ and $s = baababaababa$. This computation rule is given in [25].

2.4 Justin's Formula

Justin's formula gives a useful relation between standard words and central words generated by iterated right palindromic closure. Let A be an alphabet and let $\psi : A^* \to \mathrm{End}(A^*)$ be the morphism that maps a letter a to the morphism ψ_a defined, for $b \in A$, by

$$\psi_a(b) = \begin{cases} ab & \text{if } b \neq a, \\ a & \text{otherwise.} \end{cases}$$

For instance, if a, b, c are letters, then

$$\psi_a(bac) = abaac.$$

Composition is defined for words u, v by

$$\psi_{uv} = \psi_u \circ \psi_v,$$

that is

$$\psi_{uv}(w) = \psi_u(\psi_v(w)).$$

For instance,

$$\psi_{abc}(a) = \psi_{ab}(ca) = \psi_a(bcba) = abacaba.$$

A word of the form $\psi_u(a)$ for some word u and some letter a is an *epistandard word*. The morphisms ψ_u are pure *epistandard morphisms*. In the binary case, these morphisms are called pure Sturmian morphisms, and the words they produce are indeed the standard words. Justin's formula establishes a relation between the morphism ψ and right palindromic closure P.

Proposition 5. (Justin's Formula) *The following holds for any words u, v:*

$$P(uv) = \psi_u(P(v))P(u). \tag{3}$$

As an example, let $u = ab$, $v = ac$. Then $P(u) = aba$, $P(v) = aca$, $\psi_u(P(v)) = \psi_a(\psi_b(aca)) = \psi_a(babcba) = abaabacaba$, whereas $P(abac) = ((abaa)^{(+)}c)^{(+)} = abaabac^{(+)} = abaabacabaaba$, so indeed $P(abac) = \psi_{ab}(aca)aba$.

The formula admits several interesting special cases. First, when u is a letter, then (3) becomes

$$P(av) = \psi_a(P(v))a\,.$$

This shows that $P(av)$ is obtained from $P(v)$ by simply inserting the letter a before each letter of $P(v)$ which is not an a, and then adding a final a. For instance, since $P(ba) = bab$, one gets $P(aba) = abaaba$. Observe that $P(av)$ is also obtained from $P(v)$ by inserting the letter a *after* each non-a letter.

Another special case arises when v is just a letter. Then (3) becomes

$$P(ua) = \psi_u(a)P(u)\,. \tag{4}$$

This shows a way to compute the right palindrome closure $P(ua)$ by prefixing $P(u)$ the standard word $\psi_u(a)$. Recall that by definition $P(ua) = P(u)a\tilde{y}$, where $P(u)a = yz$ with z a maximal suffix of $P(u)a$ which is a palindrome. Since $P(u)$ and $P(ua)$ both are palindromes, one has $P(ua) = yaP(u)$ and so $\psi_u(a) = ya$.

As an example, consider the computation of $P(acbc)$. By (4), it suffices to compute $\psi_{acb}(c) = acabac$ and $P(acb) = acabaca$ to get the word

$$P(acbc) = acabacacabaca\,.$$

Finally, iteration of (4) gives, for a word $u = a_1a_2\cdots a_n$ the formula

$$P(a_1a_2\cdots a_n) = \psi_{a_1a_2\cdots a_{n-1}}(a_n)\psi_{a_1a_2\cdots a_{n-2}}(a_{n-1})\cdots\psi_{a_1a_2}(a_3)\psi_{a_1}(a_2)a_1\,.$$

For instance

$$P(acbc) = \psi_{acb}(c)\psi_{ac}(b)\psi_a(c)a = acabac\cdot acab\cdot ac\cdot a\,.$$

As an illustration of the uses of the formula, we prove the following observation.

Remark 6. A standard episturmian word w has the form $\psi_u(v)$, where u is a finite word and v is a strict standard episturmian word.

Proof. Let d be the infinite word such that $w = P(d)$. Let d' be a suffix of d such that $\mathrm{Ult}(d') = \mathrm{Alph}(d')$, and let $d = ud'$. By Justin's formula, $w = \psi_u(P(d'))$, and by construction $P(d')$ is strict.

Another remark concerns eventually periodic standard episturmian words. If w is such a word, then it is purely periodic. Indeed, by Theorem 3 in [5], one has $w = P(va^\omega)$ for some word v and some letter a, and consequently $w = \psi_v(P(a^\omega)) = \psi_v(a^\omega) = (\psi_v(a))^\omega$.

Example 7. The *Tribonacci* word is a generalization of the Fibonacci word. Finite Tribonacci words are the words t_n defined over three letters a, b, c by

$$t_{-1} = c,\ t_0 = a,\ t_1 = ab,\ t_n = t_{n-1}t_{n-2}t_{n-3}\ (n \geq 2)\,.$$

Thus

$$t_2 = abac$$
$$t_3 = abacaba$$
$$t_4 = abacabaabacab$$
$$t_5 = abacabaabacabababacabaabac$$

The infinite Tribonacci word t is the limit of the words t_n. An equivalent definition of the t_n is through the morphism

$$\psi : a \mapsto ab, \ b \mapsto ac, \ c \mapsto a.$$

Indeed, it is easy to check that $t_n = \psi^n(a)$ for $n \geq 0$. Finally, one has also

$$t = P((abc)^\omega)$$

showing that t is a strict standard episturmian word. Indeed, denote by δ_n the prefix of length n of $(abc)^\omega$ and set $u_n = P(\delta_n)$. Then it can be shown that $u_n = t_{n-1}u_{n-1}$ for $n \geq 1$. Thus $t = \lim u_n$. Also

$$u_n = t_{n-1}t_{n-2}\cdots t_0.$$

This formula has been extended to more general words in [26]. For other properties of the Tribonacci word, see [27,28] and the chapter by Allouche and Berthé in [29].

3 Sturmian Words

Sturmian words have particular properties related to their geometric interpretation. This holds especially for finite Sturmian words.

3.1 Mechanical Words

Sturmian words have a geometric interpretation as *cutting sequences* of straight lines (this word comes from [30]) and therefore are closely related to digitization and pattern recognition. An equivalent formulation is through *mechanical words* (as they are called in [2]) or as *rotation words* (this is the name given for instance in [31]).

Consider a straight line in the plane. At each intersection point with the integer grid, write the letter a if the line intersects grid vertically, and write the letter b otherwise, see Figure 1. This is the definition of Sturmian words as cutting sequences. By a "shear", that is the mapping $(x, y) \mapsto (x + y, y)$, one gets the definition as "mechanical words". These are infinite words defined, for reals $0 < \alpha < 1$ and $0 \leq \rho \leq 1$, by

$$s_{\alpha,\rho}(n) = \begin{cases} a & \text{if } \lfloor (n+1)\alpha + \rho \rfloor = \lfloor n\alpha + \rho \rfloor, \\ b & \text{otherwise.} \end{cases}$$

$$s'_{\alpha,\rho}(n) = \begin{cases} a & \text{if } \lceil (n+1)\alpha + \rho \rceil = \lceil n\alpha + \rho \rceil, \\ b & \text{otherwise.} \end{cases}$$

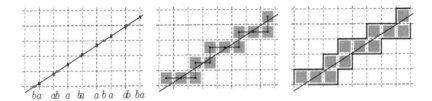

Fig. 1. A Sturmian word defined as a cutting sequence by intersection or by adjacent squares, and the upper and the lower mechanical word

for $n \geq 0$. The word $s_{\alpha,\rho}$ ($s'_{\alpha,\rho}$) is called the lower (upper) *mechanical word* with slope α and *intercept* ρ.

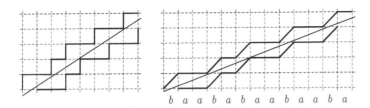

Fig. 2. "Shear" of the cutting sequence

There is an equivalent definition by rotation. Consider indeed the torus $\mathbb{T} = \mathbb{R}/\mathbb{Z}$ of reals modulo 1, and partition $\mathbb{T}$

$$I_a = [0, 1 - \alpha), \ I_b = [1 - \alpha, 1), \ I'_a = (0, 1 - \alpha], \ I'_b = (1 - \alpha, 1],$$

and let $R_\alpha : \mathbb{T} \to \mathbb{T}$ be the rotation of angle α. Then

$$s_{\alpha,\rho}(n) = \begin{cases} a & \text{if } R_\alpha^n(\rho) \in I_a, \\ b & \text{otherwise.} \end{cases}, \quad s'_{\alpha,\rho}(n) = \begin{cases} a & \text{if } R_\alpha^n(\rho) \in I'_a, \\ b & \text{otherwise.} \end{cases}$$

This is why mechanical words are also called *rotation words*. They are rational words when α is rational, and irrational words when α is irrational. It is known [2] that irrational mechanical words are exactly Sturmian words. It is also known that two Sturmian words with the same slope have the same set of factors. When $\rho = \alpha$, one has $s_{\alpha,\rho} = s'_{\alpha,\rho}$. This word is called the *characteristic* word of slope α, and is denoted by c_α. For a systematic exposition, see [32] and [33].

4   Finite Sturmian Words

In this section, all words are binary over the alphabet $A = \{a, b\}$.

A finite word is *Sturmian* if it is a factor of some infinite Sturmian word. Among finite Sturmian words, particular classes are the standard words, the central words, and the Christoffel words.

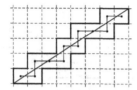 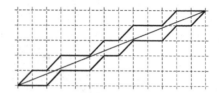

Fig. 3. The central word corresponding to the point $(8,5)$ is $x = abaababaaba$. The upper and lower Christoffel words are $bxa = babaababaabaa$ and $axb = aabaababaabab$. Two standard words are associated with them, namely $xab = abaababaabaab$ and $xba = abaababaababa$.

The mechanical words $s_{\alpha,\rho}$ and $s'_{\alpha,\rho}$ are purely periodic when α is rational. Moreover, if $\alpha = p/(p+q)$ for $p \perp q$, then $s_{\alpha,0} = w^\omega$ and $s'_{\alpha,0} = w'^\omega$ where w and w' are precisely the lower and upper Christoffel words defined by p and q. It is easily checked that for $0 \le n < p+q$,

$$\left\lfloor (n+1)\frac{p}{q} \right\rfloor = \left\lfloor n\frac{p}{q} \right\rfloor \iff np \bmod p+q < (n+1)p \bmod p+q.$$

So the lower Christoffel word is obtained simply by considering consecutive values in the sequence $np \bmod p + q$. For $p = 5$ and $q = 8$, one gets the sequence

$$0 \xrightarrow{a} 5 \xrightarrow{a} 10 \xrightarrow{b} 2 \xrightarrow{a} 7 \xrightarrow{a} 12 \xrightarrow{b} 4 \xrightarrow{a} 9 \xrightarrow{b} 1 \xrightarrow{a} 6 \xrightarrow{a} 11 \xrightarrow{b} 3 \xrightarrow{a} 8 \xrightarrow{b} 0$$

This is the construction as given by Christoffel in [34]. Another equivalent definition is by directive sequences and will be given below.

A finite word w is *balanced* if, for each pair of factors x, y of w of equal length, $\left| |x|_a - |y|_a \right| \le 1$ for the letter a. Here $|x|_a$ denotes the number of occurrences of a in x.

4.1   Standard and Central Words

A *directive sequence* $d = (d_0, d_1, \ldots, d_k)$ is a sequence of integers with $d_0 \ge 0$ and $d_i > 0$ for $i \ge 1$. The *standard word* produced by d is the word $S(d) = s_{k+1}$, where

$$s_{-1} = b, \; s_0 = a, \; s_{n+1} = s_n^{d_n} s_{n-1}, \; n \ge 0.$$

Example 8. For $d = (3,1,2,1)$, one gets $s_1 = a^3b$, $s_2 = a^3ba$, $s_3 = a^3ba^4ba^4b$, $S(d) = s_4 = a^3ba^4ba^4ba^3ba$.

The standard word produced by the empty sequence is a, the standard word produced by (0) is b.

If $k \ge 0$, the sequences $d = (d_0, d_1, \ldots, d_k, 1)$ and $d' = (d_0, d_1, \ldots, d_k+1)$ produce the same word up to the last two letters which are interchanged, because

$$S(d) = s_k^{d_k} s_{k-1} s_k, \; S(d') = s_k^{d_k} s_k s_{k-1},$$

and $s_{k-1}s_k$ and $s_k s_{k-1}$ are easily seen to be the same up to the last two letters, by induction.

A *central word* is a standard word without its two last letters: a word x is central if and only if $x = s^=$ for some standard word s.

A upper (lower) *Christoffel word* is a word of the form bxa (axb) for some central word x.

The relation between the mechanical definition and the description by the directive sequence is through the continued fraction expansion of the slope. Let again p and q be positive integers with $p \perp q$. The rational number q/p has two expansions into continued fractions, say

$$[d_0, d_1, \ldots, d_k, 1] = [d_0, d_1, \ldots, d_k + 1].$$

These are the directive sequences for the two standard words with q letters a and p letters b. For example, if $q = 5$ and $p = 8$, then $q/p = [1, 1, 1, 1, 1] = [1, 1, 1, 2]$. Also, for the word $s_4 = a^3ba^4ba^4ba^3ba$ produced by the directive sequence $d = (3, 1, 2, 1)$ given above, one has $q/p = [3, 1, 2, 1]$ with $p = |s_4|_a = 4$ and $q = |s_4|_b = 15$.

Proposition 9. *Let x be a word. Then the following are equivalent*

1. x is a central word;
2. xab is a standard word;
3. xba is a standard word;
4. bxa is an upper Christoffel word;
5. axb is a lower Christoffel word.

As a consequence, every characterization of central words translates automatically into a characterization of standard words and of Christoffel words. In particular, we may speak about the central word produced by a directive sequence, and as mentioned above, the sequences $d = (d_0, d_1, \ldots, d_k, 1)$ and $d' = (d_0, d_1, \ldots, d_k + 1)$ produce the same central word.

4.2   Characterizations of Central Words

Proposition 10. [35] *A word x is central if and only if the words axb and bxa are conjugate.*

Proposition 11. [36] *A word is central if and only if it is a palindrome prefix of a characteristic Sturmian word.*

Proposition 12. [23] *A word is central if and only if it is a binary right iterated palindrome.*

Proposition 13. [36] *A word w is central if and only if wab or wba is a standard Sturmian word.*

Proposition 14. [36] *A word w is central if and only if it is a palindrome and wab (or wba) is a product of two palindromes.*

Proposition 15. [37] *A word w is a conjugate of a standard Sturmian word if and only if it is primitive and all its conjugates are balanced.*

Proposition 16. [37] *A word w is a conjugate of a standard Sturmian word if and only if the circular word w has $k+1$ factors of length k for $0 \le k < |w|$, and this holds if and only if w is primitive and has $|w| - 1$ factors of length $|w| - 2$.*

Proposition 17. [36] *A word w is central if and only if the words awa, awb, bwa, bwb are balanced.*

In fact, a weaker condition is sufficient.

Proposition 18. [36] *A word w is central if and only if the words awb and bwa are balanced.*

Proposition 19. [23,38] *A word w is central if and only if it is a palindrome and the words wa and wb are balanced.*

Denote by π_w the minimal period of w. Then one has

Proposition 20. [38] *A word w is central if and only if it is a power of a letter or it is a palindrome and its prefix of length $\pi_w - 2$ is a right special factor of w.*

Example 21. Consider the word $w = baaabaaab$ has minimal period 4. Its prefix of length 2 is ba which is not a right special factor of w. So, according to Proposition 20, this word is not central. The conclusion follows also from Proposition 19, since $wb = baaabaaabb$ is not balanced.

The next proposition is actually a consequence of a result of [23].

Proposition 22. [39] *A word w is central if it is a power of a single letter or it satisfies the equation $w = w_1abw_2 = w_2baw_1$ with $w_1, w_2 \in a, b^*$. Moreover, in this latter case w_1 and w_2 are central words, $p = |w_1| + 2$ and $q = |w_2| + 2$ are co-prime periods of w and $\min p, q$ is the minimal period of w.*

Proposition 23. [36] *A word w is central if and only if there exist integers $p \perp q$ with $|w| = p + q - 2$ such that w has periods p, q.*

There is a duality between periods and number of letters in central words as already described in [23] and in [40]. Further results are in [24]. This duality has been developed recently in [41].

Proposition 24. [40] *A word w is central if and only if the word awb is a balanced Lyndon word.*

A *Sturmian palindrome* is a finite Sturmian word which is a palindrome. Every central word is a Sturmian palindrome but the converse is false. For instance, $baab$ is a Sturmian palindrome (it is a factor of the infinite Fibonacci word $f = abaab \cdots$) but is it not central in view of Proposition 18 since $bbaaba$ is not balanced. The following characterization holds.

Theorem 25. [37,23,5,38] *A word is a Sturmian palindrome if and only if it is a median factor of a central word.*

There are much more Sturmian palindromes than central words. The number of central words of length n is $\phi(n+2)$ since a central word of length n is described by two positive integers $p \perp q$ with $p + q = n + 2$. On the contrary, one has

Theorem 26. [38] *Denote by $h(n)$ the number of Sturmian palindromes of length n. Then*

$$h(2n) = 1 + \sum_{i=1}^{n} \phi(2i), \quad h(2n+1) = 1 + \sum_{i=1}^{n} \phi(2i+1).$$

4.3 Directive Word and Directive Sequence

Given a directive sequence $d = (d_0, d_1, \ldots)$, the word $S(d)$ produced by d is a standard word if d is finite, a characteristic word if d is infinite. Define a directive word δ by $\delta = a^{d_0} b^{d_1} a^{d_2} \cdots c^{d_n}$, where $c = a$ if n is even, and $c = b$ otherwise. The relation between directive words and directive sequences in the binary case is the following.

Proposition 27. *Let d and δ be as above. Then $S(d) = \psi_\delta(\bar{c})$ where $\bar{c}$ is the opposite letter of c and moreover $S(d)^= = P(\delta)$.*

5 Balance

Let $\ell \geq 1$ be an integer. A set X of words over an alphabet A is ℓ-*balanced* if, for each x, y in X of equal length, $\left| |x|_a - |y|_a \right| \leq \ell$ for all letters a. Here $|x|_a$ denotes the number of occurrences of a in x. A word is ℓ-balanced if the set of its factors is balanced. Binary balanced words are precisely 1-balanced words. A word w is *strongly balanced* if w is primitive and w^2 is balanced. A word w such that w^2 is balanced, without being necessarily primitive is called *cyclically balanced* in [42]. Thus a word is cyclically balanced if it is a power of some strongly balanced word. For instance, $abba$ is balanced but is not strongly balanced because the square $abbaabba$ contains both factors aa and bb. The word $ababab$ is cyclically balanced. A *finite Sturmian word* is a word which is a factor of some (infinite) Sturmian word.

Proposition 28. *A finite binary word is balanced if and only if it is a finite Sturmian word.*

Proposition 29. [43,42,44] *A finite binary word is strongly balanced if and only if it is a conjugate of some standard Sturmian word.*

For infinite words, we recall the following characterization of Sturmian words.

Proposition 30. *A binary infinite word is Sturmian if and only if it is balanced and aperiodic.*

It is easy to find balanced eventually periodic words, such as ab^ω. These are not Sturmian. We discuss this in the next section.

In fact, Sturmian words share a stronger balance property. Denote by $|x|_u$ the number of distinct occurrences of the word u as a factor in the word x, counting also overlaps. For instance, $|abbabaab|_{ba} = 2$ and $|abaababa|_{aba} = 3$. Then, one has

Theorem 31. [45] *A binary infinite word w is Sturmian if and only if for each word u,*

$$\left| |x|_u - |y|_u \right| \leq |u|$$

for each pair of factors x, y of the same length of w,

A characterization of episturmian words by a balance property like Proposition 30 does not exist. It is known that the Tribonacci word t is 2-balanced. However, when applying a well chosen pure epistandard morphism, it does not remain 2-balanced. For instance, the word $\mu(t)$ with $\mu = \psi_{aabbac}$, contains the factors $baabaaabaabaabaaabaab$ and $aacaabaabaaabaabaacaa$ of length 21. Indeed, the first is a factor of $\mu(bab)$ and the second is a factor of $\mu(aca)$. The number of b in these factors are 7 and 4, so their balance is 3. It has been proved by [46] that there exist AR-sequences which not ℓ-balanced for any ℓ.

There is a closed formula for the number of finite balanced words, that is of factors of Sturmian words.

Proposition 32. *The number of balanced binary words of length n is*

$$1 + \sum_{i=1}^{n}(n + 1 - i)\phi(i)$$

where ϕ is the Euler's totient function.

The first proof of this formula is perhaps [47]. Other proofs are in [48,36,23,49,50]. Related results also appear in [51,52]. An exact formula for the number $g_\ell(n)$ of ℓ-balance words of length n seems not to be known. It was shown already in [47] that it is exponential for $\ell \geq 2$ (whereas usual number theory shows that $g_1(n) = N^3/\pi^2 + O(n^2)$) and more exactly that

$$g_\ell(n) = \Theta\binom{\ell + 1}{\lfloor \ell/2 \rfloor}^{n/(\ell+1)}$$

which gives $g_2(n) = \Theta(3^{n/3})$. Heinis provided independently in [53] a lower bound, and Tarannikov [54] shows that

$$g_\ell(n) = \Theta(n^2(2\cos\frac{\pi}{\ell + 2})^n)$$

which is better for $\ell \geq 3$.

On the other hand, the number of factors of length n of strict episturmian words (or equivalently of Arnoux-Rauzy words) has been considered. A *bispecial factor* is a word that is both a left and a right special factor of the same Arnoux-Rauzy word.

Proposition 33. [55] *The number of factors of length n of strict episturmian words over a k-letter alphabet is*

$$k + (n-1)k(k-1) + (k-1)^2 \sum_{i=1}^{n-2}(n-i-1)b(i)$$

where $b(m)$ is the number of bispecial factors of length m of Arnoux-Rauzy words.

The number of bispecial factors is evaluated, in [55], in terms of a generalized Euclidean algorithm.

We already mentioned that episturmian words are not balanced in general. In fact, almost the opposite is true: episturmian words are never balanced, except in simple cases. More precisely, the following holds.

Theorem 34. [56] *Let x be a standard episturmian word over the alphabet $A = \{1, 2, \ldots, k\}$ with $k \geq 3$. Then x is balanced if and only if its directive word δ can be written in one of the following forms, up to a permutation of the alphabet.*

1. *$123 \cdots k1^\omega$*
2. *$1^n 23 \cdots (k-1)k^\omega$ for some $n \geq 1$*
3. *$12 \cdots \ell 1(\ell+1) \cdots (k-1)k^\omega$ for some $1 \leq \ell < k$.*

For $k = 5$, an example of the last case is 121345^ω with $\ell = 2$. All episturmian words of the theorem are eventually periodic.

6 Lexicographic Ordering

Every (total) order on an alphabet A defines a lexicographic order on (right) infinite words. Given an infinite word x, we denote by $\min(x)$ and by $\max(x)$ the minimal and the maximal word, for the lexicographic order, of the orbit of x. This is simply defined by the condition that, for each integer n, the prefix of length n of $\min(x)$ (of $\max(x)$) is the smallest (largest) word in $F_n(x)$. For Sturmian words, it is easily seen that $s_{\alpha, \rho} < s_{\alpha, \rho'}$ if and only if $\rho < \rho'$ (recall that α is irrational). Thus for the ordering $a < b$, one gets that $\min(s_{\alpha, \rho}) = s_{\alpha, 0} = ac_\alpha$ and $\max(s_{\alpha, \rho}) = bc_\alpha$, where c_α denotes the characteristic word with slope α. As an example, consider the Fibonacci word $f = abaababaabaab \cdots$. Then $\min(f) = af$ and $\max(f) = bf$.

The comparison of words for the lexicographic order is well suited for the study of balanced infinite words, and can be extended to the case of more than two letters. It will be convenient to use the following old terminology from [2]. A *Sturmian trajectory* is an infinite binary word whose (finite) factors are finite Sturmian words. Thus, Sturmian trajectories are precisely balanced binary words.

Similarly, we call *episturmian trajectory* an infinite word whose finite factors are finite episturmian words. Episturmian trajectories are called episturmian words *in the wide sense* in [57].

It is known since [2] that Sturmian trajectories can be partitioned into three classes:

1. aperiodic words: these are exactly all Sturmian words or equivalently all irrational mechanical words;
2. (purely) periodic words : these are the rational mechanical words; they are of the form w^ω, where w is a conjugate of some standard word;
3. eventually periodic but not purely periodic words. These are called *skew* words. They are not mechanical words. It has been shown that they are those suffixes of the words of the form $\mu(a^n ba^\omega)$, for some pure standard Sturmian morphism μ and some integer $n \geq 0$ which are not suffixes of $\mu(a^\omega)$.

The three classes of Sturmian trajectories can be grouped together in three manners. First, group (1) is compose of aperiodic words whereas groups (2) + (3) are eventually periodic words. Next, words in (1) + (2) are uniformly recurrent, whereas words of type (3) are not recurrent. Finally, words of type (1) + (3) are precisely the words called *fine* by Pirillo in [58] and that we will describe in a moment. First we give the following characterization.

Theorem 35. *A binary infinite word x over $\{a, b\}$ with $a < b$ is a Sturmian trajectory if and only if there is an infinite y such that $ay \leq \min(x)$ and $\max(x) \leq by$.*

This is a corollary of the next theorem, and appears also, under a different guise, in [59]. We denote by $\min(A)$ the smallest letter in the alphabet A for the given order.

Theorem 36. [57] *An infinite word x over A is an episturmian trajectory if and only if there exists an infinite word y such that $\min(A)y \leq \min(x)$ for every order over A.*

Episturmian trajectories are either episturmian words or belong to the family of so-called *episkew* words. These are exactly the episturmian trajectories which are not recurrent. It is quite interesting to note that the characterization of skew Sturmian trajectories carries over, with some complications, to episkew words. This is done in [60], see also [57].

Proposition 37. *An infinite word x with $A = \mathrm{Alph}(x)$ is episkew if and only if there is a letter a, a standard episturmian word y on $B = A \setminus \{a\}$, a finite prefix p of y and a pure epistandard morphism μ such that $zx = \mu(\tilde{p}ay)$ for some proper prefix z of $\mu(\tilde{p}a)$.*

If, in the proposition, the word y is strict, then the word x itself is called strict episkew. Observe also that in the Sturmian case the word $\tilde{p}ay$ indeed reduces to a word of the form $a^p ba^\omega$.

In the case of characteristic words or of epistandard words, one has stronger conditions.

Theorem 38. [61] *A binary word x over $A = \{a, b\}$ with $a < b$ is a characteristic Sturmian word if and only if $ax = \min(x)$ and $\max(x) = bx$.*

A result similar to Theorem 38 holds for strict episturmian (or Arnoux-Rauzy) words.

Proposition 39. [6] *An infinite word x over some alphabet A is a strict epistandard word if and only if $\min(A)x = \min(x)$ for any order on A.*

This is related to the following.

Proposition 40. [61] *An infinite word x over some alphabet A is an epistandard word if and only if $\min(A)x \leq \min(x)$ for any order on A.*

Let x be an infinite word and let $A = \mathrm{Alph}(x)$. The word x is *fine* if there exists an infinite word y such that $\min(x) = \min(A)y$ holds for any lexicographic order. As announced, we have the following.

Proposition 41. [58] *A binary word is fine if and only if it is a Sturmian word or a skew Sturmian word.*

Thus, the Sturmian trajectories which are not fine are precisely the rational mechanical words. This has been extended to episturmian trajectories.

Proposition 42. [60] *A word x is fine if and only if it is a strict epistandard word or a strict skew episturmian word.*

7 Burrows-Wheeler Transformation

The Burrows-Wheeler transformation, introduced in [62], is a reversible transformation that produces a permutation $BWT(w)$ of an input sequence w. It appears that the transform is easier to compress than the original sequence because there is some clustering effect in the transformed word. BWT is used in the BZIP2 data compression algorithm. The Burrows-Wheeler transformation has a strong relation to a transformation called the Gessel-Reutenauer transform, introduced in [63]. This connection has been described in [64]. As has been shown in [65] the Burrows-Wheeler transformation takes a very particular form when applied to standard Sturmian words. Recent results are given in the forthcoming paper [66].

The Burrows-Wheeler transformation takes as input a word w, and produces as output a permutation $BWT(w)$, obtained as follows. Let $M(w)$ be the matrix composed of all conjugates of w, ordered lexicographically. Then $BWT(w)$ is the last column of $M(w)$.

Example 43. For the input word $w = abraca$, the matrix is

$$
M(w) = \begin{bmatrix}
a & a & b & r & a & c \\
a & b & r & a & c & a \\
a & c & a & a & b & r \\
b & r & a & c & a & a \\
c & a & a & b & r & a \\
r & a & c & a & a & b
\end{bmatrix}
$$

and the output is the last column, that is $BWT(w) = caraab$.

Clearly, two words u and v are conjugate if and only if $M(u) = M(v)$. In particular, $BWT(u) = BWT(v)$. In order to make the transformation injective, the position of the input word in the matrix is added to the transform. If $u = v^m$ for some integer m and $BWT(v) = a_0 a_1 \cdots a_{n-1}$, the $BWT(u) = a_0^m a_1^m \cdots a_{n-1}^m$. In fact, the matrix $M(u)$ has every row repeated m times and every column duplicated m times.

The Burrows-Wheeler Transform is reversible: given $x = BWT(w)$ and an index i, it is possible to recover w. To do this, one first recovers the first column of $M(w)$ by ordering lexicographically the letters of the word $BWT(w)$. Next, one defines a permutation τ on the set $\{0, \ldots, n-1\}$ that maps a position in the first column of $M(w)$ to the corresponding position in x. This permutation gives the word w, when started in the position i.

Example 44. Consider $x = caraab$, and let us compute w such that $BWT(w) = x$. The matrix $M(w)$ has the form

$$M(w) = \begin{bmatrix} a & \cdots & c \\ a & \cdots & a \\ a & \cdots & r \\ b & \cdots & a \\ c & \cdots & a \\ r & \cdots & b \end{bmatrix}$$

The correspondence τ is

$$\tau = \begin{pmatrix} 0\,1\,2\,3\,4\,5 \\ 1\,3\,4\,5\,0\,2 \end{pmatrix} = \begin{pmatrix} 1\,3\,5\,4\,0 \end{pmatrix}$$

Thus

$$\begin{array}{c} 1\ 3\ 5\ 2\ 4\ 0 \\ w = a\ b\ r\ a\ c\ a \end{array}$$

The main observation concerning the relation with Sturmian words is the following remarkable theorem. Recall that a binary word is strongly balanced if and only if it is a conjugate of a standard Sturmian word.

Theorem 45. [66] *A word w over $\{a, b\}$, with $a < b$ is the power of a strongly balanced word if and only if its Burrows-Wheeler Transform is of the form $b^q a^p$. Moreover, in the matrix $M(w)$, each row is obtained from the preceding by replacing a factor ab by a factor ba, and all columns also are conjugates.*

Example 46. Consider the strongly balanced word $abaabab$. The matrix is

$$M(abaabab) = \begin{bmatrix} a\ a\ b\ a\ b\ a\ b \\ a\ b\ a\ a\ b\ a\ b \\ a\ b\ a\ b\ a\ a\ b \\ a\ b\ a\ b\ a\ b\ a \\ b\ a\ a\ b\ a\ b\ a \\ b\ a\ b\ a\ a\ b\ a \\ b\ a\ b\ a\ b\ a\ a \end{bmatrix} \tag{5}$$

Observe that the first (last) row is the lower (upper) Christoffel word, and these rows are composed of the central word bordered by a, b and b, a respectively.

The matrix $M(w)$ defined for the Burrows-Wheeler transform has also been considered in [43] in the process of giving characterizations of strongly balanced binary words. Denote by $P(w)$ the matrix of partial sums of $M(w)$ where $P(w)_{i,j}$ is defined to be the number of b in the prefix of length j of the ith row in $M(w)$. For instance, the matrix in (5) has the matrix of partial sums.

$$P(abaabab) = \begin{bmatrix} 0 & 0 & 1 & 1 & 2 & 2 & 3 \\ 0 & 1 & 1 & 1 & 2 & 2 & 3 \\ 0 & 1 & 1 & 2 & 2 & 2 & 3 \\ 0 & 1 & 1 & 2 & 2 & 3 & 3 \\ 1 & 1 & 1 & 2 & 2 & 3 & 3 \\ 1 & 1 & 2 & 2 & 2 & 3 & 3 \\ 1 & 1 & 2 & 2 & 3 & 3 & 3 \end{bmatrix}$$

They prove the following

Theorem 47. [43] *A word w over $\{a, b\}$, with $a < b$ is the power of a strongly balanced word if and only if every column in the matrix of partial sum is increasing when read from top to bottom.*

Let us mention briefly the connection of the Burrows-Wheeler and the Gessel-Reutenauer transformation [63]. The Burrows-Wheeler transformation is the inverse of the Gessel-Reutenauer transformation. Define the *standardization* associated to a word $w = a_1 \cdots a_n$ over an ordered alphabet A as the permutation σ given by

$$\sigma(i) < \sigma(j) \quad \text{iff} \quad a_i < a_j \text{ or } (a_i = a_j \text{ and } i < j)$$

Example 48. Consider the word $ccbbbcacaaabba$. After a lexicographic sort, the symbols a are at positions 1–5, symbols b at positions 6–10. The symbols c appear in position 11 to 14. This gives the permutation.

$$\begin{pmatrix} 1 & 2 & 3 & 4 & 5 & 6 & 7 & 8 & 9 & 10 & 11 & 12 & 13 & 14 \\ c & c & b & b & b & c & a & c & a & a & a & b & b & a \\ 11 & 12 & 6 & 7 & 8 & 13 & 1 & 14 & 2 & 3 & 4 & 9 & 10 & 5 \end{pmatrix}$$

After cycle decomposition, one gets

$$(1\ 11\ 4\ 7)\ (2\ 12\ 9)\ (3\ 6\ 13\ 10)\ (5\ 8\ 14)$$
$$c\quad a\quad b\quad a\qquad c\quad b\quad a\qquad b\ c\quad b\quad a\qquad b\ c\ a$$

The result is $(caba)(cba)(bcba)(bca)$

Theorem 49. [63] *The standardization σ induces a bijection between all words over A and the family of multisets of conjugacy classes of primitive words over A.*

Define a new order on finite order on words by

$$u \preceq v \quad \text{if and only if} \quad u^\omega < v^\omega \text{ or } (u^\omega = v^\omega \text{ and } |u| \le |v|)$$

For example, $aba \prec ab$ because $abaaba \cdots < ababab \cdots$. Recovering the word w from its decomposition S into conjugacy classes is done as follows: One sorts the conjugates of words in S by $\prec$. Then the word w is the sequence of last letters in this table.

Example 50. Consider the set $S = \{caba, bcba, bca, cba\}$. The conjugates of all words in S are ordered with respect to the new order $\prec$. This gives the sequence $(abac, abc, abcb, acab, acb, babc, baca, bac, bca, bcba, caba, cab, cbab, cba)$. The word composed of the last letters of the words in this sequence is $ccbbbcacaaabba$.

Conversely, to get the decomposition S from w, one sorts the word w alphabetically, then computes the letter-correspondence permutation and then outputs the permutation in cycle form, and computes the multiset.

Example 51. Starting with $ccbbbcacaaabba$, one gets the table

$$\begin{pmatrix} a & a & a & a & a & b & b & b & b & b & b & c & c & c & c \\ 1 & 2 & 3 & 4 & 5 & 6 & 7 & 8 & 9 & 10 & 11 & 12 & 13 & 14 \\ 7 & 9 & 10 & 11 & 14 & 3 & 4 & 5 & 12 & 13 & 1 & 2 & 6 & 8 \\ c & c & b & b & b & c & a & c & a & a & a & b & b & a \end{pmatrix}$$

In cycle form, one gets

$$(1\ 7\ 4\ 11)\ (2\ 9\ 12)\ (3\ 10\ 13\ 6)\ (5\ 14\ 8)$$
$$a\ b\ a\ c \qquad a\ b\ c \qquad a\ b\ c\ b \qquad a\ c\ b$$

The output is the set $S = \{caba, bca, bcba, cba\}$.

8   Sturmian Graphs

Given a standard or a central Sturmian word it appears interesting to consider, in this special case, some well known constructs, such as the (compacted) suffix tree or the suffix automaton (also called DAWG for directed acyclic word graph). A compacted version of the minimal suffix automaton has been considered by [67] for the Fibonacci word and by [68] for arbitrary central words.

The CDWAG (compact directed acyclic word graph) $G(w)$ of a word w is the minimal automaton recognizing the set of suffixes of w, after removing non-final states with out-degree 1.

The terminology DWAG stems from [69]. See also [70].

Example 52. For $w = abaababaaba$, the automaton $G(w)$ (all states are final) is given in Figure 4.

Any CDAWG is homogeneous, that is all edges leading to a state have the same label. For the description of the method of construction, we use $u[d]$ to

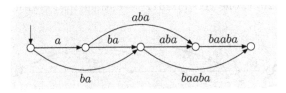

Fig. 4. The automaton $G(abaababaaba)$

denote the reversal of the standard word with directive sequence d. Thus $u[21] = abaa$ because the standard word produced by $(2,1)$ is $aaba$. We write $c[d]$ for the central word produced by the directive sequence d. We use the identity

$$c[d_0 d_1 \cdots d_n 1] = u[\varepsilon]^{d_0} u[d_0]^{d_1} u[d_0 d_1]^{d_2} \cdots u[d_0 d_1 \cdots d_{n-1}]^{d_n}.$$

The CDAWG of a central word c with directive sequence d is constructed by induction. The method goes as follows. Set $d = d'\delta 1$.

1. if $\delta \neq 1$, repeat the last edge of the graph of $d'(\delta - 1)1$.
2. otherwise (that is d ends with 11), set $d = d''\delta'11$, take the graph of $d'1$, add a new state and $1 + \delta'$ edges to this state. The common label of these fresh edges is $u[d'']$.

Example 53. In order to compute the graph of 12311, we start with $d = 11$, $c[11] = a$, and the graph

$$\circ \xrightarrow{\;a\;} \circ$$

Then, using the second rule with $d = 111$, $c[111] = a|ba$, we get

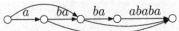

Now the first rule is applied for $d = 121$, $c[121] = a|ba|ba$. This gives

For $d = 1211$ and $c[1211] = a|ba|ba|ababa$, the second rule gives

For $d = 1221$, and setting $z = ababa$, one gets $c[1221] = a|ba|ba|z|z$ and the graph is

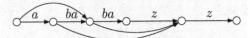

For $d = 1231$, one has $c[1231] = a|ba|ba|z|z|z$ and

Finally, for $d = 12311$, one gets $c[12311] = a|ba|ba|z|z|z|t$ with $t = bazzz$ and the graph

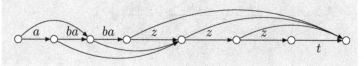

The length of the central word c defined by $d = (d_0, d_1, \ldots, d_k)$ is $|\ell_k| - 2$, where $\ell_n = |s_n| - 2$ and $\ell_{-1} = \ell_0 = 1$, $\ell_{n+1} = d_n \ell_n + \ell_{n-1}$. Let $H(c)$ be the graph obtained from the $G(c)$ by replacing each label by its length. Then $H(c)$ *counts from* 0 *to* $|c|$ in the following sense: each integer h with $0 \leq h \leq |c|$ is the sum of the weights of exactly one path in $H(c)$ starting at the initial state. In

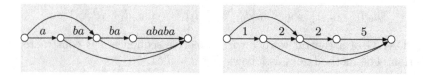

Fig. 5. The CDAWG for 1211 and the corresponding counting graph

other words, the set of weights in $H(c)$ is complete and unambiguous base for representing integer up to $|c|$, provided the representation is a path in the graph. For example, the graph on the right of Figure 5 counts up to 10.

Problem 54. What is the minimal size of a graph with out-degree at most 2 counting from 0 to n?

If the size of the labels increase exponentially, like for the Fibonacci word, then the size is $O(\log n)$. It is conjectured that the bound $O(\log n)$ always holds. This is related to the following number-theoretic conjecture (see [68] for details).

Conjecture 55 (Zaremba). There exists an integer K such that for all positive m, there exists some $i \perp m$, $i < m$ such that all partial quotients in the continued fraction expansions of i/m are bounded by K.

Acknowledgments

I thank Amy Glen, Aldo de Luca for their helpful comments, and Alessandro De Luca for sending me several preprints.

References

1. Klette, R., Rosenfeld, A.: Digital straightness—a review. Discrete Appl. Math. 139, 197–230 (2004)
2. Morse, M., Hedlund, G.A.: Symbolic dynamics II. Sturmian trajectories. Amer. J. Math. 62, 1–42 (1940)

3. Berstel, J., Boasson, L., Carton, O., Fagnot, I.: A first investigation of Sturmian trees. In: Thomas, W., Weil, P. (eds.) STACS 2007. LNCS, vol. 4393, pp. 73–84. Springer, Heidelberg (2007)
4. de Luca, A., De Luca, A.: Pseudopalindrome closure operators in free monoids. Theoret. Comput. Sci. 362(1-3), 282–300 (2006)
5. Droubay, X., Justin, J., Pirillo, G.: Epi-Sturmian words and some constructions of de Luca and Rauzy. Theoret. Comput. Sci. 255(1-2), 539–553 (2001)
6. Justin, J., Pirillo, G.: On a characteristic property of Arnoux-Rauzy sequences. Theor. Inform. Appl. 36(4), 385–388 (2002)
7. Justin, J., Pirillo, G.: Episturmian words and episturmian morphisms. Theoret. Comput. Sci. 276(1-2), 281–313 (2002)
8. Justin, J.: Episturmian words and morphisms (results and conjectures). In: Crapo, H., Senato, D. (eds.) Algebraic Combinatorics and Computer Science, pp. 533–539. Springer, Heidelberg (2001)
9. Arnoux, P., Rauzy, G.: Représentation géométrique de suites de complexité $2n+1$. Bull. Soc. Math. France 119, 199–215 (1991)
10. Coven, E.M., Hedlund, G.A.: Sequences with minimal block growth. Math. Systems Theory 7, 138–153 (1973)
11. Allouche, J.P., Baake, M., Cassaigne, J., Damanik, D.: Palindrome complexity. Theoret. Comput. Sci. 292(1), 9–31 (2003)
12. Baláži, P., Masáková, Z., Pelantová, E.: Factor versus palindromic complexity of uniformly recurrent infinite words. Theoret. Comput. Sci. 380, 266–275 (2007)
13. Droubay, X., Pirillo, G.: Palindromes and Sturmian words. Theoret. Comput. Sci. 223(1-2), 73–85 (1999)
14. Damanik, D., Zamboni, L.Q.: Combinatorial properties of Arnoux-Rauzy subshifts and applications to Schrödinger operators. Rev. Math. Phys. 15(7), 745–763 (2003)
15. Avgustinovich, S.V., Fon-Der-Flaas, D.G., Frid, A.E.: Arithmetical complexity of infinite words. In: Words, Languages and Combinatorics. Proc. 3rd Conf. Words, Languages and Combinatorics, Kyoto, March 2000, vol. III, pp. 51–62. World Scientific, Singapore (2003)
16. Cassaigne, J., Frid, A.E.: On the arithmetical complexity of Sturmian words. Theoret. Comput. Sci. 380, 304–316 (2007)
17. Avgustinovich, S.V., Cassaigne, J., Frid, A.E.: Sequences of low arithmetical complexity. Theor. Inform. Appl. 40(4), 569–582 (2006)
18. Kamae, T., Zamboni, L.Q.: Maximal pattern complexity for discrete systems. Ergodic Theory Dynam. Systems 22(4), 1201–1214 (2002)
19. Kamae, T., Rao, H., Tan, B., Xue, Y.M.: Language structure of pattern Sturmian words. Discrete Math. 306(15), 1651–1668 (2006)
20. Kamae, T., Rao, H.: Maximal pattern complexity of words over l letters. European J. Combin. 27(1), 125–137 (2006)
21. Nakashima, I., Tamura, J.I., Yasutomi, S.I.: Modified complexity and ∗-Sturmian word. Proc. Japan Acad. Ser. A Math. Sci. 75(3), 26–28 (1999)
22. Nakashima, I., Tamura, J.I., Yasutomi, S.I.: ∗-Sturmian words and complexity. J. Theor. Nombres Bordeaux 15(3), 767–804 (2003)
23. de Luca, A.: Sturmian words: structure, combinatorics, and their arithmetics. Theoret. Comput. Sci. 183(1), 45–82 (1997)
24. de Luca, A.: Combinatorics of standard Sturmian words. In: Mycielski, J., Rozenberg, G., Salomaa, A. (eds.) Structures in Logic and Computer Science. LNCS, vol. 1261, pp. 249–267. Springer, Heidelberg (1997)
25. Risley, R., Zamboni, L.Q.: A generalization of Sturmian sequences: combinatorial structure and transcendence. Acta Arith. 95, 167–184 (2000)

26. Glen, A.: Powers in a class of ⊣-strict standard episturmian words. Theoret. Comput. Sci. 380, 330–354 (2007)
27. Tan, B., Wen, Z.Y.: Some properties of the Tribonacci sequence. European J. Combin. (2007)
28. Chekhova, N., Hubert, P., Messaoudi, A.: Propriétés combinatoires, ergodiques et arithmétiques de la substitution de Tribonacci. J. Theor. Nombres Bordeaux 13(2), 371–394 (2001)
29. Lothaire, M.: Applied Combinatorics on Words. Encyclopedia of Mathematics and its Applications, vol. 105. Cambridge University Press, Cambridge (2005)
30. Series, C.: The geometry of Markoff numbers. Math. Intelligencer 7(3), 20–29 (1985)
31. Berthé, V., Ei, H., Ito, S., Rao, H.: Invertible substitutions and Sturmian words: an application to Rauzy fractals. Theor. Inform. Appl. (to appear, 2007)
32. Lothaire, M.: Algebraic Combinatorics on Words. Encyclopedia of Mathematics and its Applications, vol. 90. Cambridge University Press, Cambridge (2002)
33. Pytheas Fogg, N.: Substitutions in dynamics, arithmetics and combinatorics. In: Berthé, V., Ferenczi, S., Mauduit, C., Siegel, A. (eds.) Lecture Notes in Mathematics, vol. 1794, Springer, Heidelberg (2002)
34. Christoffel, E.B.: Observatio arithmetica. Annali di Mathematica 6, 145–152 (1875)
35. Pirillo, G.: A curious characteristic property of standard Sturmian words. In: Crapo, H., Senato, D. (eds.) Algebraic Combinatorics and Computer Science. A tribute to Gian-Carlo Rota., pp. 541–546. Springer, Heidelberg (2001)
36. de Luca, A., Mignosi, F.: Some combinatorial properties of Sturmian words. Theoret. Comput. Sci. 136(2), 361–385 (1994)
37. Borel, J.P., Reutenauer, C.: Palindromic factors of billiard words. Theoret. Comput. Sci. 340(2), 334–348 (2005)
38. de Luca, A., De Luca, A.: Combinatorial properties of Sturmian palindromes. Internat. J. Found. Comput. Sci. 17(3), 557–573 (2006)
39. Carpi, A., de Luca, A.: Codes of central Sturmian words. Theoret. Comput. Sci. 340(2), 220–239 (2005)
40. Berstel, J., de Luca, A.: Sturmian words, Lyndon words and trees. Theoret. Comput. Sci. 178(1-2), 171–203 (1997)
41. Berthé, V., de Luca, A., Reutenauer, C.: On an involution of Christoffel words and Sturmian morphisms. In: European J. Combinatorics (in press, 2007)
42. Chuan, W.F.: Moments of conjugacy classes of binary words. Theoret. Comput. Sci. 310(1-3), 273–285 (2004)
43. Jenkinson, O., Zamboni, L.Q.: Characterisations of balanced words via orderings. Theoret. Comput. Sci. 310(1-3), 247–271 (2004)
44. de Luca, A., De Luca, A.: Some characterizations of finite Sturmian words. Theoret. Comput. Sci. 356(1-2), 118–125 (2006)
45. Fagnot, I., Vuillon, L.: Generalized balances in Sturmian words. Discrete Appl. Math. 121(1-3), 83–101 (2002)
46. Cassaigne, J., Ferenczi, S., Zamboni, L.Q.: Imbalances in Arnoux-Rauzy sequences. Ann. Inst. Fourier (Grenoble) 50(4), 1265–1276 (2000)
47. Lipatov, E.P.: A classification of binary collections and properties of homogeneity classes. Problemy Kibernet 39, 67–84 (1982)
48. Mignosi, F.: On the number of factors of Sturmian words. Theoret. Comput. Sci. 82(1), 71–84 (1991)
49. Berstel, J., Pocchiola, M.: A geometric proof of the enumeration formula for Sturmian words. Internat. J. Algebra Comput. 3(3), 349–355 (1993)

50. Berstel, J., Pocchiola, M.: Random generation of finite Sturmian words. In: Proceedings of the 5th Conference on Formal Power Series and Algebraic Combinatorics (Florence, 1993), vol. 153, pp. 29–39 (1996)

51. Berenstein, C.A., Lavine, D.: On the number of digital straight line segments. IEEE Trans. Pattern Anal. Mach. Intell. 10(6), 880–887 (1988)

52. Koplowitz, J., Lindenbaum, M., Bruckstein, A.M.: The number of digital straight lines on an $n \times n$ grid. IEEE Transactions on Information Theory 36(1), 192–197 (1990)

53. Heinis, A.: On low-complexity bi-infinite words and their factors. J. Theor. Nombres Bordeaux 13(2), 421–442 (2001)

54. Tarannikov, Y.: On the bounds for the number of ℓ-balanced words. Technical report, Mech. & Math. Department, Moscow State University (2007)

55. Mignosi, F., Zamboni, L.Q.: On the number of Arnoux-Rauzy words. Boolean Calculus of Differences 101(2), 121–129 (2002)

56. Paquin, G., Vuillon, L.: A characterization of balanced episturmian sequences. Electronic J. Combinatorics 14(1) R33, pages 12 (2007)

57. Glen, A., Justin, J., Pirillo, G.: Characterizations of finite and infinite episturmian words via lexicographic orderings. European Journal of Combinatorics (2007)

58. Pirillo, G.: Morse and Hedlund's skew Sturmian words revisited. Annals Combinatorics (to appear, 2007)

59. Gan, S.: Sturmian sequences and the lexicographic world. Proc. Amer. Math. Soc. 129, electronic, 1445–1451 (2001)

60. Glen, A.: A characterization of fine words over a finite alphabet. Theoret. Comput. Sci. CANT conference, Liege, Belgium, May 8-19, 2007, 8–19 (to appear, 2007)

61. Pirillo, G.: Inequalities characterizing standard Sturmian and episturmian words. Theoret. Comput. Sci. 341, 276–292 (2005)

62. Burrows, M., Wheeler, D.J.: A block sorting data compression algorithm. Technical report, Digital System Research Center (1994)

63. Gessel, I., Reutenauer, C.: Counting permutations with given cycle structure and descent set. J. Comb. Theory A 64, 189–215 (1993)

64. Crochemore, M., Désarménien, J., Perrin, D.: A note on the Burrows-Wheeler transformation. Theoret. Comput. Sci. 332, 567–572 (2005)

65. Mantaci, S., Restivo, A., Sciortino, M.: Burrows Wheeler transform and Sturmian words. Inform. Proc. Letters 86, 241–246 (2003)

66. Mantaci, S., Restivo, A., Rosone, G., Sciortino, M.: An extension of the Burrows Wheeler transform. Theoret. Comput. Sci. (2007)

67. Rytter, W.: The structure of subword graphs and suffix trees of Fibonacci words. Theoret. Comput. Sci. 363(2), 211–223 (2006)

68. Epifanio, C., Mignosi, F., Shallit, J., Venturini, I.: On Sturmian graphs. Discrete Appl. Math 155, 1014–1030 (2007)

69. Blumer, A., Blumer, J.A., Haussler, D., Ehrenfeucht, A., Chen, M.T., Seiferas, J.I.: The smallest automaton recognizing the subwords of a text. Theoret. Comput. Sci. 40(1), 31–55 (1985) (Special issue: Eleventh international colloquium on automata, languages and programming, Antwerp, (1984))

70. Crochemore, M., Rytter, W.: Jewels of stringology. World Scientific Publishing Co. Inc, River Edge, NJ (2003)

From Tree-Based Generators to Delegation Networks

Frank Drewes

Department of Computing Science,
Umeå University, S-901 87 Umeå (Sweden)
drewes@cs.umu.se

Abstract. The first part of this paper is a brief survey on tree-based generators, including some typical examples taken from the fields of string, tree, graph, and picture generation. In the second part, an extension of the tree-based generator called delegation network is proposed. Intuitively, a delegation network is a network of tree-based generators that can "delegate" subtasks to each other. In this way, different types of tree-based generators can be combined to generate complex objects.

1 Introduction

The theory of tree languages and tree transformations is an important and lively field of theoretical computer science [GS84, NP92, GS97, FV98, CDG+02]. It is concerned with formal devices that generate, recognize or transform trees. A tree in this sense is a term, i.e., a formal expression composed of abstract operation symbols. The usefulness of devices dealing with such trees is to a large extent based on the fact that trees can be interpreted by choosing a domain $\mathbb{A}$ and associating an operation on $\mathbb{A}$ (of the appropriate arity) with each symbol. Thus, given such an interpretation, also called algebra, every tree denotes an element of $\mathbb{A}$. This means that a device generating trees provides the syntactic basis for a *tree-based generator* – a system consisting of the tree generator and an interpretation, that generates elements of $\mathbb{A}$:

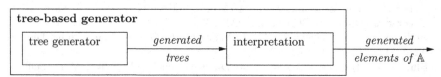

Similarly, a device that transforms trees, together with two interpretations, can be used to compute a function from $\mathbb{A}$ to $\mathbb{B}$. The tree transformation can then be seen as a symbolic algorithm [Eng80].

In this paper, we will focus on generation rather than transformation. We will first recall the formal notions needed for the definition of tree-based generators, and give some examples from the areas of string, tree, graph, and picture generation. Afterwards, a generalization called delegation network is proposed

S. Bozapalidis and G. Rahonis (Eds.): CAI 2007, LNCS 4728, pp. 48–72, 2007.
© Springer-Verlag Berlin Heidelberg 2007

and illustrated by means of an example. A first attempt to formalize the notion of delegation networks, and to prove some of their basic properties, was made in [Dre07]. However, as pointed out by Engelfriet[1], the evaluation of trees with respect to nondeterministic algebras is not defined correctly in this paper, and not all results (in particular the "Mezei&Wright-like" Theorem 5.9) hold for the nondeterministic case. Therefore, the second part of the present paper tries to formalize the notion of delegation networks in a more appropriate way, thus laying the basis for future work in this area.

The purpose of delegation networks is to be able to combine several tree-based generators (possibly of different types) with each other. Intuitively, a delegation network consists of a finite number of tree-based generators that may "delegate" parts of the generation process to each other. Roughly speaking, every delegating generator in the network is associated with a symbol $\mathbf{g} \colon \mathbb{A}_1 \times \cdots \times \mathbb{A}_k \to \mathbb{A}$ from a certain signature. The semantics of the delegation network interprets $\mathbf{g}$ as a function $g \colon \mathbb{A}_1 \times \cdots \times \mathbb{A}_k \to \mathbb{A}$.[2] Delegation means that another generator from the network, associated with a symbol $\mathbf{g}'$, can generate trees that contain occurrences of $\mathbf{g}$. If such a tree is evaluated, $\mathbf{g}$ is interpreted as g. Thus, the interpretation g' of $\mathbf{g}'$ depends on g – and as delegation networks can be cyclic, g may also depend on g'. For this reason, we choose a least fixed-point semantics for delegation networks (Definition 5).

The structure of this paper is thus as follows. In the next section, the basic notions regarding tree-based generators are recalled. In Section 3, the tree-based versions of some well-known grammatical devices are discussed. Section 4 introduces delegation networks, which are illustrated by means of an example in Section 5. Section 6 concludes the paper.

2 Tree-Based Generators

Before recalling the notion of tree-based generators, let us summarize some standard notions and notation. Throughout this paper, $\mathbb{N}$ denotes the set of natural numbers (including zero). For $n \in \mathbb{N}$, the set $\{1, \ldots, n\}$ is denoted by $[n]$. The powerset of a set A is denoted by $\wp(A)$. A function f of arity 0 is identified with the constant $f()$.

2.1 Signatures and Trees

Let S be a set of sorts. An S-sorted signature (or just signature) is a finite set Σ of symbols $\mathbf{f}$, each of which has an associated profile $\mathbb{A}_1 \times \cdots \times \mathbb{A}_k \to \mathbb{A}$, where $k \in \mathbb{N}$ and $\mathbb{A}_1, \ldots, \mathbb{A}_k, \mathbb{A} \in S$. To indicate that $\mathbf{f}$ has this profile, we also write $\mathbf{f} \colon \mathbb{A}_1 \times \cdots \times \mathbb{A}_k \to \mathbb{A}$. The number k is also called the rank of $\mathbf{f}$. Symbols of rank 0 are called constant symbols. We write the profile of a constant symbol without the arrow, i.e., $\mathbf{a} \colon \mathbb{A}$, thus saying that its profile is $\mathbb{A}$.

---

[1] Personal communication.

[2] Actually, these are nondeterministic rather than ordinary functions, but this is not important for the moment.

Given a sort A, the set of all *trees of sort A over Σ* is denoted by T_Σ^A. By definition, the sets T_Σ^A are the smallest sets of strings simultaneously satisfying the following conditions:

- For every $a\colon A$ in Σ, the string a is in T_Σ^A.
- For all $f\colon A_1 \times \cdots \times A_k \to A$ in Σ ($k > 0$) and all $t_1 \in T_\Sigma^{A_1}, \ldots, t_k \in T_\Sigma^{A_k}$, the string $f[t_1, \ldots, t_k]$ is in T_Σ^A. (Here, the square brackets and the comma are assumed to be special symbols not in Σ.)

The set T_Σ of all trees over Σ is given by $T_\Sigma = \bigcup_{A \in S} T_\Sigma^A$.

2.2 Algebras and Evaluation

Given an S-sorted signature Σ, a *Σ-algebra* (or just *algebra*) is a pair $\mathcal{A} = (dom, \sigma)$. Here, dom is a *domain mapping for S* – a function assigning to every sort $A \in S$ a set $dom(A)$ called its *domain*, and σ is the *interpretation of symbols* – a function that assigns to every function symbol $f\colon A_1 \times \cdots \times A_k \to A$ in Σ a corresponding function $\sigma(f)\colon dom(A_1) \times \cdots \times dom(A_k) \to dom(A)$. The function σ is also called a *(Σ, dom)-interpretation*.

Throughout the rest of the paper, we will use the following typographical conventions in connection with algebras, unless there is a risk of confusion: the domain $dom(A)$ assigned to a sort A is denoted by $\mathbb{A}$, and the interpretation of a symbol f is denoted by f. Using these conventions, defining an algebra means to associate a domain $\mathbb{A}$ with every sort A, and a function $f\colon \mathbb{A}_1 \times \cdots \times \mathbb{A}_k \to \mathbb{A}$ with every function symbol $f\colon A_1 \times \cdots \times A_k \to A$ in Σ.

Given a Σ-algebra $\mathcal{A}$, trees over Σ can be viewed as expressions that can be evaluated. This evaluation, denoted by $val_\mathcal{A}$, is defined recursively, as one would expect: $val_\mathcal{A}(t) = f(val_\mathcal{A}(t_1), \ldots, val_\mathcal{A}(t_k))$ for every tree $t = f[t_1, \ldots, t_k] \in T_\Sigma$. Note that the definition of T_Σ makes sure that $val_\mathcal{A}(t)$ is well defined. In the following, we may write $val(t)$ if $\mathcal{A}$ is understood.

Especially in examples, we shall frequently work with algebras over *unsorted signatures*, which is an S-sorted signature such that S is a singleton $\{A\}$. In this case, the notation $f\colon A^k \to A$ may be abbreviated as $f^{(k)}$. The (unique) domain of an algebra $\mathcal{A}$ over Σ is denoted by $dom(\mathcal{A})$.

2.3 Tree-Based Generators

A *tree language* is a subset of T_Σ^A, for some S-sorted signature Σ and a sort $A \in S$. A formal device γ defining a tree language $L(\gamma)$ is called a *tree generator*. If $L(\gamma) \subseteq T_\Sigma$, then Σ is called the *output signature* of γ. (Thus, to be picky, one should rather speak of *an* output signature of γ.)

Now, a *tree-based generator* is a pair $G = (\gamma, \mathcal{A})$ that consists of a tree generator γ and a Σ-algebra $\mathcal{A}$, where Σ is the output signature of γ. The *language generated by G* is given by $L(G) = \{val(t) \mid t \in L(\gamma)\}$.

2.4 TREEBAG

One of the advantages of the notion of tree-based generators is that it gives rise to a flexible implementation in a rather straightforward manner. The system TREEBAG [Dre06, Chapter 8] is such a system. It allows its user to interactively load instances of a variety of tree generators (and tree transducers), algebras, and so-called displays, and to establish input-output relations between them. In this way, tree-based generators can be assembled, and displays that show the resulting objects on the screen can be attached to them. The pictures in Sections 3.4 and 3.5 have been created in this way. Moreover, the delegation network discussed in Section 5 has been simulated in TREEBAG in order to create the pictures shown.

3 Examples of Tree-Based Generators

We shall now discuss a few typical classes of tree-based generators. All of them are based on tree generators with unsorted output signatures; sorted signatures will become important in the next section.

Let us start with one of the simplest meaningful cases: the tree-based version of string grammars.

3.1 String Generation

Let T be a set of symbols. We denote the set of all strings over T by T^*, and the empty string by ε.

Now, consider an unsorted signature Σ. The Σ-algebra $\mathcal{A}_{\Sigma,T}$, which allows to assemble strings by means of concatenation, is given by

- $dom(\mathcal{A}_{\Sigma,T}) = T^*$,
- $a = \mathbf{a}$ for every constant symbol in Σ which belongs to T, and
- $f(u_1, \ldots, u_k) = u_1 \cdots u_k$ for all other symbols $\mathbf{f}^{(k)} \in \Sigma$ and all strings $u_1, \ldots, u_k \in T^*$. Thus, all symbols not in T are interpreted as concatenation operators of the appropriate arity. In particular, $f = \varepsilon$ for all $\mathbf{f}^{(0)} \in \Sigma$ which are not in T.

By definition, a tree-based generator of the form $G = (\gamma, \mathcal{A}_{\Sigma,T})$ generates a string language $L(G) \subseteq T^*$. Several types of string grammars well known from traditional string language theory can be formulated in this way. In fact, historically, this was one of the motivations for developing a theory of tree languages and tree transformations (see, e.g., [Rou70, Tha73]).

As one of the easiest examples, let us see how the context-free grammar can be turned into a tree-based generator. For this, we use a tree generator known as regular tree grammar, which is defined as follows. (As a minor extension of the usual definition found in the literature, we define regular tree grammars generating trees over sorted signatures.)

Definition 1 (regular tree grammar). *Let S be a set of sorts. A regular tree grammar is a tuple $\gamma = (\Xi, \Sigma, R, \xi_{\text{ini}})$, where*

- *Ξ is an S-sorted signature of constant symbols called* nonterminals,
- *Σ is an S-sorted signature of output symbols which is disjoint with Ξ,*
- *R is a finite set of* rules *$\xi \to r$, where $\xi \in \Xi$ and $r \in T_{\Sigma \cup \Xi}$ are of the same sort, and*
- *$\xi_{\text{ini}} \in \Xi$ is the* initial nonterminal.

A derivation step $s \to_\gamma t$ (or simply $s \to t$, if γ is understood) consists of two trees $s, t \in T_{\Sigma \cup \Xi}$ such that t is obtained from s by replacing a single occurrence of a nonterminal ξ with r, for some rule $\xi \to r$ in R. The tree language generated by γ, called a regular tree language, *is*

$$L(\gamma) = \{t \in T_\Sigma \mid \xi_{\text{ini}} \to^* t\},$$

where $\to^*$ denotes the transitive and reflexive closure of the relation $\to$.

Now, let $G = (\Xi, T, R, \xi_{\text{ini}})$ be a context-free grammar consisting, as usual, of finite sets Ξ and T of nonterminal and terminal symbols, resp., a set R of context-free rules, and an initial nonterminal $\xi_{\text{ini}} \in \Xi$. We turn G into an equivalent tree-based generator, as follows.

First, we need a suitable unsorted signature Σ. For each rule $r = (\xi \to u)$ in R, let Σ contain a unique symbol $\mathbf{r} \notin T$ of rank $|u|$ (where $|u|$ denotes the length of u). In addition, Σ contains all symbols in T as symbols of rank 0.

Second, let $\gamma = (\Xi, \Sigma, R', \xi_{\text{ini}})$ be the regular tree grammar obtained by turning every rule ($r = \xi \to \mathbf{a}_1 \cdots \mathbf{a}_k$) in R (where $\mathbf{a}_1, \ldots, \mathbf{a}_k \in T$) into a rule $\xi \to \mathbf{r}[\mathbf{a}_1, \ldots, \mathbf{a}_k]$ in R'. It is an easy exercise to show that the tree-based generator $G' = (\gamma, \mathcal{A}_{\Sigma,T})$ satisfies $L(G') = L(G)$. Every tree t generated by γ', where $val(t) = u$, corresponds to an abstract syntax tree of u with respect to G.

Of course, there are several natural ways to choose Σ in the previous construction. For example, instead of including $\mathbf{r}$ for every rule $r = (\xi \to u)$, one could choose a symbol $\xi_{[k]}$ of rank $k = |u|$. Then, the trees generated by γ would correspond to the derivation trees of G. Another possibility is to include only two symbols in addition to the symbols in T, namely $\circ^{(2)}$ and $\epsilon^{(0)}$, and to use rules of the form $\xi \to \circ[\mathbf{a}_1, \circ[\cdots, \circ[\mathbf{a}_k, \epsilon]\cdots]]$ in R'.

It is also easy to show that the construction can be reversed: for every tree-based generator of the form $(\gamma, \mathcal{A}_{\Sigma,T})$, where γ is a regular tree grammar, there is a context-free grammar generating the same language. Thus, a characterization of the class of context-free languages in terms of languages generated by tree-based generators has been obtained.

3.2   Tree Generation

Trivially, trees can be generated by tree-based generators. For this, just use a tree-based generator $G = (\gamma, \mathcal{A})$, where $\mathcal{A}$ is the free term algebra over the output signature of γ. In this algebra, the interpretation of symbols is given by

$f(t_1,\ldots,t_k) = \mathtt{f}[t_1,\ldots,t_k]$, which means that *val* is the identity on T_Σ, and $L(G) = L(\gamma)$.

The situation becomes more interesting if $\mathcal{A}$ is not as simple as the free term algebra. An important example for this is the so-called YIELD algebra (or YIELD mapping, see [Mai74, Eng80, ES77, ES78, FV98, Dre06]), which formalizes the construction of trees using variable substitution. We will only discuss the variant dealing with trees over an unsorted signature. The extension to arbitrary S-sorted signatures is straightforward, but technical.

Let $X = \{x_1, x_2, \ldots\}$ be a countably infinite set of special symbols of rank 0, called variables. For $l \in \mathbb{N}$, we let X_l denote $\{x_1, \ldots, x_l\}$. For $t, t_1, \ldots, t_l \in T_{\Sigma \cup X}$,[3] we let $t[\![t_1, \ldots, t_l]\!]$ denote the tree obtained from t by simultaneously replacing all occurrences of variables x_i by t_i $(i \in [l])$.

Definition 2 (YIELD algebra). *Let Σ be an unsorted signature and let $l \geq \max\{k \in \mathbb{N} \mid \mathtt{f}^{(k)} \in \Sigma\}$. The (unsorted) derived signature $\Sigma_{Y,l}$ is given by*

$$\Sigma_{Y,l} = \{\mathtt{subst}^{(l+1)}\} \cup \{\mathtt{f}^{\langle k \rangle \, (0)} \mid \mathtt{f}^{(k)} \in \Sigma \cup X_l\}.$$

The YIELD algebra (with respect to Σ and l) is the $\Sigma_{Y,l}$-algebra $\mathcal{Y}$ such that

- *$dom(\mathcal{Y}) = T_{\Sigma \cup X}$,*
- *$\mathtt{subst}(t, t_1, \ldots, t_l) = t[\![t_1, \ldots, t_l]\!]$ for all $t, t_1 \ldots, t_l \in T_{\Sigma \cup X}$, and*
- *$\mathtt{f}^{\langle k \rangle} = f[x_1, \ldots, x_k]$ for all $\mathtt{f}^{(k)} \in \Sigma \cup X_l$.*

Tree-based generators of the form $G = (\gamma, \mathcal{Y})$, where γ is a regular tree grammar, generate the so-called IO context-free tree languages. As an example, let $\Sigma = \{\mathtt{f}^{(2)}, \mathtt{g}^{(2)}, \mathtt{a}^{(0)}\}$. We show how to generate the set of all trees of the form $t[\![t']\!]$, where t is an arbitrary tree over $\{\mathtt{f}^{(2)}, x_1^{(0)}\}$, and t' is a totally balanced tree over $\{\mathtt{g}^{(2)}, \mathtt{a}^{(0)}\}$:

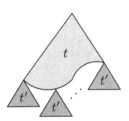

For this, use the regular tree grammar $\gamma = (\{\xi_{\mathrm{ini}}, \xi_{\mathrm{arb}}, \xi_{\mathrm{bal}}\}, \Sigma_{Y,2}, R, \xi_{\mathrm{ini}})$, where R is given as in Table 1. In an IO context-free tree grammar, these rules would be written as

$$
\begin{aligned}
\xi_{\mathrm{ini}} &\to \xi_{\mathrm{arb}}[\xi_{\mathrm{bal}}[\mathtt{a}]], \\
\xi_{\mathrm{arb}}[x_1] &\to \mathtt{f}[\xi_{\mathrm{arb}}[x_1], \xi_{\mathrm{arb}}[x_1]] \mid x_1, \\
\xi_{\mathrm{bal}}[x_1] &\to \xi_{\mathrm{bal}}[\mathtt{f}[x_1, x_1]] \mid x_1.
\end{aligned}
$$

---

[3] We use the notation $T_{\Sigma \cup X}$ to abbreviate $\bigcup_{l \in \mathbb{N}} T_{\Sigma \cup \{x_1,\ldots,x_l\}}$.

Table 1. Rules of a regular tree grammar that, in connection with the YIELD algebra $\mathcal{Y}$, mimics an IO context-free tree grammar, where '|' is used to separate alternatives. We choose $l = 2$ in Definition 2; hence, subst is of rank 3. For simplicity, we write $\mathrm{subst}[t, t_1]$ if the third subtree is uninteresting (i.e., would never be used). The right column shows the rules in a partially evaluated form, obtained by recursively replacing all subterms $t = \mathrm{subst}[\mathbf{h}^{\langle k \rangle}, t_1, \ldots, t_k]$ with $t' = \mathbf{h}[t'_1, \ldots, t'_k]$ (and all $\mathbf{h}^{\langle 0 \rangle}$ with their values $\mathbf{h}$).

Actual rules	Simplified form
$\xi_{\mathrm{ini}} \;\to\; \mathrm{subst}[\xi_{\mathrm{arb}}, \mathrm{subst}[\xi_{\mathrm{bal}}, \mathbf{a}^{\langle 0 \rangle}]]$	$\xi_{\mathrm{ini}} \;\to\; \mathrm{subst}[\xi_{\mathrm{arb}}, \mathrm{subst}[\xi_{\mathrm{bal}}, \mathbf{a}]]$
$\xi_{\mathrm{arb}} \;\to\; \mathrm{subst}[\mathbf{f}^{\langle 2 \rangle}, \xi_{\mathrm{arb}}, \xi_{\mathrm{arb}}] \mid x_1^{\langle 0 \rangle}$	$\xi_{\mathrm{arb}} \;\to\; \mathbf{f}[\xi_{\mathrm{arb}}, \xi_{\mathrm{arb}}] \mid x_1$
$\xi_{\mathrm{bal}} \;\to\; \mathrm{subst}[\xi_{\mathrm{bal}}, \mathrm{subst}[\mathbf{f}^{\langle 2 \rangle}, x_1^{\langle 0 \rangle}, x_1^{\langle 0 \rangle}]] \mid x_1^{\langle 0 \rangle}$	$\xi_{\mathrm{bal}} \;\to\; \mathrm{subst}[\xi_{\mathrm{bal}}, \mathbf{f}[x_1, x_1]] \mid x_1$

3.3 Graph Generation

The tree-based perspective has turned out to be very fruitful in the area of context-free graph grammars. The first paper investigating the generation of context-free graph languages in a tree-based manner was [BC87]; see also the surveys [Cou90, Eng97].

There are two major types of context-free graph languages: those generated by *hyperedge replacement* (HR) and those generated by *node replacement* (NR). Both have equivalent formulations in terms of tree-based generators, where the underlying tree generator is the regular tree grammar. Thus, only the algebras differ. Here, we will consider the HR case.

For simplicity, we restrict ourselves to directed unlabelled graphs. For technical reasons, these graphs are equipped with a number of distinguished nodes, so-called ports. More precisely, a *graph* is a quadruple $H = (V, E, att, port)$ consisting of

- finite sets V and E of *nodes* and *edges*, resp.,
- a mapping $att \colon E \to V^2$ assigning to every edge its attached nodes (i.e., $att(e) = (v, v')$ means that e points from v to v'), and
- a partial mapping $port \colon \mathbb{N} \to V$, the *port labelling*, which is defined on a finite subset $df(port)$ of $\mathbb{N}$.

For $i \in df(port)$, the node $port(i)$ is called the i-port of H. Note that graphs whose port labelling is the totally undefined function can be considered as ordinary graphs without ports. If a graph is of this kind, we may omit the fourth component.

We consider the following two types of operations on graphs.[4]

1. Let $H = (V, E, att, port)$ and $H' = (V', E', att', port')$ be graphs and assume, without loss of generality, that $V \cap V' = \emptyset = E \cap E'$ (otherwise, take

---

[4] Strictly speaking, these operations work on abstract graphs (i.e., isomorphism classes of graphs) rather than concrete ones. Intuitively, this means that isomorphic graphs are considered to be the same.

isomorphic copies). Let $\equiv$ be the equivalence relation on $V \cup V'$ generated by $\{(port(i), port'(i)) \mid i \in df(port) \cap df(port')\}$. Then the *parallel composition* $par(H, H')$ of H and H' is obtained by taking their union and identifying ports with the same label. Formally,

$$par(H, H') = (V'', E \cup E', att'', port''),$$

where

(a) $V'' = \{[v]_\equiv \mid v \in V \cup V'\}$ is the set of equivalence classes of $V \cup V'$ with respect to $\equiv$,

(b) $att''(e) = ([v]_\equiv, [v']_\equiv)$ for $e \in E$ with $att(e) = (v, v')$ or $e \in E'$ with $att'(e) = (v, v')$, and

(c) for all $i \in df(port) \cup df(port')$,

$$port''(i) = \begin{cases} [port(i)]_\equiv & \text{if } i \in df(port) \\ [port'(i)]_\equiv & \text{otherwise.} \end{cases}$$

2. Given a partial function $\rho \colon \mathbb{N} \to \mathbb{N}$ which is defined on a finite subset $df(\rho)$ of $\mathbb{N}$, the *port relabelling* of a graph $H = (V, E, att, port)$ with respect to ρ is given by $rel_\rho(H) = (V, E, att, port \circ \rho)$. Here, $port \circ \rho$ denotes the composition of partial functions, i.e., $(port \circ \rho)(i)$ is defined if $i \in df(\rho)$ and $\rho(i) \in df(port)$, and yields $port(\rho(i))$ in this case.

Now, for an unsorted signature Σ, a Σ-algebra $\mathcal{A}$ is an *HR Σ-algebra* if it has as its domain the set of all graphs, and Σ contains, in addition to a finite number of constant symbols,

- the symbol $\mathbf{par}^{(2)}$, which is interpreted as par, and
- finitely many symbols of the form $\mathbf{rel}_\rho^{(1)}$ (where ρ is as above), each of which is interpreted as rel_ρ.

Remark. For readers who wonder about the definition of NR algebras, it can be mentioned that the graphs in NR algebras are allowed to contain any number of distinct nodes with the same port label, i.e., *port* is turned into a binary relation rather than a partial function. Moreover, let $ports_i(H)$ denote the set of all i-ports of such a graph H. Instead of the operation par, NR algebras contain binary operations $connect_C$, where C is a binary relation on port labels. For graphs H, H', $connect_C(H, H')$ is obtained by taking their disjoint union and adding, for all $(i, j) \in C$ and $(v, v') \in ports_i(H) \times ports_j(H') \cup ports_i(H') \times ports_j(H)$, an edge from v to v'. Thus, this operation connects the disjoint components H and H' with edges according to C.

As an example, we consider an HR context-free graph grammar (i.e., a tree-based generator consisting of a regular tree grammar and an HR algebra) that generates the set of all *wheels* W_n, for $n \geq 1$. Here, $W_n = (\{v_0, v_1, \dots, v_n\}, \{e_1, \dots, e_n, e'_1, \dots, e'_n\}, att)$, where $att(e_i) = (v_0, v_i)$ (the "spokes" of the wheel) and $att(e'_i) = (v_i, v_{(i \bmod n)+1})$ (the "rim"), for $i \in [n]$. For instance, Figure 1 shows W_4.

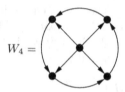

Fig. 1. The wheel with four spokes

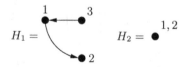

Fig. 2. Graphs to compose wheels of; the numbers indicate the ports, e.g., the bottom node of H_1 is its 2-port

Our grammar assembles such wheels from the graphs shown in Figure 2. It uses two nonterminals, ξ_{ini} and ξ, and the rules

$$\xi_{\text{ini}} \rightarrow \text{rel}_{\square}[\text{par}[\xi, H_2]]$$
$$\xi \rightarrow \text{rel}_{\begin{bmatrix} 1 \mapsto 1 \\ 2 \mapsto 4 \\ 3 \mapsto 3 \end{bmatrix}}[\text{par}[\xi, \text{rel}_{\begin{bmatrix} 2 \mapsto 1 \\ 3 \mapsto 3 \\ 4 \mapsto 2 \end{bmatrix}}[\xi]]]$$
$$\xi \rightarrow H_1.$$

Here, a subscript of the form $\begin{bmatrix} i_1 \mapsto i'_1 \\ \vdots \\ i_k \mapsto i'_k \end{bmatrix}$ denotes the partial function $\rho \colon \mathbb{N} \to \mathbb{N}$ with $df(\rho) = \{i_1, \ldots, i_k\}$ and $\rho(i_j) = i'_j$ for all $j \in [k]$. In the first rule, parallel composition with H_2 just means that the 1-port of the first argument is identified with its 2-port, thus closing the rim, whereas the application of $\text{rel}_{\square}$ removes all port labels. Of course, the second rule could be replaced by the linear rule

$$\xi \rightarrow \text{rel}_{\begin{bmatrix} 1 \mapsto 1 \\ 2 \mapsto 4 \\ 3 \mapsto 3 \end{bmatrix}}[\text{par}[\xi, \text{rel}_{\begin{bmatrix} 2 \mapsto 1 \\ 3 \mapsto 3 \\ 4 \mapsto 2 \end{bmatrix}}[H_1]]]$$

without affecting the generated language. Figure 3 indicates how to evaluate the right-hand side of the second rule if both occurrences of ξ are replaced with H_1.

Using an ET0L tree grammar (see Section 3.4) instead of a regular one, taking the same rules but putting them into separate tables, the set of all wheels with 2^n spokes would be generated, which is neither HR nor NR context free.

3.4 Generation of Line Drawings

Tree-based generation has also turned out to be very useful in the area of picture generation, because several well-known devices generating pictures can be given a

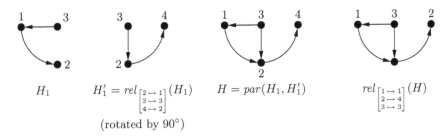

$$H_1 \qquad H_1' = rel_{\begin{smallmatrix}2 \mapsto 1\\3 \mapsto 3\\4 \mapsto 2\end{smallmatrix}}(H_1) \qquad H = par(H_1, H_1') \qquad rel_{\begin{smallmatrix}1 \mapsto 1\\2 \mapsto 4\\3 \mapsto 3\end{smallmatrix}}(H)$$

(rotated by 90°)

Fig. 3. The evaluation of $\mathtt{rel}_{\begin{smallmatrix}1 \mapsto 1\\2 \mapsto 4\\3 \mapsto 3\end{smallmatrix}} [\mathtt{par}[\mathtt{H}_1, \mathtt{rel}_{\begin{smallmatrix}2 \mapsto 1\\3 \mapsto 3\\4 \mapsto 2\end{smallmatrix}} [\mathtt{H}_1]]]$

tree-based definition.[5] Two such devices are chain-code picture grammars and L-systems with turtle interpretation. We will briefly discuss the latter, which can be seen as an extension of the former. In its original definition, the turtle mechanism interprets strings whose symbols are regarded as instructions to a plotter-like device. It originated from the "turtle" of the programming language LOGO [Ad80] and was made popular in the area of grammatical picture generation using L-systems through the book by Prusinkiewicz and Lindenmayer [PL90]; see also the later survey [PHHM97].

Let us say that a *line drawing* (in $\mathbb{R}^2$) is a pair $\Delta = (D, e)$ consisting of a finite set D of straight line segments and an end point $e \in \mathbb{R}^2$. Furthermore, choose two angles α_0 and α. Given an unsorted signature Σ containing

$$\Sigma_{\mathrm{turtle}} = \{F^{(0)}, +^{(1)}, -^{(1)}, \mathrm{enc}^{(1)}\}$$

as a subset, the *turtle Σ-algebra*[6] $\mathcal{A}$ has as its domain the set of all line drawings. The interpretation of symbols in Σ by $\mathcal{A}$ depends on α_0 and α, and is given as follows.

- F is the line drawing consisting of a single line segment extending one unit from the origin into the direction given by α_0. The end point of this line segment is the end point of F.
- The symbols $+$ and $-$ are interpreted as rotation around the origin by α and $-\alpha$ degrees, resp. Here, both the line segments and the end point are rotated.
- The symbol enc is interpreted as *encapsulation*, replacing the end point of the argument by the origin: $enc(D, e) = (D, (0, 0))$.
- Every symbol $f^{(k)} \notin \Sigma_{\mathrm{turtle}}$ is interpreted as the k-ary concatenation of line drawings: $f(\Delta_1, \ldots, \Delta_k) = (\cdots (\Delta_1 \circ \Delta_2) \circ \cdots) \circ \Delta_k$, where $\Delta \circ \Delta'$ is obtained by translating Δ' by the vector given by the end point of Δ and taking the union of the sets of line segments of both. The translated end point of Δ' becomes the end point of the resulting line drawing.

---

[5] See [Dre06] for an extensive treatment of tree-based picture generation.
[6] Slightly simplified, compared to [Dre06].

As the turtle mechanism is usually studied in connection with L-systems, let us recall the definition of ET0L tree grammars, which is the tree grammar version of ET0L systems.

Definition 3 (ET0L tree grammar). *Let S be a set of sorts. An* ET0L *tree grammar is a tuple $\gamma = (\Xi, \Sigma, R, t_0)$ consisting of*

- *(not necessarily disjoint) S-sorted signatures Ξ and Σ of nonterminals (each of rank 0) and output symbols, resp.,*
- *a finite set R of tables $R_1, \ldots, R_n$, each table being a finite set of rules as in the case of regular tree grammars, and*
- *an axiom $t_0 \in T_{\Sigma \cup \Xi}$.*

To guarantee that $\Sigma \cup \Xi$ is a well-defined signature, it is required that each nonterminal occurring in Σ has the same profile in both signatures. Moreover, in each table R_i, every nonterminal is required to occur among the left-hand sides of rules in R_i.

For trees $s, t \in T_{\Sigma \cup \Xi}$, there is a derivation step $s \Rightarrow_\gamma t$ (or just $s \Rightarrow t$) if there is a table R_i such that t can be obtained from s by simultaneously replacing every occurrence of a nonterminal ξ by r, where $\Xi \to r$ is a rule in R_i. The ET0L tree language generated by γ is given by

$$L(\gamma) = \{t \in T_\Sigma \mid t_0 \Rightarrow^* t\}.$$

A tree-based generator consisting of an ET0L tree grammar and a turtle algebra is called an ET0L turtle grammar. Such grammars have been used quite extensively to capture plant architecture by means of grammatical rules. To discuss an example, let $\gamma = (\Xi, \Sigma, \{R_1, R_2\}, \xi_{ini})$, where $\Xi = \{\xi_{ini}, \xi\}$, $\Sigma = \Xi \cup \Sigma_{turtle} \cup \{c_2^{(2)}, c_3^{(3)}\}$, and

$$R_1 = \{\xi_{ini} \to c_3[F, enc[+[c_3[F, \xi_{ini}, -[F]]]], enc[-[\xi_{ini}]]], \; F \to c_2[F, \xi_{ini}]\},$$
$$R_2 = \{\xi_{ini} \to c_3[F, enc[-[c_3[F, \xi_{ini}, +[F]]]], enc[+[\xi_{ini}]]], \; F \to c_2[F, \xi_{ini}]\}.$$

Thus, γ is a so-called DT0L tree grammar: its tables are deterministic, and all nonterminals are output symbols.

Note that derivations in γ never terminate. However, as $\Xi \subseteq \Sigma$, all trees that are derivable from ξ_{ini} are in $L(\gamma)$. Now, let $\mathcal{A}$ be the turtle Σ-algebra with $\alpha_0 = 90°$ and $\alpha = 22.5°$. An initial part of a derivation in $G = (\gamma, \mathcal{A})$ (i.e., a derivation in γ whose individual trees are interpreted using $\mathcal{A}$) is shown in Figure 4, while Figure 5 shows some randomly chosen pictures in $L(G)$. Note that the figures do not show correct relative sizes. As the pictures grow beyond any bound as derivations get longer and longer, they must be scaled in an appropriate manner.

3.5   Generation of Collages

Another well-known type of picture generator is the collage grammar, which was originally introduced in [HK91] (see also [DK99] and, for the tree-based version,

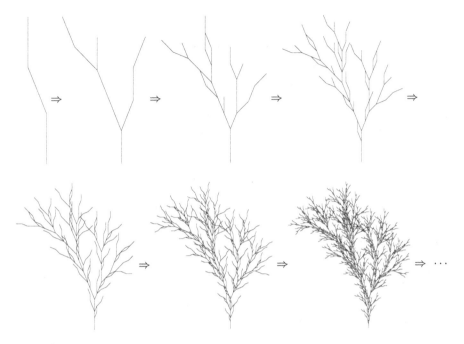

Fig. 4. An initial part of a derivation in an ET0L turtle grammar

Fig. 5. Pictures generated by an ET0L turtle grammar

[Dre06]). Choose some dimension $d \geq 1$. A *collage* in the Euclidean space $\mathbb{R}^d$ is a finite set of *parts*, each part being a nonempty and bounded subset of $\mathbb{R}^d$. Recall that an affine transformation of $\mathbb{R}^d$ is a mapping of the form $\alpha(x) = Mx + b$, where M is a $d \times d$ matrix and $b \in \mathbb{R}^d$. Such an affine transformation is applied to a part in a pointwise manner. Applying it to a collage means to apply it to each of its parts.

Now, given injective affine transformations $\alpha_1, \ldots, \alpha_k$ and a collage C, let $\langle \alpha_1 \cdots \alpha_k, C \rangle$ denote the k-ary operation f on collages given by

$$f(C_1, \ldots, C_k) = C \cup \bigcup_{i \in [k]} \alpha_i(C_i).$$

As a side remark, it may be interesting to note that, in particular, f is equal to the constant C if $k = 0$. Moreover, if $C = \emptyset$, then f is equal to the affine transformation α_1 (viewed as a unary operation on collages) if $k = 1$, and equal to the union of collages if $k = 2$ and $\alpha_1 = \alpha_2 = id$ (where id denotes the identity). For many types of tree-based collage generators, these three types of operations suffice to obtain the full generative power.

A *collage algebra* is a Σ-algebra whose domain is the set of all collages (in $\mathbb{R}^d$) and which interprets every symbol in Σ as a collage operation. Here, Σ is any unsorted signature.

Let us briefly look at an example for $d = 2$ (taken from [Dre06]). We use an EDT0L tree grammar[7] γ whose nonterminals are ξ_{ini} and ξ, and whose output signature is $\Sigma = \{\mathbf{f}^{(2)}, \mathbf{g}^{(1)}, \mathtt{snail}^{(0)}\}$. The axiom consists of the nonterminal ξ_{ini}, and the tables are

$$\{\xi_{\mathrm{ini}} \to \mathbf{f}[\xi, \xi_{\mathrm{ini}}], \ \xi \to \mathbf{g}[\xi]\}, \ \{\xi_{\mathrm{ini}} \to \mathtt{snail}, \ \xi_{\mathrm{ini}} \to \mathtt{snail}\}.$$

Instead of giving a formal definition of the collage Σ-algebra $\mathcal{A}$ used to interpret the trees in $L(\gamma)$, let us show how the rules in the first table look if they are interpreted in $\mathcal{A}$. For this, we extend $\mathcal{A}$ to $\Sigma \cup \varXi$, and interpret every constant symbol as a part whose outline resembles a snail. To be able to distinguish between the symbols, $\mathtt{snail}$ is filled with black, ξ_{ini} with white, and ξ with grey[8]. Using this interpretation of constant symbols, the two rules in the first table look like this:

Clearly, each of the rules in the second table replaces the corresponding "nonterminal snail" with the black one.

---

[7] i.e., with deterministic tables.

[8] Formally, the grey fill colour may be interpreted as a sparse area of the part, e.g., a region where the part contains only rational points.

Fig. 6. Deriving a picture made of snails

Fig. 7. Many snails

In this example, all relevant derivations consist of a number of applications of the first table, followed by one application of the second table. A short derivation of this kind is shown in Figure 6, while Figure 7 shows a picture generated by a rather long derivation.

3.6 Music Generation

Let us briefly mention that tree-based generators are even suitable for the generation of "music" – sound structures that adhere to certain basic rules of composition. A suitable algebra whose domain is the set of all musical pieces (in a certain sense) is proposed in [DH07]. It contains operations that, e.g., invert a piece, play it backwards, concatenate two pieces or create their overlay. In [DH07], a tree generator consisting of a regular tree grammar and a sequence of so-called tolerant top-down and macro tree transducers is used in connection with the music algebra in order to generate simple musical pieces. On http://www.cs.umu.se/~johanna/algebra, the implementation in TREEBAG and some generated pieces can be found.

4   Delegation Networks

A potential application area of tree-based generators concerns the generation of complex scenes involving structural, spatial, and pictorial elements. As an example, one may think of a virtual reality consisting of streets, buildings, plants, and other objects. For such an application, grammatical approaches to picture generation could be particularly useful, as they allow to generate very detailed models using simple and well-understood rules. As the previous section showed, one may indeed speak of models being generated since most of these systems actually do not generate pictures in a strict sense. Instead, they generate objects having a natural pictorial representation. For example, a collage grammar in $\mathbb{R}^3$ generates collections of three-dimensional objects that become pictures only if they are passed through a ray tracer or similar software. Thus, in principle, a collage grammar could be used to generate a virtual reality. For example, the street shown in Figure 8 has been generated in TREEBAG using a collage grammar designed by C. von Totth. Unfortunately, each of the grammatical picture generators studied in the theoretical literature (such as ET0L turtle grammars and collage grammars) is only suitable for generating a rather specific type of structures. Moreover, the devices themselves provide no support for assembling large systems from smaller components – which would certainly be needed for

Fig. 8. A street generated by a collage grammar (designed by C. von Totth)

systems comprising thousands of rules or in a context where generators are designed by a group of developers. A third disadvantage is that they provide no means for integrating nongrammatical methods in an elegant way.

In this section, we propose the delegation network, a generalization of the tree-based generator which tries to overcome these limitations by allowing to combine several generating devices. A delegation network $\mathcal{N}$ contains a finite number of so-called delegating generators, which are basically tree-based generators. However, each of them may "delegate" the generation of certain parts of the object to other delegating generators in the network. Moreover, delegation networks are based on many-sorted algebras whose predefined operations can be nondeterministic. This makes it possible to

- modularize the generation of complex objects in a meaningful way,
- combine different classes of generators, each working on a part of the problem and the domains it is appropriate for, and
- let parts of the generation task be performed by devices that are not tree based, and possibly not even grammatical at all, by viewing them as nondeterministic predefined operations.

To understand the last item, imagine that we are given some implementation of a (nongrammatical) method for generating random textures. We could then use this device in order to define a nondeterministic operation that takes a (grammatically generated) collage as input and applies nondeterministically one of the randomly generated texture patterns to it.

Conceptually, delegation is easily achieved. The tree generator component γ of a delegating generator $(\gamma, \mathcal{A})$ generates trees each of whose symbols is either interpreted by $\mathcal{A}$ or refers to another delegating generator of the network. The second case is what gives rise to delegation.

To give the formal definition of delegation networks, some preliminaries are needed. We start with the definition of nondeterministic functions.

4.1   Nondeterministic Functions and Algebras

A nondeterministic function from $\mathbb{A}_1 \times \cdots \times \mathbb{A}_k$ to $\mathbb{A}$ is a function of the form $f \colon \mathbb{A}_1 \times \cdots \times \mathbb{A}_k \to \wp(\mathbb{A})$. We identify f with the function $f' \colon \wp(\mathbb{A}_1) \times \cdots \times \wp(\mathbb{A}_k) \to \wp(\mathbb{A})$ such that, for all $A_1 \subseteq \mathbb{A}_1, \ldots, A_k \subseteq \mathbb{A}_k$,

$$f'(A_1, \ldots, A_k) = \bigcup \{ f(a_1, \ldots, a_k) \mid a_1 \in A_1, \ldots, a_k \in A_k \}.$$

Further, we write $f \colon \mathbb{A}_1 \times \cdots \times \mathbb{A}_k \twoheadrightarrow \mathbb{A}$ in order to indicate that f is a nondeterministic function, and call $\mathbb{A}_1 \times \cdots \times \mathbb{A}_k \twoheadrightarrow \mathbb{A}$ its (nondeterministic) profile. As in the deterministic case, we write $f \colon \mathbb{A}$ instead of $f \colon \twoheadrightarrow \mathbb{A}$ if $k = 0$, and identify f with the set $f() \subseteq \mathbb{A}$.

Note that a partial function $f \colon \mathbb{A}_1 \times \cdots \times \mathbb{A}_k \to \mathbb{A}$ can be regarded as the special case where $|f(a_1, \ldots, a_k)| \leq 1$ for all $(a_1, \ldots, a_k) \in \mathbb{A}_1 \times \cdots \times \mathbb{A}_k$.

The basic definitions regarding signatures, trees, and algebras carry over from the deterministic case. In an S-sorted signature Σ, profiles may from now on be nondeterministic (and a deterministic profile is regarded as a special case of a nondeterministic one). The definition of trees over Σ and the related notation are literally the same as in the deterministic case, except that $\rightarrow$ must be replaced with $\twoheadrightarrow$. Similarly, the definition of nondeterministic Σ-algebras and the related notations carry over from the deterministic case by replacing $\rightarrow$ with $\twoheadrightarrow$.

4.2 Evaluating Trees with Parameters in Nondeterministic Algebras

In the following, we want to represent complex operations by trees. For this purpose, let us reserve a set of special symbols which represent the parameters of such an operation. For every every sort A, let $\{x_1^A, x_2^A, \dots\}$ be a countably infinite set of pairwise distinct *parameter symbols* of sort A. The *parameter signature of type* $A_1 \times \cdots \times A_k$ is the signature $\{x_1^{A_1}, \dots, x_k^{A_k}\}$, containing the i-th parameter symbol $x_i^{A_i}$ of profile A_i for every $i \in [k]$. In the following, we will simply denote $x_i^{A_i}$ by x_i. To avoid confusion, we assume that parameter symbols occur only in parameter signatures, i.e., are not used as elements of ordinary signatures.

Given a nondeterministic Σ-algebra $\mathcal{A}$ and the parameter signature X of type $A_1 \times \cdots \times A_k$, we can evaluate trees $t \in T_{\Sigma \cup X}^A$. The result of this evaluation is denoted by $val_{\mathcal{A}}^X(t)$. It is the function $\varphi \colon A_1 \times \cdots \times A_k \twoheadrightarrow A$ such that $\varphi(a_1, \dots, a_k)$ is given as follows, for all $a_1 \in A_1, \dots, a_k \in A_k$:

- If $t = x_i$, then $\varphi(a_1, \dots, a_k) = \{a_i\}$.
- Otherwise, if $t = f[t_1, \dots, t_n]$ with $\varphi_i = val_{\mathcal{A}}^X(t_i)$ for all $i \in [n]$, then $\varphi(a_1, \dots, a_k) = f(\varphi_1(a_1, \dots, a_k), \dots, \varphi_n(a_1, \dots, a_k))$.

Given a set $T \subseteq T_{\Sigma \cup X}^A$ of trees rather than a single tree, we let $val_{\mathcal{A}}^X(T) = \Phi$, where $\Phi(a_1, \dots, a_k) = \bigcup_{t \in T} val_{\mathcal{A}}^X(t)(a_1, \dots, a_k)$.

During the rest of this paper, we will drop the qualifier *nondeterministic* when talking about nondeterministic Σ-algebras.

4.3 The Definition and Semantics of Delegation Networks

For the formal definition of delegation networks, one additional notion is needed. Consider an S-sorted signature Σ and a Σ-algebra $\mathcal{A} = (dom, \sigma)$. Given a domain mapping dom' for another set S' of sorts, $\mathcal{A}$ is said to be dom'-*compatible* if $dom(A) = dom'(A)$ for all $A \in S \cap S'$.

Now, the formal definition of delegation networks reads as follows.

Definition 4 (delegation network). *A delegation network is a system* $\mathcal{N} = (\Sigma, dom, G, g_0)$, *where*

- Σ *is an S-sorted signature of generator symbols, for some set S of sorts,*
- dom *is a domain mapping for S,*
- g_0 *is a constant symbol in Σ, and*
- $G = (G_g)_{g \in \Sigma}$ *is a Σ-indexed family of delegating generators $G_g = (\gamma_g, \mathcal{A}_g)$, where, for every generator symbol* $g \colon A_1 \times \cdots \times A_k \twoheadrightarrow A$,

- $\mathcal{A}_g$ is a dom-compatible Σ_g-algebra, for some signature Σ_g disjoint with Σ, and
- γ_g is a tree generator such that $L(\gamma_g) \subseteq T^A_{\Sigma_g \cup \Sigma \cup X}$, where X is the parameter signature of type $A_1 \times \cdots \times A_k$.

The semantics of $\mathcal{N}$ is obtained by constructing a Σ-algebra $\mathcal{A}_\mathcal{N}$. Intuitively, a symbol $g \in \Sigma$ is interpreted by evaluating the trees in $L(\gamma_g)$. To see what this means, let $t \in L(\gamma_g)$. According to Definition 4, every non-parameter symbol $f \colon A_1 \times \cdots \times A_k \twoheadrightarrow A$ occurring in t is either interpreted by $\mathcal{A}_g$ or a generator symbol. Thus, both cases yield an appropriate interpretation of f (using recursion if $f \in \Sigma$), which can be used to evaluate t in the way defined earlier.

However, unfortunately, the situation is not as simple as it intuitively might seem, because delegation networks can be cyclic. This invalidates the simple inductive definition the previous paragraph may have suggested. For this reason, we choose a least fixed-point semantics for delegation networks.

For this purpose, we turn the set of all functions of $f, g \colon A_1 \times \cdots \times A_k \twoheadrightarrow A$ into a complete lattice by defining $f \leq g$ if and only if $f(a_1, \ldots, a_k) \subseteq g(a_1, \ldots, a_k)$, for all $a_1 \in A_1, \ldots, a_k \in A_k$. This extends to (Σ, dom)-interpretations σ, σ', where Σ is S-sorted, in the obvious way: $\sigma \leq \sigma'$ if and only if $\sigma(f) \leq \sigma'(f)$ for all $f \in \Sigma$.

For the following definition, if $\mathcal{A} = (dom, \sigma)$ and $\mathcal{A}' = (dom', \sigma')$ are Σ- and Σ'-algebras, resp., where Σ and Σ' are disjoint and $\mathcal{A}'$ is dom-compatible, we let $\mathcal{A} \cup \mathcal{A}'$ denote the $\Sigma \cup \Sigma'$-algebra (dom'', σ'') such that $dom''(A) = dom(A)$ for all $A \in S$ and

$$\sigma''(f) = \begin{cases} \sigma(f) & \text{if } f \in \Sigma \\ \sigma'(f) & \text{otherwise.} \end{cases}$$

We are now ready to define the semantics of delegation networks.

Definition 5 (semantics of delegation networks). Let $\mathcal{N} = (\Sigma, dom, G, g_0)$ be a delegation network.

1. The operator $\text{iterate}_\mathcal{N}$ on (Σ, dom)-interpretations σ is defined as follows: $\text{iterate}_\mathcal{N}(\sigma)$ is the (Σ, dom)-interpretation σ' such that, for every symbol $g \colon A_1 \times \cdots \times A_k \twoheadrightarrow A$ in Σ,

$$\sigma'(g) = val^X_{\mathcal{A}_g \cup (dom, \sigma)}(L(\gamma_g)),$$

where X is the parameter signature of type $A_1 \times \cdots \times A_k$.
2. The least fixed point of $\text{iterate}_\mathcal{N}$ is denoted by $\sigma_\mathcal{N}$, and $\mathcal{A}_\mathcal{N} = (dom, \sigma_\mathcal{N})$. (Note that, by construction, $\text{iterate}_\mathcal{N}$ is monotonically increasing. Thus, by Tarski's fixed-point theorem, it has a least fixed point.)
3. The language generated by $\mathcal{N}$ is $L(\mathcal{N}) = \sigma_\mathcal{N}(g_0)$.

It should be noticed that, for a delegation network as above and $g \in \Sigma$, $\sigma_\mathcal{N}(g)$ is just $val^X_{\mathcal{A}_g}(L(\gamma_g))$ if no symbols from Σ appear in the trees generated by γ_g. Thus, a tree-based generator can be identified with a delegation network in which $\Sigma = \{g_0\}$, $\gamma(g_0)$ has the output signature Σ_{g_0} (i.e., does not delegate to itself), and $\mathcal{A}_{g_0}$ is a deterministic Σ_{g_0}-algebra.

From an implementation point of view, one may think of each delegating generator G_g in $\mathcal{N}$ as a nondeterministic device working as follows. Let the profile of g be $A_1 \times \cdots \times A_k \twoheadrightarrow A$, and let $a_1 \in \mathbb{A}_1, \ldots, a_k \in \mathbb{A}_k$ be arguments. We may then nondeterministically "execute" G_g in order to produce an element of the set $g(a_1, \ldots, a_k)$. For this purpose, we first run γ_g as a tree generator, which produces a tree $t \in T_{\Sigma_g \cup \Sigma \cup X}^A$. The tree t is evaluated in a bottom-up manner by nondeterministically assigning a value to each of its nodes. To each leaf carrying a variable $x_i \in X$, the value a_i is assigned. Now, consider the root node of a subtree $s \notin X$ with direct subtrees $s_1, \ldots, s_l$, and suppose that we have already assigned values $b_1, \ldots, b_l$ to its direct descendants, i.e., to the roots of $s_1, \ldots, s_l$. There are two cases.

- If $s = f[s_1, \ldots, s_l]$ with $f \in \Sigma_g$, we choose nondeterministically any element of $f(b_1, \ldots, b_l)$ (where f is the interpretation of f in $\mathcal{A}_g$) as the value of s.
- If $s = g'[s_1, \ldots, s_l]$ with $g' \in \Sigma$, we create (recursively) an instance of $G_{g'}$, apply it to the arguments $b_1, \ldots, b_l$, and consider the result to be the value assigned to the root node of s.

Without the last case, this is just the evaluation of trees with respect to $\mathcal{A}_g$, in the sense that the set of all possible results that can be obtained equals $val_{\mathcal{A}_g}^X(t)(a_1, \ldots, a_k)$. Thus, the base case is the one where $t \in T_{\Sigma_g \cup X}$, as such trees can be evaluated directly. Of course, a naive implementation may lead to an infinitely descending recursion because of the second case, if the delegation structure is cyclic. Thus, some care must be taken in an implementation.

5   An Example

Let us now consider an example. We make use of two domains, namely the set $\mathbb{C}$ of collages in $\mathbb{R}^2$ and the set $\mathbb{N}$ of natural numbers. The corresponding sorts are C and N, resp. This defines the domain mapping dom to be used throughout this section, i.e., $dom(\text{N}) = \mathbb{N}$ and $dom(\text{C}) = \mathbb{C}$. As operations on $\mathbb{N}$, we use the constant 0 and the successor function s. The operations on $\mathbb{C}$ used are of two different types. On the one hand, we use the collage operations explained in Section 3.5. One the other hand, we turn cellular automata into operations on collages. For this, let us first recall what a cellular automaton is.

The concept of cellular automata (CA) was developed by von Neumann, with contributions by Ulam, Zuse, and others, in the middle of the last century. A (two-dimensional) CA CA is a parallel device that operates on an infinite two-dimensional array of cells $cell_{ij}$ $(i, j \in \mathbb{Z})$. Geometrically, we identify the cell $cell_{ij}$ with the unit square whose lower left corner has the coordinates (i, j). The cellular automaton consists of two components. The first is a set $Q = \{0, \ldots, k\}$ of states, where $k > 0$. At each moment in time, every cell contains one of these states. The second component of CA is a transition function of the form $\Delta: Q^{3 \times 3} \to Q$, where $\Delta\left(\begin{smallmatrix} 000 \\ 000 \\ 000 \end{smallmatrix}\right) = 0$. Initially, the so-called inactive state 0 is assigned to all cells $cell_{ij}$ except $cell_{00}$, which is assigned the state 1. In each step, all cells synchronously update their states according to Δ and the states of cells in their neighbourhood. More precisely, each cell $cell_{ij}$ changes its state to $\Delta(N)$, where N is the 3×3 array of states of its neighbouring cells (including $cell_{ij}$ itself), i.e., the states of the cells $cell_{pq}$ such that $i - 1 \leq p \leq i + 1$ and $j - 1 \leq q \leq j + 1$. For example, after the first step, the state of $cell_{00}$ will be $\Delta\left(\begin{smallmatrix}000\\000\\000\end{smallmatrix}\right)$, and the state of $cell_{11}$ will be $\Delta\left(\begin{smallmatrix}000\\000\\100\end{smallmatrix}\right)$ since $cell_{11}$ has $cell_{00}$ as its lower left neighbour. Note that the number of active cells will always remain finite, because $\Delta\left(\begin{smallmatrix}000\\000\\000\end{smallmatrix}\right) = 0$.

Now, to turn a cellular automaton CA as above into an operation acting on collages, we view it as a function $CA\colon \mathbb{N} \times \mathbb{C}^k \to \mathbb{C}$, where $CA(n, C_1, \ldots, C_k)$ is obtained as follows. First, CA is executed n steps. Then, for every cell whose current state is $q \in [k]$, a copy of C_q is horizontally and vertically scaled and translated in such a way that the cell $cell_{ij}$ becomes its bounding box[9]. The resulting collage is the union of all these transformed copies of $C_1, \ldots, C_k$.

Note that one could alternatively view CA as a nondeterministic function $CA'\colon \mathbb{C}^k \to \mathbb{C}$, where $CA'(C_1, \ldots, C_k) = \{ca(n, C_1, \ldots, C_k) \mid n \in \mathbb{N}\}$. Obviously, this would not allow for as much control as the variant used here.

In the following, we use only one nontrivial cellular automaton CA, where $k = 3$. Rather than defining CA formally, let us use a delegation network $\mathcal{N}_0 = (\{g_0\colon \mathbb{C}\}, dom, (\gamma_{g_0}, \mathcal{A}_{g_0}), g_0)$ containing only one delegating generator, to show how CA behaves. The algebra $\mathcal{A}_{g_0}$ contains the operations CA, 0, s, and constant collages C_i, $1 \leq i \leq 3$. The C_i are hollow squares with different figures placed inside: a square with indented edges, a triangle, and a hollow circle, resp. The tree generator γ_{g_0} is a regular tree grammar generating the tree language $\{CA[s^n[0], C_1, C_2, C_3] \mid n \in \mathbb{N}\}$.[10] The obvious definition of γ_{g_0} is omitted here. The resulting collages for $0 \leq i \leq 6$ are shown in Figure 9.

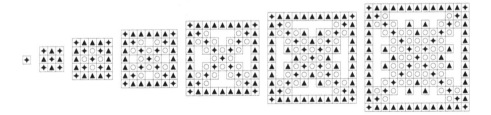

Fig. 9. Initial steps of the cellular automaton CA

Now, let us discuss a delegation network $\mathcal{N} = (\Sigma, dom, G, \mathbf{g})$ that makes use of CA in a slightly more interesting way. The signature Σ consists of the symbols $g\colon \mathbb{C}$, $ca\colon \mathbb{N} \times \mathbb{C}^3 \to \mathbb{C}$, and $ifs\colon \mathbb{C} \to \mathbb{C}$. The delegation structure in this example is such that g uses the other two, and ifs delegates to itself. All tree generators employed are regular tree grammars.

---

[9] For the purpose of this example, we may disregard collages C_q for which such a scaling is impossible.

[10] As usual, $s^0[0] = 0$ and $s^{i+1}[0] = s[s^i[0]]$ for all $i \in \mathbb{N}$.

Let us first discuss $(\gamma_{\texttt{ifs}}, \mathcal{A}_{\texttt{ifs}})$, which basically implements an iterated function system. To achieve this, the tree generator $\gamma_{\texttt{ifs}}$ generates the finite tree language $\{x_1, \texttt{ifs}[\texttt{f}[x_1, x_1, x_1]]\}$. In $\mathcal{A}_{\texttt{ifs}}$, $\texttt{f}$ is interpreted as a collage operation f of the form $\langle \alpha_1 \alpha_2 \alpha_3, \textit{line} \rangle$, where the transformations $\alpha_1, \alpha_2, \alpha_3$ and the collage $\textit{line}$ are chosen in such a way that, e.g.,

$$f(\text{♥}, \text{♥}, \text{♥}) = \text{⋎}.$$

As a consequence of the fact that $L(\gamma_{\texttt{ifs}}) = \{x_1, \texttt{ifs}[\texttt{f}[x_1, x_1, x_1]]\}$, the application of $\textit{ifs} = \sigma_{\mathcal{N}}(\texttt{ifs})$ to a collage C yields C and all collages obtained by applying $\textit{ifs}$ recursively to $f(C, C, C)$. Thus, for example, $\textit{ifs}(\text{♥})$ yields the set of collages indicated in Figure 10.

Fig. 10. Collages generated by $\textit{ifs}$, if applied to ♥ (up to scaling)

Now, let us discuss $G_{\texttt{ca}} = (\gamma_{\texttt{ca}}, \mathcal{A}_{\texttt{ca}})$. The tree generator $\gamma_{\texttt{ca}}$ is the regular tree grammar having only one nonterminal $\xi_{\text{ini}} : \texttt{C}$ and the rules

$$\xi_{\text{ini}} \to \texttt{CA}[x_1, \xi_{\text{ini}}, x_3, x_4],$$
$$\xi_{\text{ini}} \to \texttt{CA}[x_1, x_2, x_3, x_4].$$

The cellular automaton CA (i.e., the interpretation of the symbol $\texttt{CA}$) is as before. Obviously, $\gamma_{\texttt{ca}}$ generates all trees

$$\texttt{CA}[x_1, \texttt{CA}[x_1, \dots \texttt{CA}[x_1, x_2, x_3, x_4] \dots, x_3, x_4], x_3, x_4].$$

Note that the recursion takes place in the argument corresponding to state 1 of CA (i.e., where Figure 9 contains a copy of ♦), and all instances of CA perform the same number of steps, as determined by the parameter x_1.

Finally, the first component of $G_{\texttt{g}} = (\gamma_{\texttt{g}}, \mathcal{A}_{\texttt{g}})$ is the regular tree grammar with nonterminals $\xi_{\text{ini}} : \texttt{C}$ and $\xi : \texttt{N}$, and the rules

$$\xi_{\text{ini}} \to \texttt{ca}[\xi, \text{♦}, \texttt{ifs}[\text{♥}], \square],$$
$$\xi \quad \to \texttt{s}[\xi],$$
$$\xi \quad \to 0.$$

In $\mathcal{A}_{\texttt{g}}$, ♦, ♥, and $\square$ are interpreted as the collages consisting of the corresponding parts (and $\texttt{s}$ and 0 are interpreted as successor, s, and zero, 0). Thus, the purpose of $G_{\texttt{g}}$ in this example is to provide $G_{\texttt{ca}}$ with sample arguments. From the point of view of $\gamma_{\texttt{g}}$, the only argument that is not fixed, but generated in a nondeterministic manner, is the first one, determining how many steps CA

Fig. 11. Collages generated by the delegation network $\mathcal{N}$

is executed. However, actually, even the third argument is nondeterministic, as $ifs(\blacklozenge)$ is not a single collage, but the set displayed in Figure 10.

Figure 11 shows some of the collages in $L(\mathcal{N})$. They result from the trees $\mathtt{ca}[\mathtt{s}^i[0], \blacklozenge, \mathtt{ifs}[\blacklozenge], \square]$ in $L(\gamma_{\mathtt{g}})$, for $i = 0, \ldots, 3$ (top) and $i = 5$ (bottom). Among the trees in $L(\gamma_{\mathtt{ca}})$, the tree $\mathtt{CA}[x_1, \mathtt{CA}[x_1, x_2, x_3, x_4], x_3, x_4]$ has been used. The reader may find it instructive to compare this figure with Figure 9, in order to discover where the structures in Figure 9 reappear in Figure 11.

6 Conclusion and Outlook

In this paper, we have given a brief survey of tree-based generators, and have introduced the delegation network as a generalization. Future work should study both theoretical and practical issues regarding delegation networks, and provide an implementation.

Some initial results regarding delegation networks are given in [Dre07]. However, as mentioned in the introduction, evaluation in nondeterministic algebras is not correctly defined in this paper. Thus, the results of [Dre07] should be taken with care. However, under the assumption that the union $\Sigma_\cup$ of the signatures Σ_g, $g \in \Sigma$, is well defined, it is clear that a delegation network $\mathcal{N} = (\Sigma, dom, G, g_0)$ can be used to generate a tree language $T(\mathcal{N})$ by defining $T(\mathcal{N}) = L(\mathcal{N}')$, where $\mathcal{N}'$ is obtained from $\mathcal{N}$ by

1. replacing dom with dom', where $dom'(\mathtt{A}) = \mathrm{T}^\mathtt{A}_{\Sigma_\cup}$ for every sort $\mathtt{A}$, and
2. replacing each $\mathcal{A}_g$ ($g \in \Sigma$) with the free term algebra over $\Sigma_\cup$.

Now, if the algebras $\mathcal{A}_g$ are deterministic and satisfy some reasonable compatibility requirements (i.e., do not interpret common sorts or function symbols differently), the "Mezei&Wright-like" result $L(\mathcal{N}) = val_\mathcal{A}(T(\mathcal{N}))$ holds, where $\mathcal{A}$ is the union of the algebras $\mathcal{A}_g$. For the case of nondeterministic algebras, a counterexample was given in [ES78, p. 72]. In fact, looking at the notions and results in [ES77, ES78], it seems that $T(\mathcal{N})$ can be characterized by a generalized version of IO context-free tree grammars. Together with the Mezei&Wright-like result, this would yield an equivalent operational semantics for delegation networks whose algebras are deterministic (or, in other words, a characterization in terms of tree-based generators).

As delegation networks can generate tree languages, they can take the role of tree generators in delegation networks. A characterization of the tree languages $T(\mathcal{N})$ in terms of extended IO context-free tree grammars may also help to understand the resulting *delegation hierarchy* $(DEL^n(REG))_{n \in \mathbb{N}}$. Here, $DEL^n(REG)$ is the class of tree languages generated by $\mathtt{DEL}^n(REG)$, which is defined as follows: REG is the class of all regular tree grammars, and $\mathtt{DEL}^{n+1}(REG)$ is the set of all delegation networks in which each γ_g is in $DEL^n(REG)$.

Let us now discuss some more practical issues. Future plans include the implementation of a system that allows to define and execute delegation networks in a flexible and, to the extent possible, efficient manner. The flexibility of this system should preferably be similar to that of the system TREEBAG (see Section 2.4). However, in contrast to TREEBAG, whose implementation does not pay much attention to practical issues such as efficiency and usability for large examples, the development of the new system should address these points in particular.

In connection with this, it would be interesting to study the parallel and distributed execution of delegating generators. As indicated at the end of Section 4, one may view a delegating generator $(\gamma_{\mathbf{g}}, \mathcal{A}_{\mathbf{g}})$ belonging to a delegation network $\mathcal{N} = (\Sigma, dom, G, \mathbf{g}_0)$ as a device that, internally, generates a tree (in a nondeterministic fashion) and evaluates it to a (nondeterministic) function. To be able to evaluate the tree, it creates an instance I of $(\gamma_{\mathbf{g}'}, \mathcal{A}_{\mathbf{g}'})$ for every occurrence of a symbol $\mathbf{g}' \in \Sigma$ it generates (which, at least for regular and ET0L tree grammars, can be done as soon as the occurrence is generated). The function eventually returned by I is then used as the interpretation of the given occurrence of $\mathbf{g}'$. The instances of delegating generators resulting from this process are independent of each other (except for the fact that parent instances have to wait for their children during the evaluation process), it should be possible to execute them in parallel or even distribute their execution over a cluster of processors or machines.

Another question that future investigations may address is dynamic execution. Especially in the generation of graphical scenes, a derivation can often be seen as a development in time (cf. Figure 4). In terms of the discussion above, this would mean that an instance of a delegating generator does not terminate when it has reported the evaluation of its generated tree to its parent. Instead, it may perform further derivation steps, each time reporting the accordingly updated evaluation to its parent. To do this in an efficient manner, one needs to develop incremental techniques that avoid full re-evaluation in each step. This seems to be an interesting research question – as is the theoretical question of defining a suitable "dynamic semantics" for delegation networks.

Acknowledgment. I thank Joost Engelfriet for pointing out problems with the basic definitions (and, thus, results) of [Dre07] and suggesting possible remedies to me. Furthermore, I thank Carolina von Totth for allowing me to include the picture shown in Figure 8.

References

[Ad80]   Abelson, H., diSessa, A.: Turtle Geometry: The Computer as a Medium for Exploring Mathematics. MIT Press, Cambridge, MA (1980)

[BC87]   Bauderon, M., Courcelle, B.: Graph expressions and graph rewriting. Mathematical Systems Theory 20, 83–127 (1987)

[CDG+02]   Comon, H., Dauchet, M., Gilleron, R., Jacquemard, F., Lugiez, D., Tison, S., Tommasi, M.: Tree Automata Techniques and Applications (2002) available at, http://www.grappa.univ-lille3.fr/tata

[Cou90]   Courcelle, B.: Graph rewriting: an algebraic and logic approach. In: van Leeuwen, J. (ed.) Handbook of Theoretical Computer Science, vol. B, pp. 193–242. Elsevier, Amsterdam (1990)

[DH07]   Drewes, F., Högberg, J.: An algebra for tree-based music generation. In: Bozapalidis, S., Rahonis, G. (eds.) Cryptography and Coding 2007. LNCS, vol. 4728, pp. 161–173. Springer, Heidelberg (2007)

[DK99]   Drewes, F., Kreowski, H.-J.: Picture generation by collage grammars. In: Ehrig, H., Engels, G., Kreowski, H.-J., Rozenberg, G. (eds.) Handbook of Graph Grammars and Computing by Graph Transformation. Applications, Languages, and Tools, ch. 11, vol. 2, pp. 397–457. World Scientific, Singapore (1999)

[Dre06]   Drewes, F.: Grammatical Picture Generation – A Tree-Based Approach. In: Texts in Theoretical Computer Science. An EATCS, Springer, Heidelberg (2006)

[Dre07]   Drewes, F.: Delegation networks. Report UMINF 07.04, Umeå University (2007)

[Eng80]   Engelfriet, J.: Some open questions and recent results on tree transducers and tree languages. In: Book, R.V. (ed.) Formal Language Theory: Perspectives and Open Problems, pp. 241–286. Academic Press, London (1980)

[Eng97]   Engelfriet, J.: Context-free graph grammars. In: Rozenberg, G., Salomaa, A. (eds.) Handbook of Formal Languages. Beyond Words, ch. 3, vol. 3, pp. 125–213. Springer, Heidelberg (1997)

[ES77]   Engelfriet, J., Schmidt, E.M.: IO and OI. I. Journal of Computer and System Sciences 15, 328–353 (1977)

[ES78]   Engelfriet, J., Schmidt, E.M.: IO and OI. II. Journal of Computer and System Sciences 16, 67–99 (1978)

[FV98]   Fülöp, Z., Vogler, H.: Syntax-Directed Semantics: Formal Models Based on Tree Transducers. Springer, Heidelberg (1998)

[GS84]   Gécseg, F., Steinby, M.: Tree Automata. Akadémiai Kiadó, Budapest (1984)

[GS97]   Gécseg, F., Steinby, M.: Tree languages. In: Rozenberg, G., Salomaa, A. (eds.) Handbook of Formal Languages. Beyond Words, ch.1, vol. 3, pp. 1–68. Springer, Heidelberg (1997)

[HK91]   Habel, A., Kreowski, H.-J.: Collage grammars. In: Ehrig, H., Kreowski, H.-J., Rozenberg, G. (eds.) Graph Grammars and Their Application to Computer Science. LNCS, vol. 532, pp. 411–429. Springer, Heidelberg (1991)

[Mai74]   Maibaum, T.S.E.: A generalized approach to formal languages. Journal of Computer and System Sciences 8, 409–502 (1974)

[NP92]   Nivat, M., Podelski, A. (eds.): Tree Automata and Languages. Elsevier, Amsterdam (1992)

[PHHM97]   Prusinkiewicz, P., Hammel, M., Hanan, J., Měch, R.: Visual models of plant development. In: Rozenberg, G., Salomaa, A. (eds.) Handbook of Formal Languages. Beyond Words, ch. 9, vol. 3, pp. 535–597. Springer, Heidelberg (1997)

[PL90]   Prusinkiewicz, P., Lindenmayer, A.: The Algorithmic Beauty of Plants. Springer, Heidelberg (1990)

[Rou70]   Rounds, W.C.: Mappings and grammars on trees. Mathematical Systems Theory 4, 257–287 (1970)

[Tha73]   Thatcher, J.W.: Tree automata: an informal survey. In: Aho, A.V. (ed.) Currents in the Theory of Computing, pp. 143–172. Prentice-Hall, Englewood Cliffs (1973)

Bifinite Chu Spaces

Manfred Droste[1] and Guo-Qiang Zhang[2]

[1] Institute of Computer Science
Leipzig University, 04158 Leipzig, Germany
`droste@informatik.uni-leipzig.de`
[2] Department of Electrical Engineering and Computer Science
Case Western Reserve University
Cleveland, Ohio 44106, U.S.A.
`gq@case.edu`

Abstract. This paper studies colimits of sequences of finite Chu spaces and their ramifications. We consider three base categories of Chu spaces: the generic Chu spaces (**C**), the extensional Chu spaces (**E**), and the biextensional Chu spaces (**B**). The main results are

- a characterization of monics in each of the three categories;
- existence (or the lack thereof) of colimits and a characterization of finite objects in each of the corresponding categories using monomorphisms/injections (denoted as **iC**, **iE**, and **iB**, respectively);
- a formulation of bifinite Chu spaces with respect to **iC**;
- the existence of universal, homogeneous Chu spaces in this category.

Unanticipated results driving this development include the fact that:

- in **C**, a morphism (f, g) is monic iff f is injective and g is surjective while for **E** and **B**, (f, g) is monic iff f is injective (but g is not necessarily surjective);
- while colimits always exist in **iE**, it is not the case for **iC** and **iB**;
- not all finite Chu spaces (considered set-theoretically) are finite objects in their categories.

Chu spaces are a general framework for studying the dualities of objects and properties; points and open sets; terms and types, under rich mathematical contexts, with important connections to several sub-disciplines in computer science and mathematics. Traditionally, the study on Chu spaces had a "non-constructive" flavor. There was no framework in which to study constructions on Chu spaces with respect to their behavior in permitting a transition from finite to infinite in a continuous way and to study which constructs are continuous functors in a corresponding algebroidal category. The work presented here provides a basis for a constructive analysis of Chu spaces and opens the door to a more systematic investigation of such an analysis in a variety of settings.

References

1. Droste, M., Zhang, G.-Q.: Bifinite Chu spaces. In: Mossakowski, T., Montanari, U., Haveraaen, M. (eds.) CALCO 2007. LNCS, vol. 4624, pp. 179–193. Springer, Heidelberg (2007)

S. Bozapalidis and G. Rahonis (Eds.): CAI 2007, LNCS 4728, pp. 73–74, 2007.
© Springer-Verlag Berlin Heidelberg 2007

2. Plotkin, G.: A powerdomain construction. SIAM J. Comput. 5, 452–487 (1976)
3. Pratt, V.: Chu spaces. School on Category Theory and Applications, Textos Mat. SCr. B 21, 39–100 (1999)
4. Zhang, G.-Q.: Chu spaces, concept lattices, and domains. In: Proceedings of the 19th Conference on the Mathematical Foundations of Programming Semantics, Montreal, Canada, March 2003. Electronic Notes in Theoretical Computer Science, vol. 83, 17 pages (2004)

Tiling Recognizable Two-Dimensional Languages*

Dora Giammarresi

Dipartimento di Matematica. Università di Roma "Tor Vergata"
via della Ricerca Scientifica, 00133 Roma, Italy
giammarr@mat.uniroma2.it

Abstract. Tiling recognizable two-dimensional languages generalizes recognizable string languages to two dimensions and share with them several properties. Nevertheless two-dimensional recognizable languages are not closed under complement and this implies that are intrinsically non-deterministic. We introduce the notion of deterministic and unambiguous tiling system that generalizes deterministic and unambiguous automata for strings and show that, differently from the one-dimensional case, there exist other distinct classes besides deterministic, unambiguous and non-deterministic families that can be separated by means of examples and decidability properties. Finally we introduce a model of automaton, referred to as tiling automaton, defined as a scanning strategy plus a transition function given by a tiling system. Languages recognized by tiling automata are compared with ones recognized by on-line tesselation automata and four-way automata.

Keywords: Automata and Formal Languages, Two-dimensional languages, Tiling systems, Unambiguity, Determinism.

1 Introduction

Two-dimensional languages are sets of pictures or two-dimensional arrays of symbols chosen in a finite alphabet. The increasing interest for pattern recognition and image processing has motivated the research on two-dimensional (2D for short) languages, and nowadays this is a research field of great interest. Since sixties, many approaches have been presented in the literature in order to find in 2D a counterpart of what regular languages are in one dimension (1D): finite automata, grammars, logics and regular expressions. Among automata models we recall four-way automata ([6]), on-line tesselation automata ([13]) and the more recent quadrapolic automata ([8]). In this paper we mainly refer to a somehow unifying point of view presented by A. Restivo and D. Giammarresi in 1991 who defined the family REC of *recognizable picture languages* (see [10] and [11]). This definition takes as starting point a characterization of recognizable string

---

* This work was partially supported by PRIN project *Linguaggi Formali e Automi: aspetti matematici e applicativi.*

S. Bozapalidis and G. Rahonis (Eds.): CAI 2007, LNCS 4728, pp. 75–86, 2007.

languages in terms of local languages and projections (cf. [9]): the pair of a local picture language and a projection is called *tiling system*.

REC family inherits several properties from the class of regular string languages. A crucial difference lies in the fact that the definition of recognizability by tiling systems is intrinsically non-deterministic. Deterministic machine models to recognize two-dimensional languages have been considered in the literature: they always accept classes of languages smaller than the corresponding non-deterministic ones (see for example, [6,13,19]). This seems to be unavoidable when jumping from one to two dimensions. Further REC family is not closed under complement and therefore the definition of any constraint to force determinism in tiling systems should necessary result in a class smaller than REC.

In formal language theory, an intermediate notion between determinism and non-determinism is the notion of unambiguity. In an unambiguous model, we require that each accepted object admits only one successful computation. Both determinism and unambiguity correspond to the existence of a *unique* process of computation, but while determinism is a "local" notion, unambiguity is a fully "global" one. Unambiguous recognizable two-dimensional languages have been introduced in [10], and their family is referred to as UREC. Informally, a picture language belongs to UREC when it admits an unambiguous tiling system, that is if every picture has a unique counter-image in its corresponding local language; and this is an "orientation-free" notion. In [4], the proper inclusion of UREC in REC is proved.

Then we provide a definition of deterministic recognizable picture languages based on the formalism of tiling system, that generalizes 1D case.

The definition of determinism we introduce consists of a property on the tiling system (i.e. the undirected transitions of the automata in the 1D case) that leads to no backtracking in any reasonable associated "computation". Furthermore determinism is a decidable property that implies unambiguity and polynomial parsing. More in details we will define four types of determinism, one for each corner-to-corner direction of reading of a picture. Observe that this is also the case for string languages. The notion of determinism on strings is somehow an "oriented" notion. When a set of undirected transitions is given for strings, there are two notions of determinism according to the reading direction: determinism (from left-to-right) and co-determinism (from right-to-left). *Deterministic Recognizable Languages* are defined as languages that admit a deterministic tiling system along one of the four corner-to-corner directions: *DREC* denotes the class of all deterministic recognizable languages. As one would expect DREC class results to be closed under complement. We show that also DREC is properly included in UREC. Hence DREC$\subset$ UREC $\subset$ REC, differently from the 1D case where all the corresponding classes collapse. Then we further strengthen this result and show that there is a very rich hierarchy of classes between determinism and non -determinism in 2D. We exhibit some classes, denoted *Col-UREC* and *Row-UREC*, that strictly separate DREC from UREC. Recall that DREC is the class of languages that can be accepted with backtracking zero at each step of the computation while UREC languages may require backtracking linear in the size of the pictures during computation. (Remark that recall that parsing for 2D languages is a NP-complete problem [15].) As intermediate classes, Col-UREC and Row-UREC are defined in such a way to have backtracking at most linear in one dimension of the picture at each step of its computation: they are defined by means of column-unambiguous and row-unambiguous tiling systems, respectively.

Regarding decidability issue, it is easy to prove that it is decidable whether a given tiling system is deterministic while in [4] it is shown that it is undecidable whether it is unambiguous. Here we prove that for those intermediate notions of row-/ column-unambiguous tiling system such problem is still decidable.

We mention also that, in [7,20], it is given a different definition of determinism for tiling systems based on the way a tiling system is used to recognize pictures. Such definition is conceptually different and it does not reduce to conventional determinism on strings when restricting to one-row pictures.

The last part of the paper is devoted to a proposal of a model of automaton based on tiling systems. We start from the observation that a tiling system is not an effective computation device: given a tiling system and a picture, if we want to decide whether the picture belongs to the language recognized by the tiling system, we have to try to cover the picture with the given tiles, in a way that they match each others and the local symbols project to underlying symbols of the picture. All the attempts can be done following any scanning strategy: we could either start in the top-left corner and going row by row (from top to bottom) or by columns or in a spiral-like way or in many other more or less natural or strange ways of proceeding. Then in a sense, a set of tiles is the set of undirected transitions for a sort of automaton that reads the given picture along a fixed scanning strategy. We define what we call a *tiling automaton* as a generalization of finite string automaton that reads pictures and processes them by means of a transition function given as a tiling system. All the results mentioned are from [3]. Then languages recognized by tiling automata are compared with languages recognized by "classical" 2OTA and four-way automata.

2   Preliminaries

We recall some definitions about two-dimensional languages. The notations used and more details can be mainly found in [11].

A *two-dimensional string* (or a *picture*) over a finite alphabet Σ is a two-dimensional rectangular array of elements of Σ. The set of all pictures over Σ is denoted by Σ^{**} and a *two-dimensional language* over Σ is a subset of Σ^{**}.

Given a picture $p \in \Sigma^{**}$, let $p_{(i,j)}$ denote the symbol in p with coordinates (i,j), where position $(1,1)$ corresponds to to top-left corner. Moreover if p has m rows and n columns we denote $\ell_1(p) = m$, the number of rows and $\ell_2(p) = n$ the number of columns; the pair (m,n) is the *size* of p. The set of all pictures over Σ of size (m,n) is denoted by $\Sigma^{m,n}$. It will be needed to identify the symbols on the boundary of a given picture: for any picture p of size (m,n), we consider the *bordered picture* $\widehat{p}$ of size $(m+2, n+2)$ obtained by surrounding

p with a special *boundary symbol* $\# \notin \Sigma$: positions of $\hat{p}$ will be indexed in $\{0, 1, \cdots, m+1\} \times \{0, 1, \cdots, n+1\}$.

A *tile* is a picture of dimension $(2,2)$ and $B_{2,2}(p)$ is the set of all sub-blocks of size $(2,2)$ of a picture p. Given an alphabet Γ, a two-dimensional language $L \subseteq \Gamma^{**}$ is *local* if there exists a finite set Θ of tiles over $\Gamma \cup \{\#\}$ such that $L = \{p \in \Gamma^{**} | B_{2,2}(\hat{p}) \subseteq \Theta\}$ and we will write $L = L(\Theta)$.

A *tiling system* is a quadruple $(\Sigma, \Gamma, \Theta, \pi)$ where Σ and Γ are finite alphabets, Θ is a finite set of tiles over $\Gamma \cup \{\#\}$ and $\pi : \Gamma \to \Sigma$ is a projection. A two-dimensional language $L \subseteq \Sigma^{**}$ is *tiling recognizable* if there exists a tiling system $(\Sigma, \Gamma, \Theta, \pi)$ such that $L = \pi(L(\Theta))$ (extending π in the usual way). We denote by REC the family of all *tiling recognizable* picture languages.

We remark that tiling systems $(\Sigma, \Gamma, \Theta, \pi)$ for picture languages are in some sense a generalization of automata for string languages. Indeed, in the one-dimensional case, the quadruple $(\Sigma, \Gamma, \Theta, \pi)$ corresponds exactly to the graph of the automaton: Γ represents the edges set, Θ describes the edges adjacency while π gives the edges labels. A word of the local language defined by Θ corresponds to an accepting path in the graph and its projection by π gives the actual word recognized by the automaton (cf. [9]). Then, when rectangles degenerate in strings the definition of recognizability coincides with the classical one for strings.

The family REC is closed with respect to different types of operations (see [11] for all the proofs). The *column concatenation* of p and q (denoted by $p \oslash q$) and the *row concatenation* of p and q (denoted by $p \ominus q$) are partial operations, defined only if $\ell_1(p) = \ell_1(q)$ and if $\ell_2(p) = \ell_2(q)$, respectively and are given by:

$$p \oslash q = \boxed{\begin{array}{c|c} p & q \end{array}} \qquad p \ominus q = \boxed{\begin{array}{c} p \\ \hline q \end{array}}.$$

REC family is closed under row and column concatenation and their closures, under union, intersection and under rotation. All those closure properties confirm the close analogy with the one-dimensional case. The big difference regards the complement operation. In [11] it is shown that the family REC is *not* closed under complement.

Let us give some examples to which we will refer later.

Example 1. Let $L_{fc=lc}$ be the language of pictures over $\Sigma = \{a, b\}$ whose the first column is equal to the last one. Language $L_{fc=lc} \in REC$. Informally we can define a local language where information about first column symbols of a picture p is brought along horizontal direction, by means of subscripts, to match the last column of p. Tiles are defined to have always same subscripts within a row while, in the right-border tiles, subscripts and main symbols should match. Here are some left border, right border and middle tiles, respectively: $\begin{array}{|c|c|}\hline \# & z_z \\ \hline \# & t_t \\ \hline\end{array}$, $\begin{array}{|c|c|}\hline z_z & \# \\ \hline t_t & \# \\ \hline\end{array}$, and

$\begin{array}{|c|c|}\hline z_z & s_z \\ \hline t_t & r_t \\ \hline\end{array}$ with $r, s, z, t \in \Sigma$. Below it is an example of a picture $p \in L_{fc=lc}$ together with a corresponding local picture p'.

$$p = \begin{array}{|c|c|c|c|c|}\hline b & b & a & b & b \\\hline a & a & b & a & a \\\hline b & a & a & a & b \\\hline a & b & b & b & a \\\hline\end{array} \quad p' = \begin{array}{|c|c|c|c|c|}\hline b_b & b_b & a_b & b_b & b_b \\\hline a_a & a_a & b_a & a_a & a_a \\\hline b_b & a_b & a_b & a_b & b_b \\\hline a_a & b_a & b_a & b_a & a_a \\\hline\end{array}.$$

Let $L_{fc=c'}$ be the language of pictures such that the first column is equal to some i-th column, $i \neq 1$. Note that $L_{fc=c'} = L_{fc=lc} \oplus \Sigma^{**}$ and thus $L_{fc=c'} \in REC$. Similarly we can show that the languages $L_{c'=lc} = \Sigma^{**} \oplus L_{fc=lc}$, and $L_{c=c'} = \Sigma^{**} \oplus L_{fc=lc} \oplus \Sigma^{**}$ are in REC.   □

An interesting model of 2D automaton to recognize picture languages is the *two-dimensional on-line tessellation acceptor* (OTA) introduced in [13]. In a sense the OTA is an infinite array of identical finite-state automata in a two dimensional space. The computation goes by counter-diagonals starting from top-left towards bottom-right corner of the picture. A run of a OTA on a picture consists in associating a state to each position of the picture. Such state for some position (i,j) is given by the transition function and depends on symbol in that position and on the states already associated to positions $(i,j-1)$, $(i-1,j-1)$ and $(i-1,j)$ (note that an equivalent definition is possible with the state not depending on the state in the top-left corner, $(i-1,j-1)$). A deterministic version of this model is referred to as DOTA. The family of languages recognized by the two versions of the model ($\mathcal{L}(OTA)$, $\mathcal{L}(DOTA)$) are different. Despite this kind of automaton is quite difficult to manage, this is actually the machine counterpart of a tiling system: in [14], it is proved that $REC = \mathcal{L}(OTA)$.

Another model of automaton for two-dimensional languages is the *4-way automaton* (4NFA or 4DFA for the deterministic version): it is defined as an extension of the two-way automaton that recognizes strings (cf. [6]) by allowing it to move in four directions: *Left, Right, Up, Down*. It is proved that also for 4-way automata, the deterministic version of the model defines a class of languages smaller than the corresponding one defined by non-deterministic version (see [11,6]).

3   Unambiguity and Determinism in Tiling Systems

In this section we consider the notions of unambiguity and determinism by taking their definitions in the string languages case and by generalizing them to tiling systems recognizable languages. We always want that when pictures have only one row, all the definitions coincide with ones in the strings case. Recall that unambiguity prescribes only one accepting computation while determinism admits only one possible next "move" at each step of the computation. In some same the "uniqueness" is required globally for unambiguity while it should be local for determinism. Recall that in the string case deterministic , unambiguous and non-deterministic versions of the model for the computation correspond to the same class of languages, namely recognizable languages. In two dimension we have a more complex and rich situation as we shall see.

Let us consider first the notion of unambiguity. The definition of *unambiguous recognizable two-dimensional language* was first given in [10]. Informally, a tiling system is unambiguous if every picture has a unique counter-image in its corresponding local language. Remark that, by definition the notion of unambiguity lies between notion of non-determinism and determinism.

Definition 1. *A quadruple* $(\Sigma, \Gamma, \Theta, \pi)$ *is an* unambiguous tiling system *for a two-dimensional language* $L \subseteq \Sigma^{**}$ *if and only if for any picture* $x \in L$ *there exists a* unique *local picture* $y \in L(\Theta)$ *such that* $x = \pi(y)$.

An alternative definition for *unambiguous tiling system* is that function π extended to $\Gamma^{**} \to \Sigma^{**}$ is injective. We say that a two-dimensional language $L \subseteq \Sigma^{**}$ is *unambiguous* if and only if it admits an unambiguous tiling system and denote by UREC the family of all unambiguous recognizable two-dimensional languages. Obviously UREC $\subseteq$ REC.

Example 2. The language $L_{fc=lc}$ (see Example 1) of pictures p whose first column is equal to the last one, is in UREC. Indeed, we can define a tiling system as done before and this is unambiguous. This because there is only one possible counter-image for the first column of a picture p and there is a unique way to build, from this, the counter-image for the second column of p and so on up to the last column of p.

Given the definition of unambiguous tiling system, several questions naturally arise: the main one regards the problem whether all tiling recognizable languages admit unambiguous tiling systems. In [4], it is given a necessary condition for a language to be in UREC. Then, using such condition one can show that the language of pictures over a two letters alphabet that have two columns equals is not in UREC and therefore that family UREC is strictly contained in REC. In [4] it is also proved that UREC is closed under rotation and intersection operations while it is not closed under row and column concatenations and star operations. Moreover it is showed that it is undecidable whether a given tiling system is unambiguous.

When consider determinism we still have the property of a unique accepting computation but we also require that at each step of such computation we have a unique possible "choice". A recognition process is performed in linear time i.e. it does *not* have *any* backtracking at each step of the computation.

We recall that, in the string case there are two notions of determinism: (conventional) determinism and co-determinism. In fact, if the right-to-left automaton is deterministic we say that conventional automaton is co-deterministic. Translating "deterministic property" on a tiling system for strings, this should be given according to a fixed direction. Moreover recall that not all regular string languages admit automata that are both deterministic and co-deterministic.

Going to the two dimensional case there are 4 possible starting positions (the four corners) and therefore 4 possible main scanning directions (one from each corner). For example consider the direction from top-left corner towards the bottom-right one, denoted by *tl2br-direction*: any reading of a picture along this direction has the property that we can read position (x, y) only if we have already read all the positions that are above and to the left of (x, y) that is all the positions (i, j) with $i \leq x$ and $j \leq y$. Similarly we can define all the others corner-to-corner directions $tl2br, tr2bl, bl2tr, br2tl$.

Remark that, unlike the 1D case, once fixed a scanning direction there can be several reading paths on the picture p that are "compatible with" that direction.

We now define formally *Deterministic Tiling systems*.

Definition 2. *A tiling system* $(\Sigma, \Gamma, \Theta, \pi)$ *is tl2br-deterministic if for any* γ_1, γ_2, $\gamma_3 \in \Gamma \cup \{\#\}$ *and* $\sigma \in \Sigma$ *there exists at most one tile* $\begin{array}{|c|c|}\hline \gamma_1 & \gamma_2 \\\hline \gamma_3 & \gamma_4 \\\hline\end{array} \in \Theta$, *with* $\pi(\gamma_4) = \sigma$.

Similarly we define d-deterministic tiling systems for any direction d.

Example 3. Let $L_{fr=fc}$ be the language of squares over a two-letters alphabet $\Sigma = \{a, b\}$ with the first row equal to the first column. $L_{fr=fc} \in REC$: indeed we will exhibit a tiling system $\mathcal{T} = (\Sigma, \Gamma, \Theta, \pi)$ recognizing L. The tiling system $\mathcal{T}$ is such that, for any picture p, the information on each letter of the first row is brought down till the diagonal and then left towards the first column. More precisely, we use a local alphabet $\Gamma = \{x_y^z$ with $x, y \in \{a, b\}, z \in \{0, 1, 2\}\}$ and define $\pi(x_y^z) = x$. The superscript symbol 0 occurs only in positions below the diagonal, the symbol 1 occurs only on the diagonal and symbol 2 occurs only above the diagonal, while the subscript symbols correspond to information we are bringing from the first row to the first column (making a turn at the diagonal). Here below it is given an example of a picture $p \in L_{fr=fc}$ together with the corresponding local picture p' (i.e. $\pi(p') = p$).

$$p = \begin{array}{|c|c|c|c|c|}\hline a & a & b & b & a \\\hline a & b & b & a & a \\\hline b & b & a & a & b \\\hline b & b & a & a & a \\\hline a & a & a & a & b \\\hline\end{array} \qquad p' = \begin{array}{|c|c|c|c|c|}\hline a_a^1 & a_a^2 & b_b^2 & b_b^2 & a_a^2 \\\hline a_a^0 & b_a^1 & b_b^2 & a_b^2 & a_a^2 \\\hline b_b^0 & b_b^0 & a_b^1 & a_b^2 & b_a^2 \\\hline b_b^0 & b_b^0 & a_b^0 & a_b^1 & a_a^2 \\\hline a_a^0 & a_a^0 & a_a^0 & a_a^0 & b_a^1 \\\hline\end{array}$$

It is easy to see that the tiling system $\mathcal{T}$ is tl2br-deterministic. Remark that it is not br2tl-deterministic: tiles $\begin{array}{|c|c|}\hline a_a^1 & a_a^2 \\\hline a_a^0 & b_a^2 \\\hline\end{array}$, $\begin{array}{|c|c|}\hline a_b^1 & a_a^2 \\\hline a_a^0 & b_a^1 \\\hline\end{array} \in \Theta$ with $\pi(a_a^1) = \pi(a_b^1) = a$.

It is easy to show [2] that it is decidable whether a given tiling system is d-deterministic for some direction d. It suffices to verify that there are not pairs of tiles $\begin{array}{|c|c|}\hline \gamma_1 & \gamma_2 \\\hline \gamma_3 & \gamma_4 \\\hline\end{array}$, $\begin{array}{|c|c|}\hline \gamma_1 & \gamma_2 \\\hline \gamma_3 & \gamma_4' \\\hline\end{array} \in \Theta$, with $\gamma_4 \neq \gamma_4'$ and $\pi(\gamma_4) = \pi(\gamma_4')$.

A recognizable two-dimensional language L is *deterministic*, if it admits a d-deterministic tiling system for some corner-to-corner direction d. We denote by $DREC$, the class of *Deterministic Recognizable Two-dimensional Languages*.

Remark that, by definition, DREC is closed under rotation. Using this, it can be proved that DREC is closed under complement. The proof follows from the fact that DREC is the closure under rotation of $\mathcal{L}(DOTA)$ [2] and the closure of

$\mathcal{L}(DOTA)$ under complement [13]. Again from the definition, it is easy to show that deterministic recognizable languages are unambiguous (i.e. DREC⊆UREC). In [2] we prove that there are unambiguous recognizable languages that are not deterministic. The proof is given by showing that a certain language is unambiguous but not deterministic and it is quite sophisticated.

Taking as starting point this result of strict inclusion, it is very interesting to notice that it is possible to define other families, we denote *Col-UREC* and *Row-UREC*, between DREC and UREC that correspond to intermediate notions. Recall that DREC is the class of languages that can be accepted with backtracking 0 in their computations; while UREC languages may require backtracking linear in the size of the pictures during computation. Col-UREC and Row-UREC are defined in such a way to have backtracking at most linear in one dimension of the picture. They correspond to an intermediate notion between determinism and unambiguity, and hence they lie between DREC and UREC. Note that the situation is extremely more complex than in 1D where all the corresponding classes collapse.

We now define *column-* and *row-unambiguous* languages. For this, we use a different point of view for two-dimensional scanning directions: we somehow consider one dimension at each time and therefore move only along that direction. More precisely, we consider four side-to-side scanning directions namely left-to-right and vice versa, top-to-bottom and vice versa. In particular any reading of a picture p along the side-to-side direction for left-to-right, denoted by *l2r-direction*, has the property that we can read position (x, y) only if we have already read *all* the positions in the columns to the left, that is *all* the positions (i, j) with $j < y$. In other words the scanning of p proceeds column by column (despite we do not pay attention to the order of reading inside a given column). Similarly we can define all the others *side-to-side directions l2r, r2l, t2b*, and *b2t*.

Informally, we say that a tiling system is l2r-unambiguous if, when used to recognize a picture by reading it along a l2r direction, there is only one possible next local column.

Definition 3. *A tiling system* $(\Sigma, \Gamma, \Theta, \pi)$ *is* l2r-unambiguous *if for any column* $col' \in \Gamma^{m,1} \cup \{\#\}^{m,1}$, *and picture* $p \in \Sigma^{m,1}$, *there exists at most one local column* $col'' \in \Gamma^{m,1}$, *such that* $\pi(col'') = p$ *and* $B_{2,2}(p') \subseteq \Theta$ *where* $p' = \{\#\}^{1,2} \ominus (col' \oplus col'') \ominus \{\#\}^{1,2}$.

Similar properties define d-unambiguous tiling systems, for any side-to-side direction d. We say that a language is *column-unambiguous* if it is recognized by a d-unambiguous tiling system for some $d \in \{l2r, r2l\}$ and it is *row-unambiguous* if it is recognized by a d-unambiguous tiling system for some $d \in \{t2b, b2t\}$. Finally, we denote by *Col-UREC* the class of column-unambiguous languages and by *Row-UREC* the class of row-unambiguous languages.

Remark that, a column-unambiguous tiling system is such that, during the computation of a picture of size (m, n), the backtracking at each step is at most m. This is because the next local column is uniquely determined without ambiguity after backtracking of m steps at most. Same remarks hold for row-unambiguity. Moreover it is interesting to observe that we could similarly define diagonal unambiguity, requiring that the next diagonal of local symbols is uniquely determined from the previous one (for example, the counter-diagonals like OTA's transitions waves). In this case, such a diagonal unambiguity would coincide with determinism, since the local symbol in a position on the diagonal does not depend on the other local symbols on the diagonal.

In [2] there are given some necessary conditions for picture languages in Col-UREC and Row-UREC and then it is proved the following theorem.

Theorem 1. *DREC*$\subset$ *(Col-UREC*$\cap$ *Row-UREC)*$\subset$ *(Col-UREC*$\cup$ *Row-UREC)* $\subset$ *UREC* $\subset$ *REC, with all strict inclusions.*

We conclude with a decidability issue. We already mentioned that it is undecidable whether a given tiling system is unambiguous while we argued that it is decidable whether a given tiling system is corner-to-corner deterministic. It can be shown (see [2]) that it is still decidable whether a tiling system is column-/row-unambiguous.

4   Tiling Automata

In this section we briefly introduce a model of *Tiling Automaton* that is an effective computational device whose transition function is given by a tiling system. For all details see [3].

The starting observation is that in one-dimensional case (see proof in [9]) the tiling system represents the state-graph of a finite automaton with *non-oriented* edges. To get an automaton from a tiling system we assume the conventional way of reading the string (from left to right) and therefore we by choose a direction for such edges. Remark that we could also assume to read the string from right to left and get another automaton that accepts the same language but processes strings from right to left. Hence a tiling system is part of an automaton, i.e. a computation device, providing a scanning strategy that "gives the instructions" on how to use it to do the computation. When extend all these reasonings to two dimensions we have to fix a scanning strategy to read the input picture. We have much more possibilities: we have four main scanning directions from the one corner to the opposite one but also scanning strategies can not necessary follow a fixed direction.

To fix the ideas, choose the scanning strategy that, for any picture, goes row by row, from left to right and from top to bottom. We consider a *next-step function* $f(i,j,m,n)$ equal to $(i,j+1)$ if $j \leq n-1$ and equal to $(i+1,1)$ if $j = n$ and $i < m$. First position of the scanning will be the top-left corner: applying iteratively such next-step function to current position, we obtain a complete scanning sequence for the input picture. Note also that, when we have reached position (i,j), we have already visited its top-left contiguous positions (i.e. positions $(i-1,j-1)$, $(i-1,j)$, $(i,j-1)$) and so we have already "chosen" the local symbols for those positions. Now, at this step of the computation, it is possible to choose a suitable tile for the four positions (i, j), $(i - 1, j - 1)$, $(i - 1, j)$ and $(i, j - 1)$ and compute the local symbol for (i, j). Remark that for the computation, it is necessary to remember some of the local symbols associated to the positions of p already scanned. In particular when we have reached position (i, j), we need to remember the local symbols in the positions of the $(i - 1)$-th row, from the $(j - 1)$-th column to the last one, and the local symbols in the positions of the i-th row, from the first column to the $(j - 1)$-th one. We do this with a suitable data structure.

Then a tiling automaton will be defined by a tiling system plus a scanning strategy (that uses a next-step function) plus a data structure (equipped with some operations).

Other scanning strategies will define different tiling automata but the class of recognized languages will be the same, namely REC. The main difference will arise for definition of deterministic automata: depending on the chosen scanning strategy deterministic tiling automata define different classes of languages.

Here we do not give precise definitions. Informally a *scanning strategy* will be some computable next-position function with some "contiguity" and "filling" properties plus a starting position for the scanning. The scanning strategy will be "compatible" with some corner-to-corner direction.

According to a scanning strategy, a tiling system becomes a device able to effectively process a picture and decide whether it has to be accepted or not, whenever we can (easily) keep track of all information needed for the next steps of the computation. In other words for any scanning strategy, we need a proper data structure that supports operations of retrieval of the three states defined in the three contiguous positions and the update of structure itself. (In the simple example at the beginning of the section such data structure was a list).

Definition 4. *A* Tiling Automaton *(TA for short) of type tl2br is a quadruple* $\mathcal{A} = (\mathcal{T}, \mathcal{S}, D_0, \delta)$ *where* $\mathcal{T} = (\Sigma, \Gamma, \Theta, \pi)$ *is a tiling system,* $\mathcal{S}$ *is a tl2br-directed scanning strategy,* D_0 *is the initial content of a data structure that supports operations* $\text{state}_1(D)$, $\text{state}_2(D)$, $\text{state}_3(D)$, $\text{update}(D, \gamma)$, *for* $\gamma \in \Gamma \cup \{\#\}$, *and* $\delta : (\Gamma \cup \{\#\})^3 \times (\Sigma \cup \{\#\}) \rightarrow 2^{\Gamma \cup \{\#\}}$ *is a partial function such that*

$\gamma_4 \in \delta(\gamma_1, \gamma_2, \gamma_3, \sigma)$ *if the tile* $\begin{array}{|c|c|}\hline \gamma_1 & \gamma_2 \\ \hline \gamma_3 & \gamma_4 \\ \hline \end{array} \in \Theta$ *and* $\pi(\gamma_4) = \sigma$ *if* $\sigma \in \Sigma$, $\gamma_4 = \#$, *otherwise.*

Similarly, we define a tiling automaton of type d for any corner-to-corner direction d.

Definition 5. *A tiling automaton* $\mathcal{A} = (\mathcal{T}, \mathcal{S}_f, D_0, \delta)$ *is deterministic if for any* $\gamma_1, \gamma_2, \gamma_3 \in \Gamma \cup \{\#\}$ *and* $\sigma \in \Sigma \cup \{\#\}$ *there exists at most one symbol* γ_4 *such that* $\gamma_4 \in \delta(\gamma_1, \gamma_2, \gamma_3, \sigma)$.

In can be given also a definition of *unambiguous tiling automaton* (UTA) by requiring only one accepting computation for each picture in the language. The computation of a tiling automaton on a picture and the acceptance criteria are defined similarly to one-dimensional case via instantaneous configurations. All the details are of course much more involved because of dealing with the data structures and of the scanning strategy.

Let $\mathcal{L}(TA)$ denote the class of languages accepted by tiling automata. It can be shows that the recognition power of a tiling automaton is independent from the scanning strategy we choose: $\mathcal{L}(TA)$ coincides with REC family. Furthermore, we have that $\mathcal{L}(DTA)=DREC$ and $\mathcal{L}(UTA)=UREC$ as defined in the previous section. In fact deterministic tiling automata are the computational model capturing the notion of determinism as introduced for DREC family. The same remark holds for UREC. Hence $\mathcal{L}(DTA)$, $\mathcal{L}(UTA)$ inherit several properties of the classes DREC and UREC (see [2,4]).

Tiling automata can be viewed as a more general model than OTA, since the computation done by a OTA can be simulated by a tiling automaton of a certain type. Nevertheless OTA and tiling automata have the same recognition power, that is they recognize the same class of languages (namely REC). In particular it can be shown that any OTA can be simulated by a tiling automaton of type $tl2br$. On the contrary when restricted to their deterministic counterparts, DOTA are less powerful than deterministic tiling automata.

We finish by considering 4-way automata. Observe that the tiling automaton is a model conceptually different from 4-way automaton: while the next movement of a 4FA is determined from the pair (state,symbol), in a tiling automaton the direction of the computation is fixed in advance. Furthermore 4-way automaton can visit the same position many times, while this is forbidden to tiling automata. And in fact $\mathcal{L}(4NFA)$ is strictly contained in $\mathcal{L}(TA)=REC$. When restricting to determinism, the two models diverge: it can be shown that $\mathcal{L}(DTA)$ is incomparable with $\mathcal{L}(4DFA)$, still remaining inside $\mathcal{L}(UTA)$.

Acknowledgments

I am grateful to my friends and co-authors Marcella Anselmo, Maria Madonia and Antonio Restivo for all the results discussed in this paper.

References

1. Anselmo, M., Giammarresi, D., Madonia, M.: New Operators and Regular Expressions for two-dimensional languages over one-letter alphabet. Theoretical Computer Science 340(2), 408–431 (2005)
2. Anselmo, M., Giammarresi, D., Madonia, M.: From determinism to non-determinism in recognizable two-dimensional languages. In: Harju, T., Karhumäki, J., Lepistö, A. (eds.) DLT 2007. LNCS, vol. 4588, Springer, Heidelberg (2007)
3. Anselmo, M., Giammarresi, D., Madonia, M.: Tiling Automaton: a Computational Model for Recognizable Two-dimensional Languages. In: Holub, J., Žd'árek, J. (eds.) CIAA 2007. LNCS, vol. 4783, Springer, Heidelberg (2007)
4. Anselmo, M., Giammarresi, D., Madonia, M., Restivo, A.: Unambiguous Recognizable Two-dimensional Languages. RAIRO: Theoretical Informatics and Applications 40(2), 227–294 (2006)

5. Anselmo, M., Madonia, M.: Deterministic Two-dimensional Languages over One-letter Alphabet. In: Bozapalidis, S., Rahonis, G. (eds.) CAI 2007. LNCS, vol. 4783, Springer, Heidelberg (2007)
6. Blum, M., Hewitt, C.: Automata on a two-dimensional tape. IEEE Symposium on Switching and Automata Theory, pp. 155–160 (1967)
7. Borchert, B., Reinhardt, K.: Deterministically and Sudoku-Deterministically Recognizable Picture Languages: http://www-fs.informatik.uni-tuebingen. de/borchert/papers/borchert/papers/Borchert-Reinhardt_2006_Sudoku.pdf
8. Bozapalidis, S., Grammatikopoulou, A.: Recognizable picture series, Journal of Automata, Languages and Combinatorics, special vol. Weighted Automata (2004)
9. Eilenberg, S.: Automata, Languages and Machines, vol. A. Academic Press, London (1974)
10. Giammarresi, D., Restivo, A.: Recognizable picture languages. Int. Journal Pattern Recognition and Artificial Intelligence 6(2&3), 241–256 (1992)
11. Giammarresi, D., Restivo, A.: Two-dimensional languages. In: Rozenberg, G., et al. (eds.) Handbook of Formal Languages, vol. III, pp. 215–268. Springer, Heidelberg (1997)
12. Giammarresi, D., Restivo, A., Seibert, S., Thomas, W.: Monadic second order logic over pictures and recognizability by tiling systems. Information and Computation 125, 32–45 (1996)
13. Inoue, K., Nakamura, A.: Some properties of two-dimensional on-line tessellation acceptors. Information Sciences 13, 95–121 (1977)
14. Inoue, K., Takanami, I.: A characterization of recognizable picture languages. In: Nakamura, A., Saoudi, A., Inoue, K., Wang, P.S.P., Nivat, M. (eds.) ICPIA 1992. LNCS, vol. 654, Springer, Heidelberg (1992)
15. Lindgren, K., Moore, C., Nordahl, M.: Complexity of two-dimensional patterns. Journal of Statistical Physics 91(5-6), 909–951 (1998)
16. Matz, O.: Regular expressions and Context-free Grammars for picture languages. In: Reischuk, R., Morvan, M. (eds.) STACS 97. LNCS, vol. 1200, pp. 283–294. Springer, Heidelberg (1997)
17. Matz, O.: On piecewise testable, starfree, and recognizable picture languages. In: Nivat, M. (ed.) Foundations of Software Science and Computation Structures, vol. 1378, Springer, Berlin (1998)
18. Matz, O., Thomas, W.: The Monadic Quantifier Alternation Hierarchy over Graphs is Infinite. In: IEEE 1997. IEEE Symposium on Logic in Computer Science, LICS, pp. 236–244. IEEE Computer Society Press, Los Alamitos (1997)
19. Potthoff, A., Seibert, S., Thomas, W.: Nondeterminism versus determinism of finite automata over directed acyclic graphs. Bull. Belgian Math. Soc. 1, 285–298 (1994)
20. Reinhardt, K.: On some recognizable picture-languages. In: Brim, L., Gruska, J., Zlatuška, J. (eds.) MFCS 1998. LNCS, vol. 1450, pp. 760–770. Springer, Heidelberg (1998)

Algebraic Methods in Quantum Informatics

Jozef Gruska

Faculty of informatics, Masaryk university
Botanická 68a, 60200 Brno, Czech Republik*

Abstract. The paper presents and discuses several important problems and challenges of quantum informatics at which algebraic methods have been, or expect to be, very useful, or even of the key importance.

Some of these problems came up quite naturally, at a very straightforward quantumization of the classical informatics concepts, models and problems. However, the most attractive/important ones come up when one starts to dig deeper into the quantum world, into its phenomena, processes, laws and limitations. Among them are problems that belong to grand challenges not only of informatics (and physics), but of all current science. To get involved in exploration of such challenges is therefore much desirable for informatics community with algebraic expertise.

1   Quantum Informatics

Quantum informatics is an area of science that has emerged through a marriage of arguable two most important areas of science of 20th century, quantum physics and informatics. It has three main goals: (a) to develop a scientific basis to study laws and limitations of (quantum) information processing world; (b) to develop a scientific bases for the emerging quantum information processing and communication technology; (c) to develop new, information processing based, scientific tools to understand (quantum) physical world. To achieve its goals, quantum informatics uses all available methods of informatics (and mathematics) and physics, and algebraic methods play by that a prominent role.

All that is closely related to our view that the main scientific goal of physics is to study concepts, phenomena, processes, laws and limitations of the physical world and that the main scientific goal of informatics is to study concepts, phenomena, processes, laws and limitation of the information world (see also Calude and Gruska, 2007). Investigation of the relations between these two worlds, between their fundamental concepts and also of the question how much are these two worlds actually only two sides of a single world, belong to the most fundamental problems of science.

* Support of the grants GACR 201/07/0603 and MSM00211622419 as well as VEGA 1/3105/06 is to be acknowledged.

2   Basics of Quantum Information Processing

In order to make this paper quite self-contained, let us first introduce, very briefly, basic concepts of quantum information processing. For details see Gruska (1999, 2006), Chuang and Nielsen (2000).

The starting point is an understanding that the mathematical concept of *Hilbert space* well corresponds to the physical concept of *quantum system*.

Hilbert space H_n is an n-dimensional complex vector space with

$$\text{the } (scalar) \text{ inner product } \langle\psi|\phi\rangle = \sum_{i=1}^{n} \phi_i\psi_i^* \text{ of vectors } |\phi\rangle = \begin{vmatrix} \phi_1 \\ \phi_2 \\ \vdots \\ \phi_n \end{vmatrix}, |\psi\rangle = \begin{vmatrix} \psi_1 \\ \psi_2 \\ \vdots \\ \psi_n \end{vmatrix},$$

where the *norm of a vector* $||\phi|| = \sqrt{|\langle\phi|\phi\rangle|}$ and *metrics* $\text{dist}(\phi, \psi) = ||\phi - \psi||$ can be defined. This allows to introduce on H_n such concepts as continuity.

Elements (vectors) of a Hilbert space $\mathcal{H}_n$ are usually called n-dimensional *pure states*, or, more physically, n-level systems.

Two quantum states $|\phi\rangle$ and $|\psi\rangle$ are called *orthogonal* if their inner product is zero, that is if $\langle\phi|\psi\rangle = 0$. The concept of states orthogonality is very important because two pure quantum states are physically perfectly distinguishable if and only if they are orthogonal.

By an *orthogonal decomposition* $\mathcal{H}_1 \oplus \mathcal{H}_2 \oplus \ldots \oplus \mathcal{H}_k$ of a Hilbert space $\mathcal{H}$ we understand such a set of mutually orthogonal subspaces $\mathcal{H}_1, \ldots, \mathcal{H}_k$ of $\mathcal{H}$ that each state of $\mathcal{H}$ can be in a unique way decomposed as a superposition of states from subspaces $\mathcal{H}_1, \ldots, \mathcal{H}_k$. In every Hilbert space there are so-called *orthogonal bases* - they are bases all states of which are mutually orthogonal and of the norm one. In the following we consider only orthonormal bases.

Dirac introduced a very handy notation, so-called bra-ket notation, to deal with probability amplitudes, quantum states and linear functionals $f : H \to \mathbf{C}$ of any Hilbert space $\mathcal{H}$.

If $\psi, \phi \in H$, then

$\langle\psi|\phi\rangle$ — the *inner (scalar or Hermitian) product* of ψ and ϕ — is a probability amplitude of a transfer from the state ϕ to ψ;

$|\phi\rangle$ — the *ket-vector* — is an equivalent to ϕ

$\langle\psi|$ — the *bra-vector* — is a linear functional on H such that $\langle\psi|(|\phi\rangle) = \langle\psi|\phi\rangle$;

$|\psi\rangle\langle\phi|$ — the *outer product* of $|\psi\rangle$ and $|\phi\rangle$ — is a (density) matrix.

Example. For states $\phi = (\phi_1, \ldots, \phi_n)$ and $\psi = (\psi_1, \ldots, \psi_n)$ we have

$$|\phi\rangle = \begin{pmatrix} \phi_1 \\ \ldots \\ \phi_n \end{pmatrix}, \langle\phi| = (\phi_1^*, \ldots, \phi_n^*); \langle\phi|\psi\rangle = \sum_{i=1}^{n} \phi_i^*\psi_i; |\phi\rangle\langle\psi| = \begin{pmatrix} \phi_1\psi_1^* & \cdots & \phi_1\psi_n^* \\ \vdots & \ddots & \vdots \\ \phi_n\psi_1^* & \cdots & \phi_n\psi_n^* \end{pmatrix}.$$

It is natural to see each evolution step in a quantum system as a transition from one state to another. Such a transition has to satisfy three principles:

P1 To each transfer from a quantum state ϕ to a state ψ a complex number $\langle\psi|\phi\rangle$ is associated, so-called *probability amplitude* of the transfer, such that $|\langle\psi|\phi\rangle|^2$ is the *probability* that the transfer takes place.

P2 If a transfer from a quantum state ϕ to a quantum state ψ can be decomposed into two subsequent transfers, $\psi \leftarrow \phi' \leftarrow \phi$, then the resulting amplitude of the transfer is the *product* of amplitudes of the sub-transfers $\langle\psi|\phi\rangle = \langle\psi|\phi'\rangle\langle\phi'|\phi\rangle$.

P3 If the transfer from a state ϕ to a state ψ has two independent alternatives, with amplitudes α and β,

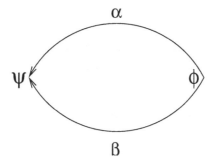

then the overall amplitude of the transfer is the sum $\alpha + \beta$ of amplitudes of two sub-transfers (and therefore $\alpha + \beta = 0$ if $\alpha = -\beta$).

In general, an evolution of a quantum system, that is in the state $\psi(t)$ in time t, is described by so-called *Schrödinger linear equation*

$$i\hbar\frac{\partial\psi(t)}{\partial t} = H(t)\psi(t),$$

where $H(t)$ is a quantum analogue of a Hamiltonian (in time t), of the classical system, and $\hbar$ is the Planck constant. From that it follows that the evolution (computation) of a quantum system is performed by a *unitary operator* $A = e^{iH}$, for the case the Hamiltonian H is constant in time, and a step of such an evolution, when the evolution is considered in discrete time, can then be seen as a multiplication of a *unitary matrix*[1] A with a vector $|\psi\rangle$, i.e. as $A|\psi\rangle$.

In case an orthonormal basis $\{\beta_i\}_{i=1}^n$ is chosen in $\mathcal{H}_n$, any state $|\phi\rangle \in \mathcal{H}_n$ can be uniquely expressed in the form

$$|\phi\rangle = \sum_{i=1}^n a_i|\beta_i\rangle, \quad \sum_{i=1}^n |a_i|^2 = 1,$$

---

[1] A matrix A is called *unitary* if $A \cdot A^\dagger = A^\dagger \cdot A = I$, where $A^\dagger$ is conjugate transpose of A.

where $a_i = \langle \beta_i | \phi \rangle$, and $|a_i|^2$ are *probabilities* that *if the state $|\phi\rangle$ is measured with respect to the basis $\{\beta_i\}_{i=1}^n$, then the state $|\phi\rangle$ collapses* (is projected) into the state $|\beta_i\rangle$. The classical "outcome" of a measurement of the state $|\phi\rangle$, with respect to the basis $\{\beta_i\}_{i=1}^n$, is then the index i of that state $|\beta_i\rangle$ into which the state $|\phi\rangle$ collapses.

A more general way to define quantum projective measurement is throuh observables - Hermitian matrices. Each such matrix A can be decomposed in a unique way in the form $A = \sum_{i=1}^k \lambda_i P_i$, where λ_i are different eigenvalues of A and each P_i is the projection operator into the subspace of eigenvectors coresponding to the eigenvalue λ_i. The result of such a measurement of a state $|\psi\rangle$ is then a (random) projection to one of such subspaces.

The most general form of quantum measurement is so-called Positve Operator Valued Measurement (POVM). Such measurement on a Hilbert space $\mathcal{H}$ actually only formally captures the impact of a projective measurement performed on a larger Hilbert space.

A *qubit* is a quantum system whose state

$$|\phi\rangle = \alpha|0\rangle + \beta|1\rangle, \quad \text{where} \quad \alpha, \beta \in \mathbf{C} \quad \text{are such that} \quad |\alpha|^2 + |\beta|^2 = 1,$$

lies in the two dimensional Hilbert space in which $\{|0\rangle, |1\rangle\}$ is a (*standard*) basis.

Example. Representation of qubits by
(a) electron in a Hydrogen atom   (b) a spin-$\frac{1}{2}$ particle

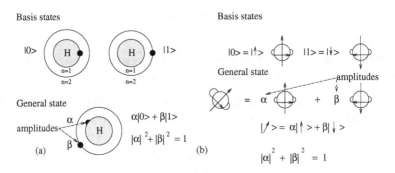

Fig. 1. Qubit representations by energy levels of an electron in a hydrogen atom and by a spin-$\frac{1}{2}$ particle. The condition $|\alpha|^2 + |\beta|^2 = 1$ is a legal one if $|\alpha|^2$ and $|\beta|^2$ are to be the probabilities of being in one of two basis states (of electrons or spin-$\frac{1}{2}$'s particles).

Another important concept, very specific indeed in the quantum case, is that of a composed quantum system. In case a quantum system S_1 (S_2) is represented by a Hilbert space $\mathcal{H}_n$ ($\mathcal{H}_m$), with an orthogonal basis $\{\alpha_i\}_{i=1}^n$ ($\{\beta_j\}_{j=1}^m$), then the *composed quantum system of S_1 and S_2* is represented by the Hilbert space $\mathcal{H}_{nm} = \mathcal{H}_n \otimes \mathcal{H}_m$, the tensor product of Hilbert spaces $\mathcal{H}_n$ and $\mathcal{H}_m$ - and one of its basis consists of all tensor products

$$\{\alpha_i \otimes \beta_j\}_{i,j=1}^{n,m}$$

of vectors of the bases of the two Hilbert (sub)spaces $\mathcal{H}_n$ and $\mathcal{H}_m$.

Example. A quantum n-qubit register can be seen as being composed of n qubit systems and as one of its basis we have all n-fold tensor products of the basis vectors/states $\{|0\rangle, |1\rangle\}$ of H_2, that is of the vectors/states

$$|b_1\rangle \otimes |b_2\rangle \otimes \ldots \otimes |b_n\rangle$$

that are usually denoted as

$$|b_1 b_2 \ldots b_n\rangle.$$

All such vectors form a (computational, or standard) basis of the Hilbert space $\mathcal{H}_{2^n}$ of n qubits.

In the classical world, each state of a composed system is composed from the states of subsystems. This is not the case in the quantum world.

States of a bipartite quantum system $S_1 \otimes S_2$, represented by tensor product of underlying Hilbert spaces $\mathcal{H}_n$ and $\mathcal{H}_m$, that cannot be decomposed as tensor products of the states of the underlying subsystems are called *entangled states*.

Example. So-called the EPR-state

$$\frac{1}{\sqrt{2}}(|00\rangle + |11\rangle)$$

is an important example of an entangled state of a two-qubit register.

Another important concept is that of a mixed state. Any probability distribution $\{(p_i, |\phi_i\rangle)\}_{i=1}^{k}$ on pure states is called a *mixed state*. To each such a mixed state a ρ so-called *density operator* is assigned that is defined as

$$\rho = \sum_{i=1}^{k} p_i |\phi_i\rangle\langle\phi_i|.$$

One interpretation of a mixed state $\{(p_i, |\phi_i\rangle)\}_{i=1}^{k}$ is that a source produces the state $|\phi_i\rangle$ with probability p_i. Any matrix representing a density operator, in a basis, is called *density matrix*. Density matrices are quantum generalization of classical probability distributions. Another important fact is that two mixed states with the same density matrix are physically undistinguishable.

A *mixed state (density matrix)* ρ *of* $\mathcal{H}$ is called entangled if ρ cannot be written in the form

$$\rho = \sum_{i=1}^{k} p_i \rho_{A,i} \otimes \rho_{B,i}$$

where $\rho_{A,i}$ ($\rho_{B,i}$) are density matrices in $\mathcal{H}_A$ (in $\mathcal{H}_B$) and $\sum_{i=1}^{k} p_i = 1$, $p_i > 0$.

The existence of entangled states has important implications. Indeed, if two particles, no matter how much space-separated they are, are in the entangled EPR-state

$$\frac{1}{\sqrt{2}}(|00\rangle + |11\rangle),$$

then a measurement of any one of them, in the standard basis, makes the EPR-state to collapse, with the same probability, to one of the states

$$|00\rangle \quad \text{or} \quad |11\rangle,$$

that is to a mixed state, and therefore the measurement (even instantaneous) of the other particle produces the same classical outcome.

Measurements of entangled states therefore produce non-local correlations. However, the non-local correlations the entangled states exhibit do not allow superluminal transmission of information and therefore do not contradict the relativity theory. Since measurement of entangled states creates non-local correlations and therefore entangled states used to be seen as strange features of the quantum mechanics as of the theory of quantum world. Nowadays, entangled states are seen as important resource of quantum information processing, what will be discussed in more details later.

3   Why Quantum Mechanics Is as It Is?

In order to discuss powerful, though often subtle, role of algebra in Quantum Information Processing and Communication (QIPC), let us start with a very foundational question/problem.

It is understood/believed, at least by some, that QIPC in general, and quantum informatics in particular, has a potential to contribute significantly to the old debates concerning various interpretations of quantum mechanics theory and concerning their mutual relations. Perhaps the main effort of QIPC people in this area has concentrated recently on such natural and foundational question as *Why quantum mechanics* (is as it is), with the goal to find *natural* (information-theoretic, or even information-processing casted) axioms of quantum mechanics - axioms that would have clear physical meaning and would deal with possibility or impossibility of various information processing phenomena and processes.

Quantum computational complexity has already been used to show why various modifications (or *fantasy versions*) of quantum mechanics are much too powerful and this way we can gain additional insight why quantum mechanics is as it is.

Another interesting idea came from Brassard and Fuchs. They asked whether we can built quantum physics from the following two axioms: (a) unconditionally secure quantum key distribution is possible; (b) unconditionally secure bit commitment is not possible; and, perhaps, from few other very simple axioms.

Clifton, Bub and Halverson (2002) took their challenge and showed that one can derive quantum mechanics from the following three axioms: no superluminal communication, no broadcasting and no unconditionally secure bit commitment, though they seem to need (as pointed out by Smolin (2004)), an assumption that quantum mechanics has to be formulated in the C^*-algebra terms. Actually, they showed that "the above constrains force any theory formulated in C^*-algebraic terms to incorporate a non-commuting algebra of observables for individual systems, kinematic independence for the algebras of space-like separated systems and the possibility of entanglement between space-like separated systems".

It is worth to observe that C^*-algebras play an important role in quantum mechanics and quantum information processing in several other ways.

4   Quantum (Finite) Automata

For all basic models of classical automata: finite automata, push-down automata, Turing machines, RAM and cellular automata, we have nowadays their quantum versions. In the following we only briefly present basic models and results for quantum finite automata. For details, and references see Gruska (1999, 2000, 2006a).

Importance of various quantum versions of the classical finite automata follows also from the following basic observation.

Quantum computations operate in the quantum world. However, for results of quantum computations to be useful, they have to get an input (an output) from (into) the classical world. Therefore, quantum computation has to operate under the classical control. How can the outcomes of quantum processes get into the classical world? Only by measurements.

Another motivation for the study of quantum finite automata goes as follows: It is still questionable whether we can design a powerful universal quantum computer. Exploration of the power of quantum finite automata is therefore a very natural goal. Moreover, exploration of the classical finite automata brought beautiful, powerful and very useful theory, with surprisingly many important applications practically in all areas of informatics, and it is therefore natural to expect that it could be so also in the quantum case.

Let us now demonstrate how we come to a natural quantum model of (finite) automata.

Input will be a binary string $\#w_1 \ldots w_n\$$, $|w| = n$, $w_i \in \Sigma$ - an input alphabet. The set of *states* will be $Q = Q_a \cup Q_r \cup Q_n$, with mutually disjoint subsets of accepting, rejecting and not-terminating states. A *configuration* is a pair (q, i) - a state $q \in Q$ and a position, $0 \leq i \leq n + 1$, on the input tape; a *set of configurations* (for an input w) is $C(Q, w) = \{(q, i) \mid q \in Q, 0 \leq i \leq |w| + 1\}$. The *underlying Hilbert space* is $H_{|C(Q,w)|}$. A *transition mapping* is defined by

$$\delta(q, i) = \sum_{q' \in Q, 1 \leq j \leq n} \alpha_{q',j} |(q', j)\rangle,$$

with an additional requirement that the *evolution* induced by δ has to be unitary.

The measurement mapping is defined by a projection into one of the mutually disjoint and orthogonal subspaces:

$$E_a = l(\{(q, i) \mid q \in Q_a\}), E_r = l(\{(q, i) \mid q \in Q_r\}), E_l = l(\{(q, i) \mid q \in Q_n\}).$$

The following two acceptance modes are of special importance for quantum finite automata:

- MM-mode (*many measurements* mode) - a measurement is peformed after each step of computation;
- MO-mode (*measurement once* mode) - only one measurement is performed - at the very end of computation.

Definitions when a word, or a language, is accepted by a quantum automaton are done in a similar way as in the case of probabilistic automata. We can therefore talk about *unbounded error acceptance* and, especially, about *bounded error acceptance, and also about an acceptance with a cut-point.*

Of the main importance seems to be the following basic model of so called one-way quantum finite automata (1QFA).

A *one-way (real-time) quantum finite automaton* (1QFA) $\mathcal{A}$ is given by: Σ — the input alphabet; Q — the set of states; q_0 – the initial state; $Q_a \subseteq Q, Q_r \subseteq Q$ — sets of accepting and rejecting states and the transition mapping

$$\delta : Q \times \Gamma \times Q \to \mathbf{C}_{[0,1]},$$

where $\Gamma = \Sigma \cup \{\#, \$\}$ and symbols $\#, \$$ are endmarker.

The evolution (computation) of $\mathcal{A}$ is performed on the Hilbert space $H_{|Q|}$, with the basis states $\{|q\rangle \,|\, q \in Q\}$, using unitary operators $V_\sigma, \sigma \in \Gamma$, defined, for basis states $\{|q\rangle \,|\, q \in Q\}$, by

$$V_\sigma |q\rangle = \sum_{q' \in Q} \delta(q, \sigma, q')|q'\rangle.$$

For *measurement* the *computational observable* is used that corresponds to the following orthogonal decomposition of $l_2(Q)$:

$$l_2(Q) = E_a \oplus E_r \oplus E_n,$$

where

$$E_a = \operatorname{span}\{|q\rangle \,|\, q \in Q_a\};$$

$$E_r = \operatorname{span}\{|q\rangle \,|\, q \in Q_r\};$$

E_n is the orthogonal complement of $E_a \oplus E_r$.

1QFA with measurement once (many) mode of acceptance accept exactly the class of group languages (a special proper subclass of regular languages).[2]

An open problem is to characterize, in a nice way, the family of languages accepted by 1QFA with MM-mode of acceptance. It is known that this class is a proper subclass of the class of regular languages and it is closed under complement, inverse homomorphism and quotient, but it is not closed under homomorphism and binary Boolean operations. Moreover, no really nice characterisation of this class is known.

Of interest are also main results concerning succinctness of 1QFA with MM-mode of acceptance. In some cases, (sequential) quantum one-way finite automata can be, likely due to the parallelism in their evolution, exponentially more succinct than classical deterministic finite automata (DFA). However, in

---

[2] Group languages are defined as languages accepted by group automata; as languages syntactical monoids of which are groups; or as languages that are accepted by permutation automata (in which each transition function $\delta(., a)$ is a permutation).

some cases, quantum one-way finite automata can be, likely due to their requirement on the unitarity of their evolution, exponentially larger, with respect to the number of states, as the corresponding DFA.

A very special model of quantum finite automata is that of so-called 1.5QFA. They are QFA heads of which can duplicate and can "move only in one (left-to-right) direction", but " a head does not have to move at each step, at some steps it can stay idle".

1.5QFA can accept non-regular languages, with respect to the unbounded error acceptance. Their power follows also from the result, by Amano and Iwama, that the emptiness problem is undecidable for 1.5QFA. An interesting old open problem is whether every regular language is accepted by a 1.5QFA.

Another interesting model of quantum automata is that of two-way quantum automata (2QFA) - a natural quantum version of the classical two-way finite automata.

A 2QFA $\mathcal{A}$ is specified by a finite (input) alphabet Σ, a finite set of states Q, an initial state q_0, sets $Q_a \subset Q$ and $Q_r \subset Q$ of accepting and rejecting states, respectively, with $Q_a \cap Q_r = \emptyset$, and the transition function

$$\delta : Q \times \Gamma \times Q \times \{\leftarrow, \downarrow, \rightarrow\} \longrightarrow \mathbf{C}_{[0,1]},$$

where $\Gamma = \Sigma \cup \{\#, \$\}$ is the tape alphabet of $\mathcal{A}$ and $\#$ and $\$$ are endmarkers, not in Σ, which satisfies the following conditions (of well-formedness) for any $q_1, q_2 \in Q$, $\sigma, \sigma_1, \sigma_2 \in \Gamma$, $d \in \{\leftarrow, \downarrow, \rightarrow\}$ (to ensure unitarity of evolution):

1. **Local probability and orthogonality condition.**
$$\sum_{q',d} \delta^*(q_1, \sigma, q', d)\delta(q_2, \sigma, q', d) = \begin{cases} 1, \text{ if } q_1 = q_2; \\ 0, \text{ otherwise.} \end{cases}$$

2. **Separability condition I.**
$$\sum_{q'} \delta^*(q_1, \sigma_1, q', \rightarrow)\delta(q_2, \sigma_2, q', \downarrow) + \sum_{q'} \delta^*(q_1, \sigma_1, q', \downarrow)\delta(q_2, \sigma_2, q', \leftarrow) = 0.$$

3. **Separability condition II.**
$$\sum_{q'} \delta^*(q_1, \sigma_1, q', \rightarrow)\delta(q_2, \sigma_2, q', \leftarrow) = 0.$$

Of importance is a special class of 2QFA, so-called *unidirectional*, or *simple*, 2QFA, in which for each pair of states, q and q', a probability amplitude is assigned that the automaton moves from the state q to the state q'. Moreover, to each state q a head movement $D(q)$ — to right, to left or no movement — is defined with the interpretation that if an automaton comes to a state q, then the head always moves in the direction $D(q)$.

It is quite straightforward to show that the class of languages accepted by simple 2QFA contains all regular languages and also some non-regular (even non context free) languages as $\{0^i 1^i \mid i \geq 0\}$.

One problem with 2QFA is that this model actually is not that of finite memory - to simulate such an automaton one needs an auxiliary memory proportional to the logarithm of the length of the (classical) input. This disadvantage does not have the classical/quantum model of two-way quantum automata discussed next.

The intuitive idea of having quantum action performed as a response to the classical input is well captured by a surprisingly simple and powerful model of two-way quantum finite automata with classical and quantum states (2QCFA) - due to Ambainis and Watrous.

A 2QCFA has a classical initial state q_0 and also a quantum initial state $|\phi_0\rangle$ in a quantum register. The evolution of a 2QCFA is specified by a mapping Θ that assigns to each classical state q and a tape symbol σ an action $\Theta(q, \sigma)$. One possibility is that

$$\Theta(q, \sigma) = (q', d, U),$$

where q' is a new state, d is the next movement of the head (to left, no movement or to right) on the classical tape, and U is a unitary operator to be performed on the current state of the quantum register.

The second possibility is that

$$\Theta(q, \sigma) = (M, m_1, q_1, d_1, m_2, q_2, d_2, \ldots, m_k, q_k, d_k),$$

where M is a projection measurement, $m_1, \ldots, m_k$ are its possible classical outcomes, and for each measurement outcome m_i a new state q_i and a new movement d_i of the head is determined. In such a case, therefore, the state transmission and the head movement are probabilistic.

Ambainis and Watrous have shown that 2QCFA with only a single qubit of quantum memory are already very powerful. Such 2QCFA can accept, with bounded error, the language of palindromes over the alphabet $\{0, 1\}$, which cannot be accepted by probabilistic 2FA at all, and also the language $\{0^i 1^i \,|\, i \geq 0\}$, in polynomial time — this language can be accepted by probabilistic 2FA, but only in exponential time.

For each model of quantum automata of interest is the task to find simplest classical model that can simulate such a quantum model. Along these lines, Rao and Vinay (2007) have shown that a large and natural class of 2QCFA, with one-sided error acceptance, can be simulated efficiently by 2-way weighted finite automata, with respect to Cortes-Mohri definition of language acceptance.

Another model of quantum automata was introduced by Bertoni, Mereghetti and Palano (2003) - they are one-way quantum finite automata with a regular language classical control (1QFACC).

1QFACC are actually the usual 1QFA that work in the MM-mode, but the measurement that is used after each move is defined by an arbitrary, though fixed, Hermitian observable and its classical outcomes (eigenvalues) are seen as elements of a special (control) alphabet Λ. With each 1QFACC $\mathcal{A}$ a regular (control) language $L \subseteq \Lambda^*$ is associated and an input word is accepted iff the corresponding word of eigenvalues obtained by measurements is in L.

Mereghetti and Palano (2006) have shown that 1QFACC accept, with respect to the isolated cut point, exactly regular languages and that for some regular languages 1QFACC are more succinct than the corresponding classical DFA.

In general, in case of quantum finite automata, the main problems to deal with are: (a) to explore power of particular automata models; (b) to compare power of various models of quantum and classical automata, especially probabilistic ones (c) to explore succinctness of quantum models comparing with the classical ones.

Main methods to explore quantum versions of the classical finite automata, their power and succinctness, are those used for the study of classical automata, especially probabilistic automata: syntactical monoids, formal power series, ...

More in the spirit of the algebraic theory of classical automata has been the approach due to Gudder (2000). He has introduced a *quantum state machine* (QSM). A QSM is defined as $\mathcal{M} = \langle Q, q_0, \delta \rangle$, where Q is a set of states, q_0 is the initial state and $\delta : Q \times Q \to \mathbf{C}$ is a transition function satisfying the well-formedness condition $\sum_q \delta(q_1, q)^* \delta(q_2, q) = \delta_{q_1, q_2}$, where δ_{q_1, q_2} is the Kronecker delta. The well-formedness condition is equivalent to the requirement that the corresponding evolution is unitary.

A so-called *q-state machine* (qSM) was defined by Guder as $\mathcal{M} = \langle H, \phi_0, U \rangle$, where H is a Hilbert space, ϕ_0 an (initial) state of H and U a unitary transformation on H.

Each QSM can be seen as a qSM. To each qSM correspond many QSM, that can be obtained by choosing a proper orthonormal basis that includes the state ϕ_0.

Both concepts of QSM and qSM have been extended by adding sets of accepting states and by considering also inputs.

Recognizable functions defined by 1QFA working in MO-mode and also by classical probabilistic one-way finite automata have been shown by Bozapalidis (2003) to be a proper subclasse of so-called B-recognizable functions.[3] B-recognizable functions have been shown to have various nice property and as ones for which cut-theorem also holds. This is another attempt to study more powerful systems, at least in some sense, than quantum ones - at least from a certain algebraic point of view.

Two another interesting generalizations of 1QFA with MO-mode of acceptance are also due to Bozapalidis (2000). As an extention to tree are (bottom-up) quantum tree automata (QTA). They are linearized tree automata whose transitions are $k^n \times k$ matrices having a certain "uniformity" property. The class of functions QTA specify (compute) is called the class of quantum recognizable (QR) functions. The functions QTA compute assign probabilities to tree. The class pf function QTA specify is convex and closed under product, complement and right derivation.

A related concept is that of quantum Γ-algebra (QA), where Γ is a ranked alphabet. Such an algebraic structure is equal, from the computational power point of view, with quantum tree automata. Quantum γ-algebra is a Hermitian (inner-product) Γ-algebra, whose functions preserve Hermitian product, endowed with a projection.

Cellular automata networks of finite automata is other very important model of classical computation. One can even say that in the quantum case this could be the most important model for local interactions performed in (quantum)

---

[3] A function $f : \Sigma^* \to \mathbf{C}$ is called *B-recognizable* if for some bounded Σ-module $\mathcal{A}$ (A bounded Σ-module is a Hermitian space $\mathcal{A}$ on which the monoid Σ^* acts linearly and whose reachability function $h_{\mathcal{A}} : \Sigma^* \to \mathcal{A}$ is bounded) of finite dimension, and for some linear form $\phi : \mathcal{A} \to \mathbf{C}$ it holds that $f(w) = \phi(h_{\mathcal{A}}(w))$, for $w \in \Sigma^*$.

nature. There have been many approaches published to define quantum cellular automata, but one can hardly say that a fully satisfactory model has already been found. The problem is how to make sure that global evolution, that has to be specified by local actions, will be unitary - in case of infinite networks of finite automata. C^*-algebras play by that also an important role.

5 Quantum Computation Primitives - Universality, Optimization

Design of sufficiently powerful quantum computers is one of the major goals of quantum information processing research. The principal difficulties that are caused by decoherence, a destructive influence of the environment, and also the principal difficulties that are behind design of entangling gates (gates that can map a product state into an entangled state) make this task so difficult that it is even unclear we can really overcome these difficulties. It is therefore of large importance to search for various small universal sets $\mathcal{G}$ of computation primitives, as well for programming (compiling) techniques that allow to implement efficiently any unitary operation through a circuit consisting only from the gates from $\mathcal{G}$, or to design even some other computational primitives than unitary gates (for details and references see Gruska, 2005).

It has been shown that the following sets of gates[4] $\mathcal{G}_1 = \{\Lambda_1(\sigma_z^{\frac{1}{2}}), H\}$ and $\mathcal{G}_2 = \{CNOT, H, \sigma_z^{\frac{1}{4}}\}$ are universal, in the sense that any unitary can be approximated arbitrarily well by circuits consisting of the gates from one of the sets $\mathcal{G}_1$ or $\mathcal{G}_2$. Another important, and this time fully universal, set (though infinite) of gates is the one consisting of the CNOT gate and all one-qubit gates. Using such gates any unitary can be exactly implemented. In case we care only for implementation (approximation) of real unitary matrices, there is a very simple universal set, due to Shi (2003), consisting of the Toffoli gate TOF and the Hadamard gate H.

The problem of designing a circuit (as efficient as possible) to implement a given unitary operation is also very non-trivial. It has been shown that the number of one qubit gates and CNOT gates needed to implement a unitary for n qubits is $\theta(4^n)$.

Concerning compilation, it has been shown, using an idea borrowed from the QR-decomposition in linear algebra, and using so-called Given's rotation matrices, that each n-qubit unitary can be implemented using $\mathcal{O}(n^3 4^n)$ one-qubit and CNOT gates. Another compilation technique, based on cosine-sine

---

[4] The basic gates that will be used in the following: Hadamard gate $H = \frac{1}{\sqrt{2}}\begin{pmatrix} 1 & 1 \\ 1 & -1 \end{pmatrix}$, Pauli gates $\sigma_x = \begin{pmatrix} 0 & 1 \\ 1 & 0 \end{pmatrix}$, $\sigma_y = \begin{pmatrix} 0 & -i \\ i & 0 \end{pmatrix}$, $\sigma_z = \begin{pmatrix} 1 & 0 \\ 0 & -1 \end{pmatrix}$, CNOT $= \begin{pmatrix} 1 & 0 & 0 & 0 \\ 0 & 1 & 0 & 0 \\ 0 & 0 & 0 & 1 \\ 0 & 0 & 1 & 0 \end{pmatrix}$, $\Lambda_1(\sigma_z^{\frac{1}{2}}) = \begin{pmatrix} \mathbf{I}_4 & \mathbf{O}_4 \\ \mathbf{O}_4 & \sigma_z^{\frac{1}{2}} \end{pmatrix}$, Toffoli gate TOF $= \begin{pmatrix} \mathbf{I}_4 & \mathbf{O}_4 \\ \mathbf{O}_4 & \sigma_x \end{pmatrix}$, where $\mathbf{I}_4$ and $\mathbf{O}_4$ are unit- and zero-matrices of degree 4.

decomposition (CSD) of unitary matrices, has been developed and later improved by R. T. Tucci. An important general and recursive method of decomposition of any unitary matrix into one- and two-qubit unitary matrices, based on the Cartan decomposition of the Lie group $su(2^n)$, is due to Khaneja and Glaser. Using these methods it has been shown, for example, that there is a universal circuit for realization of an arbitrary two-qubit gate with 3 CNOT gates and 19 elementary one-qubit rotation gates. However, already for 3-qubit gates the simplest known universal circuit has 40 CNOT gates and 98 one-qubit gates.

At the circuit optimization it is of interest to consider the following three concepts of equivalence of states:

- Two states $|\phi\rangle$ and $|\psi\rangle$ are called identical if $|\phi\rangle = |\psi\rangle$;
- Two states $|\phi\rangle$, $|\psi\rangle$ are *equivalent up to a global phase* if $|\phi\rangle = e^{i\theta}|\psi\rangle$, where $\theta \in \mathbf{R}$.
- Two states $|\phi\rangle$, $|\psi\rangle$ are *equivalent up to a relative phase* if $|\phi\rangle$ can be mapped into $|\psi\rangle$ by a unitary diagonal matrix with the diagonal $(e^{i\theta_0}, e^{i\theta_1} \dots, e^{i\theta_k})$.

As an illustration of importance of the above concepts let us consider the Toffoli gate. This gate can be exactly implemented by a circuit with 6 CNOT gates and 8 one-qubit gates.

On the other hand, the following circuit, see Figure 5, where $R_y(\theta) = e^{-i\theta\sigma_y/2}$, with only 3 CNOT gates and 4 elementary rotations is equivalent up to the relative phase to Toffoli circuit.

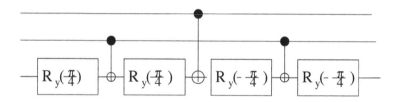

Another recent surprising outcome is that that any unitary can be implemented by a circuit consisting only of measurement gates and auxiliary ancilla qubits. There is a variety of results along these lines. One such minimal universal set of measurement gates and ancilla qubits has been determined by Perdrix (2007). It consists of one two-qubit observable $\sigma_z \otimes \sigma_x$, two one-qubit observables σ_z and $\frac{1}{\sqrt{2}}(\sigma_x - \sigma_y)$ and one one-qubit auxiliary ancilla. An obvious lower bound is one 2-qubit observable and one 1-qubit ancilla. It is an open problem which of the above bounds can be improved.

6   Quantum Circuits That Can be Simulated Classically

Another important area of the research in quantum informatics is to find out for which algorithmic problems quantum algorithms cannot (can) be essentially more efficient than classical ones.

To this area belongs the problem for which type of quantum circuits we can have an efficient classical simulation.

There are two remarkable results along these lines. An old result, due to Gottesman and Knill, that says that so-called Clifford circuits, composed from the gates CNOT, Hadamard, $\sqrt{\sigma_z}$ and projection measurements, with respect to the computational basis, can be simulate on classical computers in polynomial time. A more recent result is due to Markov and Shi (2002), who showed that quantum circuits can be simulated classically in time polynomial in the number of gates and exponential in the tree-width of the correponding circuit graph. This implies that quantum circuits with logarithmic tree-width can be simulated in polynomial time on classical computers. This result has been used recently by Aharonov et al. (2006) to show that Quantum Fourier Transform over $\mathbf{Z}_q$ can be simulated in polynomial time - a very surprising outcome.

The results just mentioned demonstrate that in order to get deeper results in this area one needs to use deep mathematical, especially algebraic, methods and results.

The above results also only underline the fact that one of the key problems of QIPC is still to get a better understanding in what really lies the power of quantum information processing and communication.

7   Quantum Algorithms Design Challenges

Attempts to solve variations of so-called *Hidden Subgroup Problem* (HSP), brought a variety of interesting and important results concerning the design of quantum algorithms. Actually, almost all speedups in quantum computing that seem to be exponential have been obtained by solving some instances of HSP.

HSP is defined as follows: Given is an (efficiently computable) function f : $G \rightarrow R$, where G is a group and R is a finite set, and a *promise* that there exists a subgroup $G_0 \leq G$ such that

- f is constant on each left coset of G (with respect to G_0);
- f is distinct on different cossets of G (with respect to G_0).

(and in this sense G_0 *is hidden by* f).

The *task* is to find a generating set for G_0 (in polynomial time (in $polylog|G|$) in the number of calls to the oracle for f and in the overall polynomial time).[5]

Four well known algorithmic problems can be reformulated as follows as special cases of HSP.

Deutsch's problem, $G = \mathbf{Z}_2$, $f : \{0,1\} \rightarrow \{0,1\}$, $x - y \in G_0 \Leftrightarrow f(x) = f(y)$.
Decide whether $G_0 = \{0\}$ (and f is balanced) or $G_0 = \{0,1\}$ (and f is constant).

---

[5] One way to solve the problem is to show that in polynomial number of oracle calls (or time) the coset states corresponding to different candidate subgroups will have exponentially small inner product and are therefore distinguishable.

Simon's problem, $G = \mathbf{Z}_2^n$, $f : G \to R$. $x - y \in G_0 \Leftrightarrow f(x) = f(y)$, $G_0 = \{0^{(n)}, s\}$, $s \in \mathbf{Z}_2^n$. Decide whether $G_0 = \{0^{(n)}\}$ or $G_0 = \{0^{(n)}, s\}$, with an $s \neq 0^{(n)}$ (and in the second case find s).

Order-finding problem, $G = \mathbf{Z}$, $a \in \mathbf{N}$, $f(x) = a^x$, $x - y \in G_0 \Leftrightarrow f(x) = f(y)$, $G_0 = \{rk \mid k \in \mathbf{Z}$ for the smallest r such that $a^r = 1.\}$ Find r.

Discrete logarithm problem, $G = \mathbf{Z}_r \times \mathbf{Z}_r$, $a^r = 1$, $b = a^m$, $a, b \in \mathbf{N}$, $f(x, y) = a^x b^y$, $f(x_1, y_1) = f(x_2, y_2) \Leftrightarrow (x_1, y_1) - (x_2, y_2) \in G_0$. $G_0 = \{(-km, m) \mid k \in \mathbf{Z}_r\}$. Find G_0 (or m).

Another important problem that can be casted as a HSP is the Graph Isomorphism Problem to be discussed in more details later.

Quantum version of the HSP assumes the access to the oracle $f(|g\rangle|0\rangle) = |g\rangle|f(g)\rangle$ for $g \in G$. If this oracle is applied to a superposition of all group elements (basis states), we have

$$U_f(\frac{1}{\sqrt{|G|}} \sum_{g \in G} |g\rangle|0\rangle) = \frac{1}{\sqrt{|G|}} \sum_{g \in G} |g\rangle|f(g)\rangle$$

and therefore, if the second register is measured, we get a mixed state

$$\rho_H = \frac{|G_0|}{|G|} \sum_{g \in \{\text{coset representatives}\}} |gG_0\rangle\langle gG_0|,$$

where $|gG_0\rangle$ are so-called *quantum (left) coset states*

$$|gG_0\rangle = \frac{1}{\sqrt{|G_0|}} \sum_{h \in G_0} |gh\rangle.$$

The quantum coset states, just introduced, play an important role in dealing with HSP. They are important because they encode the hidden subgroup. One of the key questions is how much information one can obtained from quantum coset states using POVM. In case of Abelian groups, a POVM operating on one coset states exists that can extract polynomial amount of information about the hidden subgroup. This measurement is effectively implementable using Quantum Fourier Transform over finite groups and this holds also for some non-Abelian groups. However, Moore et al. (2005) have shown that for solving the graph isomorphism problem, POVM can obtain only exponential small amount information for one or two coset states. They also showed that entangled measurements on at least $\Omega(n \lg n)$ states are needed to get useful information in case of the graph isomorphism problem.

Kitaev showed that there is a quantum polynomial time algorithm to solve the HSP in the case of Abelian groups.

The above result also implies that factoring and discrete logarithm computation can be done in quantum polynomial time - famous results of P. Shor. It is an open problem whether the HSP can be solved in quantum polynomial time also for all non-Abelian groups. A positive solution of this problem would imply, for example, that also the graph isomorphism problem, another famous and very important algorithmic problem, known to be in **NP**, but not known to be in **P** and also unlikely to be **NP**-complete, can be solved in quantum polynomial time.

A lot of research concentrates on attacking HSP for non-Abelian groups. For example, it has been shown (by Bacon et al. (2005) and others) that HSP is solvable in quantum polynomial time for several types of non-Abelian groups: for *dihedral groups*, for *almost Abelian* groups, for *near Hamiltonian* groups, for *Heisenberg groups* $\mathbf{Z}_p^2 \rtimes \mathbf{Z}_p$ (for which classical query complexity is exponential).

A series of negative results towards a standard approach to solving HSP for non-Abelian groups have also appeared, They indicate that to solve HSP for non-Abelian groups some new approach will likely be needed and not the one based on Quantum Fourier Transform. It has been even shown that in order to solve HSP for non-Abelian groups in general, techniques for efficient measurements across multiple copies of the Hidden subgroup state have to be developed. Indeed, it has been shown that the HSP becomes a problem of optimal state discrimination. It has also been shown (Bacon, 2006), that so-called Glebsch-Gordan transform over Heisenberg groups is an important new primitive for solving algorithmic problems in general and to solve HSP for Heisenberg groups in particular. (The key step by that was a demonstration that HSP and the *Hidden Subgroup Conjugacy Problem* are polynomially equivalent for the Heisenberg group.) On the other hand, it has been shown, by Ettinger et al. (2004), that quantum query complexity of HSP is polynomial.

In the above context, of importance are the following negative results of Hallgren et al. (2006): they showed that for sufficiently non-Abelian groups the HSP is hard for quantum computers in the sense that any quantum algorithm using the coset state framework requires exponential time unless it makes highly entangled measurements of $\Omega(\lg |G|)$ registers. The problem is that highly entangled measurements seem to be very hard to implement. *Quantum sieves* is one way to carry out efficiently such measurements and that was used by Kuperberg (2005), who also used the Glebsh-Gordan transform, for solving HSP for dihedral groups, with subexponential algorithm.[6] However, Moore et all. (2006) have shown that no such approach yields an effective algorithm for the HSP on symmetric groups.

Graph Isomorphism Problem (GIP) reduces to the HSP over symmetric groups, Since GIP is such a prominent case of HSP for non-Abelian groups, an intensive study of this problem brought also a variety of interesting and important results. It has been shown that the graph isomorphism problem belong to a similar complexity class **NP ∩ co-AM**, as integer factorization (**NP ∩ co-NP**), and also that HSP represents a systematic way to approach this problem. (Another known option, see Hallgren et al. (2006), not much explored yet, is to create a uniform superposition of all graphs isomorphic to a given graph.) It has been shown that in order to solve a

---

[6] He created a measurement for the dihedral group operating on $2^{\mathcal{O}(\sqrt{\lg |G|})}$ coset states that also takes $2^{\mathcal{O}(\sqrt{\lg |G|})}$ time to implement - also in some other cases efficiently implementable measurements on a constant number (1 or 2) of coset states were shown.

HSP relevant to graph isomorphism one needs to develop techniques how to implement efficiently measurement of $\mathcal{O}(\lg |G|)$ registers containing the quantum coset states. Child and Wocjan (2005) explored the hidden shift approach (to be discussed later) to the graph isomorphism and showed that $o(n \lg n)$ of the *hidden shift states* contain only exponentially little information about the isomorphism.

Negative results concerning HSP for non-Abelian groups have also positive impacts. Indeed, Moore at al. (2007) designed a simple function that is believed to be one-way and secure against even quantum attacks, because inverting of this function reduces to solving the HSP over general linear groups (which is at least as hard as HSP over symmetric groups). It is the function f_V parametrized over a set $V = \{v_1, v_2, \ldots, v_n\}$ of randomly chosen elements from $\mathbf{F}_q^n$, where q is a small prime, and for an invertible $n \times n$ matrix M over $\mathbf{F}_q$,

$$f_V(M) = \{Mv \,|\, v \in V\}$$

and f_V returns the resulting set of vectors as an unordered set.

Closely related to the HSP is the *Hidden Shift Problem* (HSHP) - another interesting/important problem. In the HSHP, given are two functions f and g for which there is a *shift s* such that $f(x) = g(x + s)$ for each x - the task is to find s. van Dam et al. (2002) solved HSHP for several types of functions; one of them is the *Shifted Legendre Symbol Problem*: Given is a function $\left(\frac{x+s}{p}\right)$ as an oracle, for a prime p, find s. The HSHP have been also solved using Quantum Fourier Transform. The *Hidden coset problem*, see van Dam (2002), is another important problem - and a generalization of both HSP and HSHP. Dealing with these problems demonstrated the ability of the Quantum Fourier Transform to capture subgroup and shift structure. Making use of specific properties of special groups is one way to approach the HSP and its variants.

Since non-Abelian groups are so hard to handle in case of the HSP, it is natural to explore which important group-theoretic algorithmic problems are solvable in quantum, but not classical, polynomial time for (some) non-Abelian groups. Along these lines, Watrous (2000) showed quantum polynomial time algorithm to compute orders of solvable groups - to this problem one can reduce several other problems for solvable groups, such as membership testing, testing equality of subgroups and so on. Algorithms work in the setting of black-box groups (where elements are uniquely encoded by strings of the same length and group operations are peformed by an oracle at unit cost). (Observe that solvable groups of order ≤ 60 are Abelian.)

8   Quantum Entanglement

The concept of quantum entanglement is primarily concerned with the states of multipartite systems.

For a bipartite quantum system, represented by the Hilbert space $\mathcal{H} = \mathcal{H}_A \otimes \mathcal{H}_B$, a pure state $|\Phi\rangle$ is called an *entangled pure state* if it cannot be decomposed into a tensor product of a state from $\mathcal{H}_A$ and a state from $\mathcal{H}_B$. A mixed state (density matrix) ρ of $\mathcal{H}$ is called an *entangled mixed state* if ρ cannot be written in the form

$$\rho = \sum_{i=1}^{k} p_i \rho_{A,i} \otimes \rho_{B,i}$$

where $\rho_{A,i}$ ($\rho_{B,i}$) are density matrices in $\mathcal{H}_A$ (in $\mathcal{H}_B$) and $\sum_{i=1}^{k} p_i = 1$, $p_i > 0$. In other words, ρ is entangled if it is not a mixture of tensor products of density matrices of subsystems.

Less formally, quantum entanglement is a subtle nonlocal and non-classical correlation among the subsystems of a quantum system. Entanglement can also be characterized and quantified as a feature of quantum system that cannot be created through local quantum operations and classical communications among the parts.

Quantum entanglement as a resource allows: (a) to perform processes that are classically impossible; (b) to speed-up some (quantum) algorithms; (c) to make communications more efficient; (d) to generate classical cryptographic keys in unconditionally secure way; (e) to make transmission of quantum information in unconditionally secure way; (f) to enlarge capacities of (classical) channels; (g) to act as catalyst for various operations that are otherwise impossible. For details see Gruska (1999, 2003), Horodeckis (2007).

Fundamental properties of quantum entanglement can be summarized as follows: (a) entanglement is an observable phenomenon that does not depend on a physical representation; (b) entanglement enables and is consumed by a variety of tasks; (c) entanglement obeys a set of as yet not fully understood principles of behaviour; (d) entanglement is shared according to strict laws and limitations; (e) entanglement cannot be increased by local actions and classical communications; (f) entanglement is a precious resource that is very difficult to create, to store and to transmit.

There are many basic problems concerning entanglement, especially concerning multipartite entanglement, that need to be solved and where algebraic methods are of large importance: (a) How to detect entanglement of multipartite systems? (b) How many inequivalent types of entanglement are there in multipartite systems? (c) Which types of entanglement we have in multipartite systems? (For example, what are the properties of bound entangled mixed states (from which one cannot distill pure entanglement)?) (d) What are the proper axioms for measures of entanglement of multipartite systems? (e) Which measures of entanglement of multipartite systems are useful and what are the relations between them? (f) What are the laws and limitations of entanglement sharing among various parties?

A variety of techniques have been developed to get deeper insights into the laws and limitations of entanglement. For example: an investigation of the states and operators transformations, reducibilities and equivalences; entanglement invariants; entanglement monotones; measures of entanglement, as well as dynamics and frequency of entanglement.

Example - *quantum invariants*. Any quantity of quantum states that is invariant with respect to local unitary transformations is called *entanglement invariant*.

There are two basic reasons for studying quantum invariants: (a) any good measure of entanglement has to be invariant under local unitary transformations - it has to be a expressed in terms of entanglement invariants; (b) the invariants of multipartite states give the finest discrimination between different types of entanglement.

A general theory of invariants of mixed multipartite states has been introduced by Rains (1997) and Grassl et al. (1998). Invariants of entanglement can be of various types. Of a special interest are so called *polynomial invariants* – polynomial functions of amplitudes of quantum state.

For example, for pure states

$$|\psi\rangle = \sum_{ij}^{n} \alpha_{ij}|e_i\rangle|e_j\rangle,$$

with bases $\{e_i\}$ and $\sum_{ij}^{m} \alpha_{ij}\alpha_{ij}^* = 1$, of an $n \times n$ bipartite system $A \otimes B$, the reduced density matrix has the form

$$\rho_0 = AA^\dagger, \text{ where } A = \{\alpha_{ij}\},$$

and of importance are the following polynomial invariants

$$I_k = Tr(AA^*)^{k+1}, k = 0, 1, \ldots, n.$$

In case $|\psi\rangle = \sum_{i=1}^{n} \sqrt{\lambda_i}|e_i\rangle|e_i\rangle$,

$$I_k = \sum_{i=1}^{n} \lambda_i^{k+1}.$$

It has also been realized that entanglement has many faces. One can see it also as a bridging notion between QIPC science and so different fields as condense-matter physics, quantum gravity and so on. There are also various approaches to generalize this concept. One of the recent ones, see Barnum et al. (2003), is based on the idea that quantum entanglement may be directly defined through expectation values of preferred observables - without reference to preferred subsystem decomposition. Such a framework allows for non-trivial entanglement to exist within a single indecomposable quantum system,....

9   Quantum and Other Non-Localities

The fact that measurement of entangled states creates non-local correlations, what can be seen, surprisingly, also as an important information processing resource, is perhaps the most important impact of entangled states. It is therefore natural that an attention in QIPC starts to orient not only on the study of non-locality quantum entanglement induces, but also other potential types of non-locality. The key observation here is the fact that quantum nonlocality is non-signaling and therefore does not contradict relativity theory. Recently, therefore, attempts started to explore whether there are, within the current physical framework, stronger non-signaling non-localities than the ones quantum mechanics allows. Let us now discuss such potential non-localities, their motivations, impacts, but also weaknesses.

The behaviour of a bipartite quantum state under measurement can be described by a conditional probability distribution $P_{ab|xy}$, where x and y denote the chosen bases for measurement and a with b are corresponding measurement outputs.

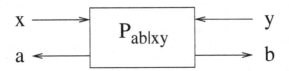

John Bell was the first to recognize that there are measurement bases such that the resulting behaviour is not local, i.e. cannot be explained by shared classical information.

Bell also showed the existence of inequalities, that motivated definition of so-called *Bell inequalities*, that cannot be violated by any local system, but are violated when some entangled states are measured.

Non-locality exhibited by the measurement of the EPR state can be seen as the implementation of the following *EPR-box*

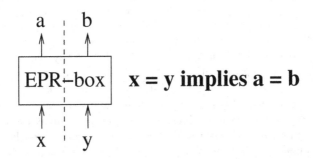

Also non-locality exhibited by the following *PR-box* does not allow superluminal communication and therefore does not contradict special relativity.

Here we denote input measurements and outcomes by binary values. For the PR-box it holds

$$\text{Prob}[a = b|(x,y) \neq (1,1)] = 1$$
$$\text{Prob}[a = b|(x,y) = (1,1)] = 0$$

The idea of PR-boxes arises in the following setting. Let us have two parties, A and B, and let each of these parties X perform two measurements on a quantum state with two outcomes m_0^X and m_1^X, with 0 and 1 as potential values.

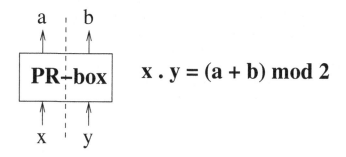

$$x \cdot y = (a + b) \bmod 2$$

Let us denote a bound on correlations between two such measurements as

$$B = \sum_{x,y \in \{0,1\}} Prob(m_x^A \oplus m_y^B = x \cdot y).$$

Famous Bell/CHCS inequality says that $B \leq 3$ in any classical hidden variable theory (that is a theory with shared random variables). So-called Cirel'son's bound (Cirel'son, 1980), says that the maximum for B in quantum mechanics is $2 + \sqrt{2}$.

Popescu and Rohrlich (1997) developed a model in which the maximal possible bound, 4, is achievable.

PR-boxes are, more exactly they would be, very powerful. Indeed, PR-boxes are non-local, yet they are causal; using one PR-box one can simulate measurement of the EPR-state; using PR-boxes one can make bit commitment and 1/2-oblivious transfer unconditionally secure; having PR-boxes one could simulate any secret multiparty computation and solve any multipartite communication problem by communicating a single bit - what is not to believe. From this result, obtained using quantum communication complexity tools, by van Dam (2000), it follows that PR-boxes cannot exist (for details and references see Scarani (2006)).

PR-boxes gave rise to the following basic questions: (a) Why are the correlations achievable by quantum mechanics not maximal among those that preserve causality? (b) How well the correlations of PR-boxes can be approximated by devices that follow laws of physics?

It has been shown, see Brassard et al. (2005), that the availability of apriori shared entanglement allows to approximate PR-boxes with success probability $\cos^2 \frac{\pi}{8} = 0.854$ and that in any physical world in which it is possible, without communication, to approximate PR-boxes with probability greater than $\frac{3+\sqrt{6}}{6} \approx$ 90.8%, every Boolean function could be, probabilistcally, computed using only one bit of communication.

PR-boxes are an important tool to study non-locality. However, they are not a universal building block of non-local correlations. Also their relation to entangled states is far from simple. One PR-box is enough to simulate (measurement on) maximally entangled two-qubit state, but not for every not-maximally entangled state. Moreover, there are tasks that can be done with n EPR-states, but require 2^n PR-boxes (see again Scarani (2006) for references and more details).

Remark. For almost 40 years it has been assumed that maximaly entangled states are the most non-local quantum states. However, it has recently emerged (see Methot and Scarani, 2006) that, for all known measures of non-locality, non-maximally entangled states are in general more non-local than maximally entangled states. An understanding has therefore developed that non-locality and entanglement are two different concepts and resources!

10   Bell Inequalities

Another important tool, and a beautiful research area by itself, to study non-locality, where algebraic and geometric tools dominate, are so-called Bell inequalities, see Gisin (2007) for a survey of the key open problems. One can even say that Bell inequalities are at the heart of the study of non-locality.

Technically, Bell inequalities are relations between conditional probabilities that are valid under locality assumption. They got of importance due to the fact that there are quantum states whose measurements produce correlations that violates some Bell inequalities - so called *Bell theorem*. This fact has changed essentially our view of the physical world. At the same time, there are beautiful and difficult problems related to Bell inequalities that require to use sophisticated algebraic and geometric methods.

Let $Pr(a, b, c, \ldots \mid x, y, z, \ldots)$ denote the conditional probability that parties $A, B, C, \ldots$ produce the outputs $a, b, c, \ldots$ when they receive inputs $x, y, z, \ldots$. Typically, parties perform measurements $x, y, z, \ldots$ with outcomes $a, b, c, \ldots$ Measurements are performed *under the assumption of non-locality*. That means that there is a probability distribution $Pr(\lambda)$ such that

$$Pr(a, b, c, \ldots \mid x, y, z, \ldots) = \sum_{\lambda} Pr(\lambda) \cdot Pr(a \mid x, \lambda) \cdot Pr(b \mid y, \lambda) \cdot Pr(c \mid z, \lambda) \cdot \ldots$$

The set of such correlations is convex, with finitely many vertices – a *polytope* bounded by *hyperplanes*.

All local correlations lie on one side of the hyperplanes and therefore they necessary satisfy inequalities of the type

$$\sum_{a,b,c,\ldots,x,y,z,\ldots} \alpha^{x,y,z,\ldots}_{a,b,c,\ldots} Pr(a, b, c, \ldots x, y, z, \ldots) \leq \gamma$$

called usually as *Bell inequalities*.

Importance of Bell inequalities stems from the fact that a quantum state ρ is said to be non-local iff there are measurements on ρ that produces a correlation that violates a Bell inequality.

Perhaps the most famous is so-called CHCS-inequality

$$E(x = 0, y = 0) + E(x = 0, y = 1) \tag{1}$$

$$+ E(x = 1, y = 0) - E(x = 1, y = 1) \leq 2 \tag{2}$$

where

$$E(x, y) = Pr(a = b \mid x, y) - Pr(a \neq b | x, y)$$

Some of the numerous open problems concerning Bell inequalities (see Gisin (2007)):

- Is there a finite set of inequalities such that no other inequality is relevant with respect to that set? (An inequality is relevant with respect to a given set of inequalities if there is a quantum state violating it, but not violating any of the inequalities in the set.)
- Why are almost all known Bell inequalities for more than two outcomes maximally violated by states that are not maximally entangled?
- Is there a bound entangled state that violates some Bell inequalities?
- Given a multiparty quantum state ρ, how can we know whether ρ is non-local, i.e. whether there is a Bell inequality and measurements such that quantum physics predicts a violation of the inequality?

11   Grand Challenges of Quantum Informatics

Finally, let us introduce several grand challenges quantum informatics should deal with where algebraic and combinatorial methods can be expected to play a very important role.

(a) To "un-reveal secret of secrets" - Is our universe polynomial or exponential place (see Aaronson (2005))? (b) To get a better understanding of quantum mechanics using QIPCC concepts and tools; (c) To understand power of non-locality that goes beyond quantum mechanics; (d) To find out how far we can get beyond quantum mechanics; (e) To find out whether QM is an approximation of a cosmological non-local theory; (f) To get a deeper understanding of multipartite entanglement.

References

1. Aaronson, S.: Are quantum states exponentially long vectors? quant-ph/0507242 (2005)
2. Aharonov, D., Landau, Z., Makowsky, J.: The quantum FFT can be classically simulated. quant-ph/0611156 (2006)
3. Ambainis, A., Watrous, J.: Two-way finite automata with quantum and classical states. quant-ph/9911009 (1999)
4. Bacon, D.: How a Glebsh-Gordan transform helps to solve the Heisenberg Hidden Subgroup problem. quant-ph/0612107 (2006)
5. Bacon, D., Childs, A., van Dam, W.: Optimal measurements for the dihedral hidden subgroup problem. quant-ph/0501044 (2005)
6. Barnum, H., Knill, E., Ortiz, G., Somma, R., Viola, L.: A subsystem-independent generalization of entanglement. quant-ph/0305023 (2003)
7. Bertoni, S., Mereghetti, K., Palano, B.: Quantum computing: 1-way quantum automata. In: Ésik, Z., Fülöp, Z. (eds.) DLT 2003. LNCS, vol. 2710, pp. 1–20. Springer, Heidelberg (2003)

8. Bozapalidis, S.: Quantum recognizabla tree functions. In: Proceedings of the Conference on Unconventional models of computation - UMC'2K, pp. 25–47 (2000)
9. Bozapalidis, S.: Extending stochastic and quantum functions. Theory of computing systems 36(3), 183–197 (2003)
10. Brassard, G., Buhrman, H., Linden, N., Méthot, A.A., Tapp, A., Unger, F.: A limit on non-locality in any world in which communication complexity is not trivial. quant-ph/0508042 (2005)
11. Cerf, N., Gisin, N., Masar, S., Popescu, S.: Quantum entanglement can be simulated without communication. Physical Review Letter, 94:220403 (2005)
12. Cirel'son, B.S.: Quantum generalization's of Bell's inequality. Letters in Mathematical Physics 4(2), 93–100 (1980)
13. Clifton, R., Bub, J., Halvorson, H.: Characterizing quantum theory in terms of information-theoretic constraints. quant-ph/0211089 (2002)
14. Ettinger, M., Hoyer, P., Knill, E.: The quantum query complexity of the hidden subgroup problem is polynomial. quant-ph/0401083 (2004)
15. Gisin, N.: Bell inequalities: many questions, a few answers. quant-ph/0702021 (2007)
16. J. Gruska. *Quantum computing*. McGraw-Hill, 1999-2005, see also additions and updatings of the book on `http://www.mcgraw-hill.co.uk/gruska`
17. Gruska, J.: Descriptional complexity issues in quantum computing. Automata, Languages and Combinatorics 5(3), 198–218 (2000)
18. Gruska, J.: Quantum entanglement as a new quantum information processing resource. New Generation Computing 21, 279–295 (2003)
19. Gruska, J.: General Theory of information transfer and combinatorics, Universal sets of quantum information processing primitives and optimal use of such primitives, pp. 356–377. Springer-Heidelberg (2005)
20. Gruska, J.: From informatics to quantum informatics. In: Proceedings of the Fourth IFIP International Conference on Theoretical Computer Science- TCS 2006 at the World Computer Congress, pp. 17–46 (2006)
21. Gruska, J.: Recent advances in Formal Languages and Applications. Studies in Computational Intelligence 25, 81–117 (2006)
22. Gudder, S.: Basic properties of quantum automata. email: sgudder@cs.du.edu (2000)
23. Hallgren, S., Moore, C., Roetteler, M., Russel, A., Sen, P.: Limitations of quantum coset states for graph isomorphism. In: Proceedings of 38th STOC, pp. 604–617 (2006)
24. Horodecki, R., Horodecki, P., Horodecki, M., Horodecki, K.: Quantum entanglement. quant-ph/0702225 (2007)
25. Kuperberg, G.: A subexponential-time quantum algorithm for the dihedral hidden subgroup problem. New Generation Computing 35, 170–188 (2005), see also quant-ph/0302112.
26. Markov, I., Shi, Y.: Simulating quantum computation by contracting tensor. quant-ph/0511069 (2002)
27. Mereghetti, C., Palano, B.: Quantum finite automata with control language. Tech. rep. (2006)
28. Methot, A.A., Scarani, V.: An anomality of non-locality. quant-ph/0601210 (2000)
29. Moore, C., Russell, A., Sniady, P.: On the impossibility of a quantum sieve algorithm for graph isomorphism: unconditional result. quant-ph/0612089 (2006)
30. Moore, C., Russell, A., Vazirani, U.: A classical one-way function to confound quantum adversaries. quant-ph/0701115 (2007)

31. Nielsen, M.A., Chuang, I.I.: Quantum information processing. Cambridge University Press, Cambridge (2000)
32. Rao, M.V.P., Vinay, V.: Quantum finite automata and weighted automata (2007)
33. Perdrix, S.: Towards minimal resources of measurement-based quantum computation. quant-ph/070.0202 (2007)
34. Popescu, S., Rohrlich, D.: Causality and non-locality as axioms for quantum mechanics. quant-ph/9709026 (1997)
35. Scarani, V.: Feats, features and failures of the PR-box. quant-ph/0603017 (2006)
36. Shi, Y.: Both Toffoli and controlled-NOT need little help to do universal computation. quant-ph/0205115 (2002)
37. Short, T., Gisin, N., Popescu, S.: The physics of no-bit commitment generalized quantum non-locality versus oblivious transfer. quant-ph/0504134 (2005)
38. Smolin, J.A.: Can quantum cryptography imply quantum mechanics. quant-ph/0310067 (2003)
39. Tucci, R.R.: A rudimentary quantum compiler. quant-ph/9902062 (1999)
40. van Dam, W.: Implausible consequences of superstrong nonlocality. quant-ph/0501159 (2005)
41. Hallgren, S., van Dim, W., Ip, L.: Quantum algorithms for some hidden shift problems. quant-ph/0211140 (2002)
42. Watrous, J.: Quantum algorithms for solvable groups. quant-ph/0011023 (2000)

Recognizable vs. Regular Picture Languages

Oliver Matz

Institut für Informatik, Universität Kiel, Germany
matz@ti.informatik.uni-kiel.de

1 Introduction

The class of regular word languages plays a central role in formal language theory. Considerable effort has been made to transfer definitions and applications from word languages to their two-dimensional analog, the *picture languages*, where one considers (two-dimensional) matrices rather than (one-dimensional) words.

One may ask for a "natural" adaption of the class of regular word languages for pictures. The class of *recognizable* picture languages is usually considered the correct answer, chiefly because it is equivalently characterized by several formalisms such as different variants of tiling-systems, non-deterministic on-line tessellation automata, Wang-systems, existential monadic second-order logic, and doubly ranked monoids. This class has been studied intensively by several authors, e.g. [12, 13, 30, 14, 15, 7, 18, 11, 10, 16, 17, 25, 26, 20, 21, 29, 22, 6].

The number of different characterizations indeed indicates that the recognizable picture languages form a robust and therefore somewhat natural class to study. The fact that the one-dimensional restriction of each of the above formalisms characterizes regular word languages indicates that this class is a promising candidate for the two-dimensional equivalent of regular word languages. We discuss picture language recognizability in Section 2.

However, there is also some indication that the class of recognizable picture languages is too large. For example, for every linear bounded automaton $\mathfrak{A}$ (a Turing machine that never exceeds its input area), the set of pictures that (in the straightforward way) encode a run of $\mathfrak{A}$, is tiling-recognizable. This implies in particular that the emptiness problem is undecidable for tiling-systems. Besides, the membership problem is NP-complete ([28]). These two facts show that recognizable picture languages are computationally hard.

The mentioned characterizations of recognizable picture languages only adapt a certain aspect of regular word languages, which I would like to name "decoration + locality". If a different aspect is taken as the starting point, one obtains a different two-dimensional adaption.

Such a potential starting point are regular expressions, see e.g. [9, 10]. This concept also allows for a straightforward adaption to two dimensions, the *regular* picture languages. Every regular picture language is recognizable, but the converse is not true. In Section 3, we collect some indication that, nonetheless, the concept of regular picture languages is a different straightforward adaption of regular word languages.

S. Bozapalidis and G. Rahonis (Eds.): CAI 2007, LNCS 4728, pp. 112–121, 2007.

2   Picture Language Recognizability = Decoration + Locality

2.1   Definitions

Throughout the paper, Σ denotes a finite alphabet and is fixed unless stated otherwise. A *picture (over Σ)* is a matrix over Σ. A *picture language (over Σ)* is a set of pictures over Σ.

Let # be a fresh symbol not in Σ. A picture language L over Σ is *local* iff there exists a set Δ of 2×2-pictures such that L contains exactly those pictures p for which Δ contains all 2×2-sub-blocks of the picture which results by surrounding p with the boundary symbol #.

Let Γ be another alphabet. A mapping $\pi : \Gamma \to \Sigma$ can be extended to pictures and to picture languages as usual.

A picture language L over Σ is a *projection* of a picture language M over Γ iff there exists a mapping $\pi : \Gamma \to \Sigma$ such that $\pi(M) = L$.

A picture language L is *recognizable* iff it is the projection of a local picture language. The class of recognizable picture languages over Σ is denoted by REC(Σ) (or REC for short, if the alphabet is fixed).

2.2   Characterizations of Recognizable Picture Languages

Tiling Systems and Variations. The ingredients Σ, Γ, Δ, and π from the above definition together form a *tiling system*. In some sense, a tiling-system is a natural analog of non-deterministic finite automata to two-dimensions. It is possible to modify the definition in order to make this analogy more evident: We may assume w.l.o.g. that Γ is of the form $\Sigma \times Q$ for some finite set Q, and that the projection mapping π is such that $\pi(a, q) = a$ for every $a \in \Sigma$ and every $q \in Q$.

In this setting, the elements of Q could be called "states", but I prefer to name them "decorations" because the word "state" originally alludes to the intuition of information that is internally stored in a device and lost after that device has done another step of computation. Such a situation is not given for tiling systems.

There are several different ways to modify the definition of tiling systems without affecting the defined class of picture languages. For example, one can trade the size of the set Q of decorations against the block sizes in Δ: you can buy smaller blocks, but you pay by increasing the number of decorations (see [11]). This trade can be continued to separate the vertical from the horizontal transitions, resulting in the definition of *domino systems* (see [17,18]), where the set Δ contains 1×2- as well as 2×1-blocks (rather than 2×2-blocks).

The above trade "decorations against transition size" can be done in the opposite direction, too: If we allow the Δ to specify larger sub-blocks of the decorated picture, then we can reduce the number of decorations to two. For technical reasons it is more convenient to specify the forbidden sub-blocks of decorated pictures rather than the allowed ones as above. The necessary construction has been carried out in [21] in the setting of existential monadic second-oder logic.

On-Line Tessellation Automata. We have sketched different versions of tiling systems above. They have in common that their operation on an input picture p can be imagined as follows: Firstly, all input cells of p are decorated simultaneously, and then the decorated input is checked for local compatibility with the transition relation Δ.

The automation model of *on-line tessellation automata* is very similar, but it sequentializes the decoration and combines the two phases. Such an automaton assigns a decoration to every input cell in a pre-determined order, namely starting in the top-left corner, proceeding diagonal-wise, and finishing in the bottom right corner. Whenever a cell is decorated, the decorations of the previously visited top or left neighbor (if present) are available.

It is immediate that for a non-deterministic automaton model, it does not mean any loss of generality to pre-determine this order of decoration assignment, so these automata indeed capture the class of recognizable picture languages. The definition of *deterministic* on-line tessellation automata is straightforward, but they are less powerful than non-deterministic ones.

Tiling Automata – Tiling Systems with Scanning Strategy. On-line tessalation automata fix a specific order in which decorations are assigned to input cell, namely diagonal-wise from top-left to bottom-right. Recently, Anselmo, Giammarresi, and Madonia (see [8]), have suggested a promising approach to generalize this concept of a scanning strategy of a tiling system, resulting in the definition of *tiling automata*.

For the *non-deterministic* case, fixing a scanning strategy does not limit the expressive power. However, the presence of a scanning strategy allows for a straightforward definition of determinism.

The various ways to introduce determinism into the concept of "decoration + locality" are current research interest, see [26, 5, 3, 2, 4].

Wang Systems. Another approach to adapt the intuition of "decoration + locality" to picture languages is described in [25] and called *Wang system*. The essential difference to tiling systems is that a Wang system does not decorate the input cells but rather the edges between them. The transitions specify 5-tuples that can be visualized in the form

$$
\begin{array}{c}
q_1 \\
q_2 \ a \ q_3 \\
q_4
\end{array}
$$

with the obvious meaning. [25] shows how to translate tiling systems into Wang systems and vice versa.

Quadrapolic Automata and Doubly Ranked Monoids. In [6], the authors define another concept of non-deterministic automata for pictures. These automata, too, are based on the "decoration + locality" intuition as their behaviors can be obtained as projections of local picture languages. Their definition is the basis for a purely algebraic characterization of REC in terms of so-called *doubly ranked monoids*.

Existential Monadic Second-Order Logic. A very important result in the theory of formal languages is that the class of regular word languages is characterized by the monadic second-oder logic (MSO) in the signature with the successor relation, and even by the existential fragment thereof (EMSO).

The definitions of both MSO and its fragment EMSO can be adapted to the two-dimensional case of pictures (or rather picture models) in a straightforward way by introducing two distinct successor relation symbols for vertical and for horizontal successors, respectively.

In the fundamental paper [11], the authors show that the class of recognizable picture languages is not closed under complement, so it cannot be characterized by MSO. In the same paper, it is shown that REC *is* characterized by EMSO.

The proof of this characterization can be seen as another application of the intuition of "decoration + locality": An existential quantification prefix over, say, k sets of picture positions introduces a k-bit decoration, and the following first-order (i.e., set-quantification-free) formula specifies a local constraint. However, the proof is much more intricate because like in the word language case, first-order formulas can express also properties that are not entirely local but involve a limited form of counting: the first-order formulas characterize the locally threshold testable (word or picture) languages.

Closure Properties. The well-known Kleene-Theorem states that the class of word languages defined by NFA is the same as the class of word languages defined by regular expressions (or, more conveniently, their corresponding closure properties—the difference is not essential). It is natural to ask how the situation is for picture languages.

For this purpose we start with the definition of two partial concatenations on the set of pictures. For two pictures p, r over alphabet Σ, their *column concatenation* $p \oplus r$ (or *row concatenation* $p \ominus r$, respectively) is defined iff p and r have equal height (or width, respectively) by the result of juxtaposing r to the right (or to the bottom, respectively) of p.

As usual, these partial concatenations may be lifted to total operations on picture languages. These operations may be iterated as follows: For a picture language L, the *column closure* $L^{\oplus+}$ (or the *row closure* $L^{\ominus+}$, respectively) is defined as the smallest picture language that is a superset of L and is closed under column concatenation (or row concatenation, respectively).

The smallest class of picture languages over Σ which contains all singletons and is closed under row- and column-concatenation and -closure, as well as under union and intersection, is denoted $\cap-\mathrm{REG}(\Sigma)$. As shown in [9], $\mathrm{REC}(\Sigma)$ has all of these closure properties but is not characterized by them, i.e., $\cap-\mathrm{REG}(\Sigma) \subsetneq \mathrm{REC}(\Sigma)$, even in case Σ is a singleton. In other words, if we consider tiling systems the "right" adaption of NFA to two dimensions, then the Kleene-Theorem does not carry over, even if the intersection is allowed on the side of regular expressions.

By definition, REC is also closed under projection, meaning: for every language $L \in \mathrm{REC}(\Gamma)$ and every alphabet mapping $\pi : \Gamma \to \Sigma$, we have $\pi(L) \in \mathrm{REC}(\Sigma)$.

On the other hand, it is easy to see that every local picture language is the projection of the intersection of picture languages in REG (see [29] for a detailed analysis). From these two observations we may conclude:

Proposition 1. *The finite-alphabet-indexed family* $(\mathrm{REC}(\Sigma))_\Sigma$ *of picture language classes is the smallest[1] family* $(\mathcal{L}(\Sigma))_\Sigma$ *such that*

- *for each finite alphabet Σ, the class $\mathcal{L}(\Sigma)$ contains all finite picture languages over Σ and is closed under row- and column-concatenation and -closure as well as under union and intersection, and*
- *for all finite alphabets Σ_1, Σ_2, every projection mapping $\pi : \Sigma_1 \to \Sigma_2$ and every picture language $L \in \mathcal{L}(\Sigma_1)$ we have $\pi(L) \in \mathcal{L}(\Sigma_2)$.*

The above proposition is sometimes stated as: "REC is the smallest class of picture languages closed under row- and column-concatenation and -closure as well as under union, intersection and projection." This is somewhat imprecise because for the characterization it is essential to have a potentially unlimited number of decoration symbols available, i.e., to pass to an arbitrarily large alphabet. The needed closure under projection is *not* a closure property of the class REC over a fixed alphabet, but a closure property of the family of classes of picture languages over all finite alphabets.

The above proposition is also often considered a Kleene-like theorem. I do not subscribe to that point of view because the value and beauty of the original Kleene-Theorem is that it shows that the assembling character of regular expressions allows for equal expressive power as the (seemingly more flexible) computational character of NFA. Allowing the intersection in regular expressions already disturbs that assembling character, but additionally allowing arbitrarily large auxiliary alphabets together with projection completely ruins that character as it introduces decorations.

3   Regular Picture Languages

The class of *regular picture languages (over Σ)* (denoted $\mathrm{REG}(\Sigma)$ or simply REG) is the smallest class of picture languages over Σ that contains all singleton languages and that is closed under row- and column-concatenation and -closure, as well as under union.

The survey [10] contains a lot of separation results about regular picture languages and related classes. For example, the class REG is not closed under intersection if the alphabet contains at least two symbols ([10]). In particular, it is a proper subset of $\cap-\mathrm{REG}$, which is in turn a proper subset of REC.

---

[1] A family $(X(\Sigma))_\Sigma$ is *smaller* than another family $(Y(\Sigma))_\Sigma$ iff for all finite alphabets Σ we have $X(\Sigma) \subseteq Y(\Sigma)$.

In [23], a non-deterministic machine model is suggested. It is called *picture position pushdown automata (PPPA)* and characterizes REG. Such a PPPA proceeds stepwise, scanning each cell exactly once. It visits the input cells in the same order as a regular expression would assemble the input picture: for a column concatenation (or row concatenation, resp.), the left (or top, resp.) part is scanned before the right (or bottom, resp.) part. In order to control this process, the PPPA uses two distinct stacks of limited height on which it stores picture positions together with states.

Another observation in [23] is that the partialness of the row- and column-concatenation can, on the language level, be overcome. Intuitively, this means that every regular picture language can be assembled without exploiting the partialness of the concatenation: whatever picture has been assembled by a subexpression must contribute to the resulting language.

For the case of a singleton alphabet, REG is characterized by the finite unions of Cartesian products of ultimately periodic subsets of $\mathbb{N}$, just like the class of regular languages over a singleton alphabet is characterized by ultimately periodic subsets of $\mathbb{N}$.

A standard example for a non-regular picture language is the set of all squares, because the set $\{(n, n) \mid n \in \mathbb{N}\}$ is not a finite union of Cartesian products of ultimately periodic subsets of $\mathbb{N}$. The simplicity of this non-regular language might be an indication that the class REG is too weak for practical purposes.

3.1   Concatenation Alternation Hierarchy

The investigation of regular picture languages naturally leads to a hierarchy therein, which I would like to call *concatenation alternation hierarchy*. It is defined as follows:

Definition 1. *For every picture language L, the* transposition *of L is the set of all transpositions of elements of L.*

For $n \geq 1$ and a finite alphabet Σ, the class $\mathbf{r}\text{REG}n(\Sigma)$ of picture languages over Σ is inductively defined as follows:

- *$\mathbf{r}\text{REG}1(\Sigma)$ is the set of regular word languages over Σ.*
- *For every $n \geq 1$, the class $\mathbf{r}\text{REG}n+1(\Sigma)$ is the smallest class of picture languages that contains all transpositions of picture languages in $\mathbf{r}\text{REG}n(\Sigma)$ and that is closed under union, row concatenation, and row closure.*

Intuitively, a picture language is in the n-th level $\mathbf{r}\text{REG}n(\Sigma)$ of that hierarchy iff it can be defined by at most n alternations of row- and column-operations. Here, we count a sequence of consecutive applications of concatenations, iterations, and unions as just *one* operation as long as its concatenations and iterations are in the same direction (either row- or column).

The proofs in [19] show that, for every n, a picture language is in the n-th level $\mathbf{r}\text{REG}n(\Sigma)$ of that hierarchy iff it is definable by a PPPA with maximal stack height n.

As to my knowledge, it is open whether the concatenation alternation hierarchy is infinite or whether it collapses to a fixed level.

4   Generalizations of Regular Picture Languages

Different ways have been suggested to generalize regular operations. The first obvious way is to add intersection or even complement to the closure properties. This increases the defined class of picture languages iff the alphabet is non-trivial. In particular, the set of all squares cannot be obtained this way either. In case of a non-trivial alphabet, adding the complement allows to define non-recognizable picture languages, see [20]. In any case, intersection spoils the assembling character and thereby the beauty of this concept.

A second way to generalize regular operations is to allow more complex concatenations. As suggested in [1], one may consider a partial 4-ary concatenation that assembles four pictures p_1, p_2, p_3, p_4 to one picture $\begin{array}{|c|c|} \hline p_1 & p_2 \\ \hline p_3 & p_4 \\ \hline \end{array}$ and is defined iff, firstly, p_1 and p_2 as well as p_3 and p_4 have matching heights, and, secondly, p_1 and p_3 as well as p_2 and p_4 have matching widths. In the same spirit one may even consider 9-ary concatenations etc.

Some preliminary results on this subject and lots of examples can be found in [1] for the case of a singleton alphabet.

A third obvious way to make regular expressions more powerful is to allow the iteration to range over more complex compositions of concatenations. This approach is presented in [18, 19] introducing the class $\mathrm{REG}^{\mathrm{OP}}$. In these papers, too, quite a lot of the results deal with the case of a singleton alphabet. I would like to report a specific one from [18] here. For a function $f : \mathbb{N} \to \mathbb{N}$, let L_f be the set of all pictures of size $m \times f(m)$ for some m. For example, if f is the identity function, then L_f is the set of squares.

Proposition 2. *([18]) If f is a polynomial with non-negative integer coefficients, then $L_f \in \mathrm{REG}^{\mathrm{OP}}$. Conversely, if f is a function with $L_f \in \mathrm{REG}^{\mathrm{OP}}$, then f is bounded by a polynomial.*

The degree of the polynomial depends on the nesting depth of the iteration.

Proposition 2 complements a result of [7] stating the corresponding fact for the class REC and the singly exponential functions. The corresponding fact is also true for the class REG and the constants functions. In particular, we have $\mathrm{REG} \subsetneq \mathrm{REG}^{\mathrm{OP}} \subsetneq \mathrm{REC}$, even for a singleton alphabet.

Also [24] provides a fact similar to Proposition 2, this time for the k-th level of the MSO quantifier alternation hierarchy and the k-fold exponential functions.

5   Conclusion

We started by asking which is the "correct" adaption of "regularity" to two dimensions. Some reasonable criteria for the jury are:

Canonicity. The definition should adapt one or more definitions of regular word languages in a straightforward way.

Robustness. Like for word languages, there should be several conceptually different characterizations.

Closure properties. Like for word languages, there should be convenient closure properties.

Simplicity. Like for word languages, it should be easy to tell for a given candidate language whether it belongs to the class.

Computational complexity. Like for word languages, standard decision problems should be computationally very easy.

Two candidate picture languages classes are REC (recognizable picture languages) and REG (regular picture languages).

My personal conclusion is that this competition ends in a tie. Let us investigate each of the five above criteria.

Concerning the **canonicity**, both classes do very well as their definition naturally adapts a concept from the word languages: NFA (with "decoration + locality" intuition) for REC, and that of assembling regular expressions in case of REG.

Concerning the **robustness**, both classes perform satisfactory: REC is characterized by tiling-systems, an adaption of non-deterministic finite automata. Different flavors of their definition are possible, but all are based on the intuition of "decoration + locality". REC provides also a logical ([11]) and an algebraic characterization ([6]), but no assembling regular expressions and no satisfactory characterization by closure properties. REG has no logical nor algebraic characterization (except for the case of a singleton alphabet), but it has a characterization by closure properties. I admit that the definition of its automaton model, the PPPA, lacks the beauty and succinctness of tiling systems, but its stepwise operation reflects the normal intuition of NFA better. Both candidates lack a deterministic automaton model. Concluding, REC performs a little better than REG in this category.

Concerning the **closure properties**, both classes perform well as they are closed under union, (iterated) row-/column-concatenation, and projection. REC additionally provides (the certainly precious) closure under intersection, whereas REG does not. On the other hand, for the case of a singleton-alphabet, REG provides closure even under complement, whereas REC does not ([22]). Again, REC performs only slightly better than REG in this category.

Concerning the **simplicity**, REG is my winner because I had made several wrong conjectures about recognizability of specific classes (see e.g. [26, 27]). Besides, the set of top rows of a recognizable picture language forms a context-sensitive word language, and every context-sensitive word language can be obtained this way ([16]). This indicates also that recognizable picture languages are "as complex" as context-sensitive word languages. In contrast to that, the set of top rows of a regular picture language is regular ([23]).

On the other hand, one might object that REG is *too* simple for practical applications as there are very simple non-regular picture languages.

Concerning the **computational complexity**, REG is the clear winner since the membership problem for tiling systems is NP-complete ([28]), and the emptiness problem for tiling systems is undecidable ([9]), whereas both of these problems are in P for regular expressions.

Despite this tie, the class of recognizable picture languages has gained much more attention. I think the class of regular picture language deserves more interest in the future. It is a promising candidate for practical applications such as two-dimensional pattern matching. Potential starting points for further investigation include deterministic automaton models in the spirit of PPPA as well as regular expressions with enhanced concatenations (as suggested at the end of [1]) or enhanced iteration of standard concatenations (as suggested in [19]).

References

1. Anselmo, M., Giammarresi, D., Madonia, M.: New operations and regular expressions for two-dimensional languages over one-letter alphabet. Theor. Comput. Sci. 340(1), 408–431 (2005)
2. Anselmo, M., Giammarresi, D., Madonia, M.: From determinism to nondeterminism in recognizable two-dimensional languages. In: Developments in Language Theory. Springer, Heidelberg (to appear, 2007)
3. Anselmo, M., Giammarresi, D., Madonia, M., Restivo, A.: Unambigiuos recognizable two-dimensional languages. Inf. Theor. Appl. 40, 277–293 (2006)
4. Anselmo, M., Madonia, M.: Deterministic two-dimensional languages over one-letter alphabet. In: Bozapalidis, S., Rahonis, G. (eds.) CAI 2007. LNCS, vol. 4728, pp. 147–159. Springer, Heidelberg (2007)
5. Borchert, B., Reinhardt, K.: Deterministically and sudoku-deterministically recognizable picture languages (2006), http://tobias-lib.ub.uni-tuebingen.de/volltexte/2006/2503/
6. Bozapalidis, S., Grammatikopoulou, A.: Recognizable picture series. Journal of Automata, Languages and Combinatorics 10(2/3), 159–183 (2005)
7. Giammarresi, D.: Two-dimensional languages and recognizable functions. In: Rozenberg, G., Salomaa, A. (eds.) Developments in Language Theory, Proceedings of the conference, Turku (Finnland), 1993, pp. 290–301. world scientific, Singapore (1994)
8. Giammarresi, D.: Tiling recognizable two-dimensional languages. Springer, Heidelberg (2007) included in these proceedings
9. Giammarresi, D., Restivo, A.: Recognizable picture languages. In: Proceedings First International Colloqium on Parallel Image Processing 1991. International Journal Pattern Recognition and Artificial Intelligence, vol. 6, pp. 241–256 (1992)
10. Giammarresi, D., Restivo, A.: Two-dimensional languages. In: Rozenberg, G., Salomaa, A. (eds.) Handbook of Formal Language Theory, vol. III, pp. 215–268. Springer, New York (1996)
11. Giammarresi, D., Restivo, A., Seibert, S., Thomas, W.: Monadic second-order logic and recognizability by tiling systems. Information and Computation 125, 32–45 (1996)
12. Inoue, K., Nakamura, A.: Nonclosure properties of two-dimensional on-line tessellation acceptors and one-way parallel/sequential array acceptors. Transaction of IECE of Japan 6, 475–476 (1977)
13. Inoue, K., Nakamura, A.: Some properties of two-dimensional on-line tessellation acceptors. Information Sciences 13, 95–121 (1977)
14. Inoue, K., Takanami, I.: A survey of two-dimensional automata theory. In: Dassow, J., Kelemen, J. (eds.) Proceedings 5th International Meeting of Young Computer Scientists. 5th International Meeting of Young Computer Scientists. LNCS, vol. 381, pp. 72–91. Springer, Heidelberg (1990)

15. Inoue, K., Takanami, I.: A characterization of recognizable picture languages. In: Dassow, J., Kelemen, J. (eds.) Machines, Languages, and Complexity. LNCS, vol. 381, pp. 133–143. Springer, Heidelberg (1992)
16. Latteux, M., Simplot, D.: Context-sensitive string languages and recognizable picture languages. Information and Computation 138, 160–169 (1997)
17. Latteux, M., Simplot, D.: Recognizable picture languages and domino tiling. Theoretical Computer Science 178(1-2), 275–283 (1997)
18. Matz, O.: Klassifizierung von Bildsprachen mit rationalen Ausdrücken, Grammatiken und Logik-Formeln. Diploma thesis, Christian-Albrechts-Universität Kiel (in German) (1995)
19. Matz, O.: Regular expressions and context-free grammars for picture languages. In: Reischuk, R., Morvan, M. (eds.) STACS 97. LNCS, vol. 1200, pp. 283–294. Springer, Heidelberg (1997)
20. Matz, O.: On piecewise testable, starfree, and recognizable picture languages. In: Nivat, M. (ed.) ETAPS 1998 and FOSSACS 1998. LNCS, vol. 1378, pp. 203–210. Springer, Heidelberg (1998)
21. Matz, O.: One quantifier will do in existential monadic second-order logic over pictures. In: Brim, L., Gruska, J., Zlatuška, J. (eds.) MFCS 1998. LNCS, vol. 1450, pp. 751–759. Springer, Heidelberg (1998)
22. Matz, O.: Dot-depth, monadic quantifier alternation, and first-order closure over grids and pictures. Theor. Comput. Sci. 270(1-2), 1–70 (2002)
23. Matz, O.: A Kleene theorem for regular picture languages. Technical Report 0703, Christian-Albrechts-Universität Kiel (2007)
24. Matz, O., Thomas, W.: The monadic quantifier alternation hierarchy over graphs is infinite. In: Twelfth Annual IEEE Symposium on Logic in Computer Science, Warsaw, Poland, pp. 236–244. IEEE Computer Society Press, Los Alamitos (1997)
25. Prophetis de, L., Varricchio, S.: Recognizability of rectangular pictures by wang systems. Journal of Automata, Languages and Combinatorics 2, 269–288 (1997)
26. Reinhardt, K.: On some recognizable picture-languages. In: Brim, L., Gruska, J., Zlatuška, J. (eds.) MFCS 1998. LNCS, vol. 1450, pp. 760–770. Springer, Heidelberg (1998)
27. Reinhardt, K.: The #a = #b pictures are recognizable. In: Symposium on Theoretical Aspects of Computer Science, pp. 527–538 (2001)
28. Schweikardt, N.: The monadic quantifier alternation hierarchy over grids and pictures. In: Nielsen, M. (ed.) CSL 1997. LNCS, vol. 1414, pp. 441–460. Springer, Heidelberg (1998)
29. Simplot, D.: A characterization of recognizable picture languages by tilings by finite sets. Theoretical Computer Science 218(2), 297–323 (1999)
30. Siromoney, R.: Advances in array languages. In: Ehrig, H., Nagl, M., Rosenfeld, A., Rozenberg, G. (eds.) Graph-Grammars and Their Application to Computer Science. LNCS, vol. 291, pp. 549–563. Springer, Heidelberg (1987)

From Algebraic Graph Transformation to Adhesive HLR Categories and Systems

Ulrike Prange and Hartmut Ehrig

Technical University of Berlin, Germany
{uprange,ehrig}@cs.tu-berlin.de

Abstract. In this paper, we present an overview of algebraic graph transformation in the double pushout approach. Basic results concerning independence, parallelism, concurrency, embedding, critical pairs and confluence are introduced. As a generalization, the categorical framework of adhesive high-level replacement systems is introduced which allows to instantiate the rich theory to several interesting classes of high-level structures.

1 Introduction to Graph Transformation

Combining the important concepts of graphs, grammars and rewriting, the research area of graph grammars or graph transformation is a discipline of computer science which dates back to the 1970s. Methods, techniques, and results from the area of graph transformation have already been studied and applied in many fields of computer science, such as formal language theory, pattern recognition and generation, compiler construction, software engineering, the modeling of concurrent and distributed systems, database design and theory, logical and functional programming, artificial intelligence, and visual modeling. A detailed presentation of various graph grammar approaches and application areas of graph transformation is given in the handbooks [1, 2, 3].

This wide applicability is due to the fact that graphs are a very natural way of explaining complex situations on an intuitive level. Hence, they are used in computer science almost everywhere, for example for data and control flow diagrams, for entity relationship and UML diagrams, for Petri nets, for visualization of software and hardware architectures, for evolution diagrams of nondeterministic processes, for SADT diagrams, and for many more purposes.

The main idea of graph transformation is the rule-based modification of graphs, as shown in Fig. 1. The core of a rule or production p is a pair of graphs (L, R), called the left-hand side L and the right-hand side R. Applying the rule $p = (L, R)$ means finding a match of L in the source graph G and replacing L by R, leading to the target graph H of the graph transformation. The main technical problems are how to delete L from G and how to connect R with the remaining context leading to the target graph H. In fact, there are several different solutions how to handle these problems, leading to several different graph transformation approaches.

S. Bozapalidis and G. Rahonis (Eds.): CAI 2007, LNCS 4728, pp. 122–146, 2007.
© Springer-Verlag Berlin Heidelberg 2007

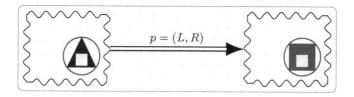

Fig. 1. Rule-based modification of graphs

The algebraic graph transformation approach is based on pushout constructions, where pushouts are used to model the gluing of two graphs along a common
subgraph. Intuitively, we use this common subgraph and add all other nodes and
edges from both graphs. In the algebraic approach, two gluing constructions are
used to model a graph transformation step. For this reason, this approach is also
known as the double-pushout (DPO) approach.

Roughly speaking, a production is given by $p = (L, K, R)$, where L and R are
the left- and right-hand side graphs and K is the common interface of L and
R, i.e. their intersection. The left-hand side L represents the preconditions of
the rule, while the right-hand side R describes the postconditions. K describes
a graph part which has to exist to apply the rule, but which is not changed.
$L \backslash K$ describes the part which is to be deleted, and $R \backslash K$ describes the part to
be created.

A direct graph transformation with a production p is defined by first finding a
match m of the left-hand side L in the current host graph G and then constructing the pushouts (1) and (2) in Fig. 2. For the construction of the first pushout,
however, a gluing condition has to be satisfied, which allows us to construct D
such that G is the gluing of L and D via K. The second pushout means that H
is the gluing of R and D via K. This means that a direct graph transformation
$G \Rightarrow H$ in Fig. 2 consists of two gluing constructions, which are pushouts in the
category of graphs and graph morphisms.

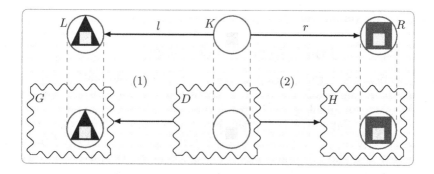

Fig. 2. DPO graph transformation

The algebraic approach to graph transformation is not restricted to (standard) graphs, but has been generalized to a large variety of different types of graphs and other kinds of high-level structures, such as labeled graphs, typed graphs, hypergraphs, attributed graphs, Petri nets, and algebraic specifications. This extension from graphs to high-level structures – in contrast to strings and trees, considered as low-level structures – was initiated in [4, 5] leading to the theory of high-level replacement (HLR) systems. In [6, 7], the concept of high-level replacement systems was joined to that of *adhesive categories* introduced by Lack and Sobociński in [8], leading to the concept of adhesive HLR categories and systems. There are several interesting instantiations of adhesive HLR systems, including not only graph and typed graph transformation systems, but also hypergraph, Petri net, algebraic specification, and typed attributed graph transformation systems.

In addition to pushouts, which correspond to the gluing of graphs, adhesive HLR categories are based on pullbacks, corresponding to the intersection and homomorphic preimages of graphs. The basic axioms of adhesive HLR categories require construction and basic compatibility properties for pushouts and pullbacks. These properties (and a few additional ones) allow to prove several interesting results concerning transformations.

In Section 2, we introduce algebraic graph transformation based on the double pushout approach and present the main results for transformations together with illustrating examples. The categorical framework of adhesive HLR systems is introduced in Section 3. For a more detailed presentation including all the proofs and further results we refer to our book [7].

2   Algebraic Graph Transformation – The Double Pushout Approach

In this section, we introduce graph transformation in the double pushout approach and give an overview of important results. We present the main results with illustrative examples, but give only an intuitive idea of some of the new notions used in these results. A formal definition of these notions and also the proofs of these results are given in [7].

2.1   Graph and Typed Graph Transformation

In this section, we introduce graph and typed graph transformation systems, or (typed) graph transformation systems, for short. In the following, we always use an abbreviated terminology of this kind to handle both cases simultaneously.

A graph has nodes, and edges, which link two nodes. We consider directed graphs, i.e. every edge has a distinguished start node (its source) and end node (its target). We allow parallel edges, as well as loops. Graphs are related by (total) graph morphisms, which map the nodes and edges of a graph to those of another one, preserving the source and target of each edge.

Definition 1 (Graph). *A graph* $G = (V, E, s, t)$ *consists of a set* V *of nodes (also called vertices), a set* E *of edges, and two functions* $s, t : E \to V$, *the source and target functions.*

Given graphs G_1, G_2 *with* $G_i = (V_i, E_i, s_i, t_i)$ *for* $i = 1, 2$, *a graph morphism* $f : G_1 \to G_2$, $f = (f_V, f_E)$ *consists of two functions* $f_V : V_1 \to V_2$ *and* $f_E : E_1 \to E_2$ *that preserve the source and target functions, i.e.* $f_V \circ s_1 = s_2 \circ f_E$ *and* $f_V \circ t_1 = t_2 \circ f_E$.

If f_V *and* f_E *are both injective (bijective) then* f *is called an injective (isomorphic) graph morphism.*

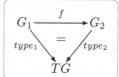

Graphs and graph morphisms form the category **Graphs** *of graphs.*

A type graph defines a set of types, which can be used to assign a type to the nodes and edges of a graph. The typing itself is done by a graph morphism between the graph and the type graph.

Definition 2 (Typed graph). *A type graph is a distinguished graph* $TG = (V_{TG}, E_{TG}, s_{TG}, t_{TG})$. V_{TG} *and* E_{TG} *are called the vertex and the edge type alphabets, respectively.*

A tuple $G^T = (G, type)$ *of a graph* G *together with a graph morphism* $type : G \to TG$ *is then called a* typed graph.

Given typed graphs $G_1^T = (G_1, type_1)$ *and* $G_2^T = (G_2, type_2)$, *a typed graph morphism* $f : G_1^T \to G_2^T$ *is a graph morphism* $f : G_1 \to G_2$ *such that* $type_2 \circ f = type_1$.

Typed graphs and typed graph morphisms form the category **Graphs$_{TG}$** *of typed graphs over the type graph* TG.

For simplicity, in the following we use the notation G for both graphs and typed graphs.

Example 1. In the following, we model a variant of Dijkstra's algorithm for mutual exclusion (see [9]). Given two processes that compete for a resource used by both of them, the aim of the algorithm is to ensure that once one process is using the resource the other has to wait and cannot access it.

There is a global variable *turn* that assigns the resource to any of the processes initially. Each process i has a flag $f(i)$ with possible values 0, 1, 2, initially set to 0, and a state that is initially *non-active*. If the process wants to access the resource, its state changes to *active* and the flag value is set to 1. If the variable *turn* has assigned the resource already to the requesting process, the flag can be set to 2, which indicates that the process is accessing the resource. Then the process uses the resource and is in its critical section. Meanwhile, no other process can access the resource, because the turn variable cannot be changed in this stage of the process. After the critical section has been exited, the flag is set back to 0 and the state to *non-active*. Otherwise, if the resource is assigned to a nonactive process, it can be reassigned and then accessed analogously by the requesting process.

The type graph TG is given in Fig. 3. Each process is typed by P, a resource is typed by R, and T denotes the turn. If the flag of a process is set to 0, we do not depict it in the graph. The flag values 1 and 2 are shown by nodes typed with $F1$ or $F2$, respectively, with a link from the corresponding process to the node and a link to the required resource.

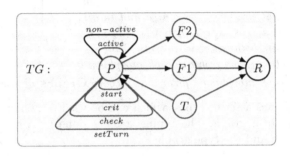

Fig. 3. Example type graph

A typed graph S is given in Fig. 4, containing two nonactive processes that can compete for one resource, where the graph morphism $type : S \to TG$ is given by the labels of the nodes and edges.   □

(Typed) graph transformation is based on (typed) graph productions, which describe a general way how to transform (typed) graphs. The application of a (typed) graph production to a (typed) graph is called a direct (typed) graph transformation. This is based on the concept of pushouts which is motivated to be a gluing construction in the introduction.

Definition 3 (Graph production and transformation). *A (typed) graph production $p = (L \xleftarrow{l} K \xrightarrow{r} R)$ consists of (typed) graphs L, K, and R, called the left-hand side, gluing graph, and the right-hand side respectively, and two injective (typed) graph morphisms l and r.*

Given p, a (typed) graph G, and a (typed) graph morphism $m : L \to G$, called match, a direct (typed) graph transformation $G \xRightarrow{p,m} H$ from G to a (typed) graph H is given by the pushouts (1) and (2), where the (typed) graph morphism n is called comatch.

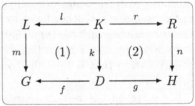

A sequence $G_0 \Rightarrow G_1 \Rightarrow \ldots \Rightarrow G_n$ of direct (typed) graph transformations is called a (typed) graph transformation and is denoted by $G_0 \xRightarrow{*} G_n$. For $n = 0$, we have the identical (typed) graph transformation $G_0 \xRightarrow{id} G_0$. Moreover, for $n = 0$ we allow also graph isomorphisms $G_0 \cong G_0'$, because pushouts and hence also direct graph transformations are only unique up to isomorphism.

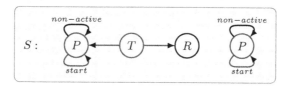

Fig. 4. Example typed graph

Example 2. For our mutual exclusion example, we have five typed graph productions shown in Fig. 5, where all morphisms are inclusions. The typed graph production *setFlag* allows a nonactive process to indicate a request for the resource by setting its flag to 1. The typed graph production *setTurn1* allows the turn to be changed to an active process if the other process, which has the turn, is nonactive. If the turn is already assigned to the active process, then the turn remains in *setTurn2*. Thereafter, in the typed graph production *enter*, the process enters its critical section. Finally, the process exits the critical section with the typed graph production *exit* and another process may get the turn and access the resource.

We can apply the typed graph production *setFlag* to the typed graph S given in Fig. 4 with a match m, leading to the direct typed graph transformation $S \overset{setFlag,m}{\Longrightarrow} G_1$ shown in Fig. 6.

If we apply the typed graph productions *setFlag*, *setTurn1*, *enter*, *setFlag*, and *exit* to S, then we obtain the typed graph transformation $S \overset{*}{\Rightarrow} G$ shown in Fig. 7. □

Now we analyze under what conditions a (typed) graph production $p = (L \leftarrow K \rightarrow R)$ can be applied to a (typed) graph G via a match m. In general, the existence of a context graph D that leads to a pushout (1) is required. This allows us to construct a direct (typed) graph transformation $G \overset{p,m}{\Longrightarrow} H$, where, in a second step, the (typed) graph H is constructed as the gluing of D and R via K leading to a pushout (2). Note that the construction of D and H is unique up to isomorphism.

Definition 4 (Gluing condition). *A (typed) graph production* $p = (L \overset{l}{\leftarrow} K \overset{r}{\rightarrow} R)$ *is applicable to a (typed) graph G via the match m if the following condition holds:*

p and m satisfy the gluing condition *if all identification points and all dangling points are also gluing points, i.e. $IP \cup DP \subseteq GP$, where*

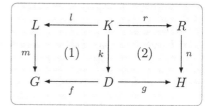

- *the* gluing points GP *are those nodes and edges in L that are not deleted by p, i.e. $GP = l_V(V_K) \cup l_E(E_K) = l(K)$,*

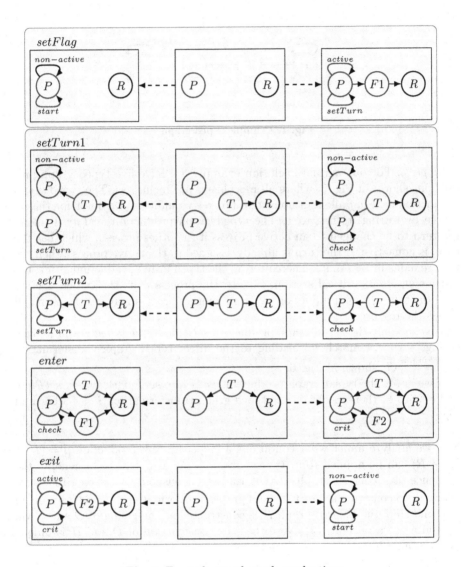

Fig. 5. Example typed graph productions

- the identification points IP are those nodes and edges in L that are identified by m, i.e. $IP = \{v \in V_L \mid \exists w \in V_L, w \neq v : m_V(v) = m_V(w)\} \cup \{e \in E_L \mid \exists f \in E_L, f \neq e : m_E(e) = m_E(f)\}$,
- the dangling points DP are those nodes in L whose images under m are the source or target of an edge in G that does not belong to $m(L)$, i.e. $DP = \{v \in V_L \mid \exists e \in E_G \backslash m_E(E_L) : s_G(e) = m_V(v) \text{ or } t_G(e) = m_V(v)\}$.

Example 3. For the direct typed graph transformation in Fig. 6, we analyze the gluing, identification, and dangling points:

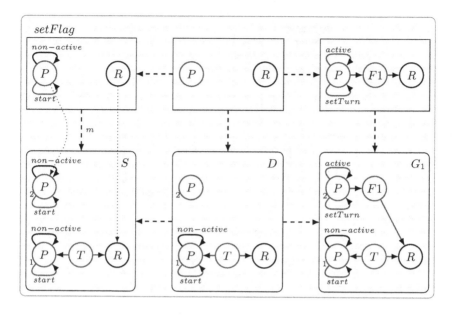

Fig. 6. Example direct typed graph transformation

- $GP = l(K)$, which means that the gluing points in L are both nodes.
- $IP = \varnothing$, since m does not identify any nodes or edges.
- The resource node is the only dangling point: in S, there is an edge from the turn node T (which has no preimage in L) to the resource node R, but there is no edge from or to the upper process node P that is not already in L.

This means that $IP \cup DP \subseteq GP$, and the gluing condition is satisfied by m and $setFlag$.

In contrast, the typed graph production *deleteProcess* given in the top row of Fig. 8 is not applicable to S with the match m'. We have:

- $GP = l(K)$, which means that there are no gluing points in L.
- $IP = \varnothing$, since m' does not identify any nodes or edges.
- The process node in L is a dangling point: in S, there are two loops at this node, which have no preimages in L.

This means that $DP \not\subseteq GP$, and the gluing condition is not satisfied by m' and *deleteProcess*.   □

Now we shall define (typed) graph transformation systems and (typed) graph grammars. The language of a (typed) graph grammar consists of those (typed) graphs that can be derived from the start graph.

Definition 5 (Graph transformation system and graph grammar). *A graph transformation system* $GTS = (P)$ *consists of a set of graph productions P.*

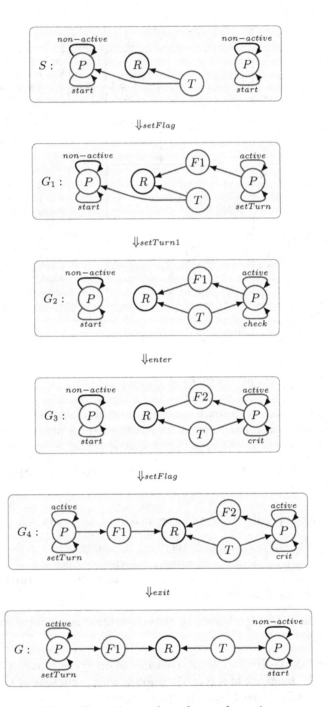

Fig. 7. Example typed graph transformation

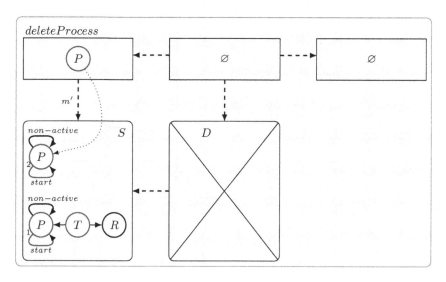

Fig. 8. Example of non-applicability

A typed graph transformation system $GTS = (TG, P)$ *consists of a type graph TG and a set of typed graph productions P.*

A (typed) graph grammar $GG = (GTS, S)$ consists of a (typed) graph transformation system GTS and a (typed) start graph S.

The (typed) graph language L of a (typed) graph grammar GG is defined by

$$L = \{G \mid \exists \ (typed) \ graph \ transformation \ S \overset{*}{\Rightarrow} G\}.$$

Example 4. Combining the type graph in Fig. 3, the typed graph productions in Fig. 5 and the start graph S in Fig. 4 we have the typed graph grammar $MutualExclusion = (TG, P, S)$ with $P = \{setFlag, setTurn1, setTurn2, enter, exit\}$.

To show that this typed graph grammar indeed ensures mutual exclusion, the whole derivation graph is depicted in Fig. 9. The nodes – which stand for the graphs in the typed graph language – show, in an abbreviated notation, the state of the processes. On the left-hand side of each node, the state of the first process is shown, and also its flag value and if the turn is assigned to that process. Analogously, this information for the second process is depicted on the right-hand side. The marked nodes are those nodes where the resource is actually accessed by a process – and only one process can access it at any one time. □

2.2 Overview of Results for (Typed) Graph Transformations

In the following, we present important results for (typed) graph transformations, namely the

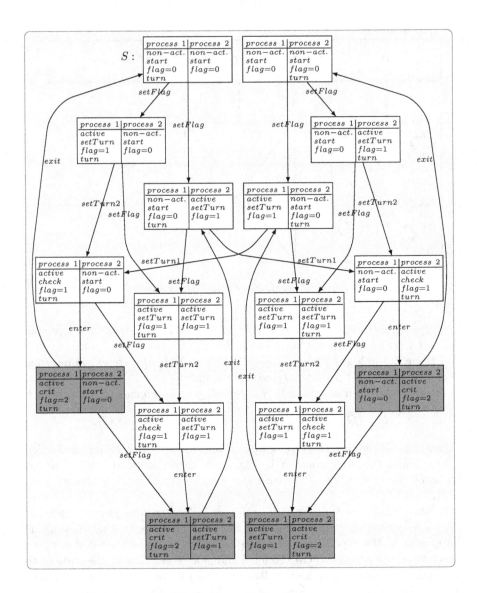

Fig. 9. Example language

- Local Church–Rosser and Parallelism Theorem,
- Concurrency Theorem,
- Embedding and Extension Theorem,
- Critical Pairs and Local Confluence Theorem,
- Graph Constraints and Application Conditions.

Local Church–Rosser and Parallelism Theorem

The first theorem is concerned with parallel and sequential independence of direct (typed) graph transformations. We study under what conditions two direct (typed) graph transformations applied to the same (typed) graph can be applied in arbitrary order, leading to the same result. This leads to the Local Church–Rosser Theorem. Moreover, the corresponding (typed) graph productions can be applied in parallel in this case, leading to the Parallelism Theorem.

Two direct (typed) graph transformations $G \overset{p_1,m_1}{\Longrightarrow} H_1$ and $G \overset{p_2,m_2}{\Longrightarrow} H_2$ are parallel independent, if p_1 does not delete anything p_2 uses, and vice versa. This means that all nodes and edges in the intersection of the two matches are gluing items with respect to both transformations, i.e.

$$m_1(L_1) \cap m_2(L_2) \subseteq m_1(l_1(K_1)) \cap m_2(l_2(K_2)).$$

Analogously, two direct (typed) graph transformations $G \overset{p_1,m_1}{\Longrightarrow} H_1 \overset{p_2,m'_2}{\Longrightarrow} G'$ are sequentially independent, if p_1 does not create something p_2 uses, and p_2 does not delete something p_1 uses or creates. This means that all nodes and edges in the intersection of the comatch $n_1 : R_1 \to H_1$ and the match m_2 are gluing items with respect to both transformations, i.e.

$$n_1(R_1) \cap m_2(L_2) \subseteq n_1(r_1(K_1)) \cap m_2(l_2(K_2)).$$

With this notion of independence, we are able to formulate the Local Church–Rosser and Parallelism Theorem.

Theorem 1 (Local Church–Rosser and Parallelism Theorem). *Given two parallel independent direct (typed) graph transformations $G \overset{p_1,m_1}{\Longrightarrow} H_1$ and $G \overset{p_2,m_2}{\Longrightarrow} H_2$, there is a (typed) graph G' together with direct (typed) graph transformations $H_1 \overset{p_2,m'_2}{\Longrightarrow} G'$ and $H_2 \overset{p_1,m'_1}{\Longrightarrow} G'$ such that $G \overset{p_1,m_1}{\Longrightarrow} H_1 \overset{p_2,m'_2}{\Longrightarrow} G'$ and $G \overset{p_2,m_2}{\Longrightarrow} H_2 \overset{p_1,m'_1}{\Longrightarrow} G'$ are sequentially independent.*

Given two sequentially independent direct (typed) graph transformations $G \overset{p_1,m_1}{\Longrightarrow} H_1 \overset{p_2,m'_2}{\Longrightarrow} G'$, there are a (typed) graph H_2 and direct (typed) graph transformations $G \overset{p_2,m_2}{\Longrightarrow} H_2 \overset{p_1,m'_1}{\Longrightarrow} G'$ such that $G \overset{p_1,m_1}{\Longrightarrow} H_1$ and $G \overset{p_2,m_2}{\Longrightarrow} H_2$ are parallel independent.

In any case of independence, there is a parallel (typed) graph transformation $G \Rightarrow G'$ via the parallel (typed) graph production $p_1 + p_2$, which is the disjoint union of the (typed) graph productions p_1 and p_2. Vice versa, the parallel (typed) graph transformation $G \Rightarrow G'$ can be sequentialized both ways.

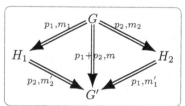

Example 5. We apply the typed graph production *setFlag* twice to the start graph S, first with the match m, and the second time with a different match m'

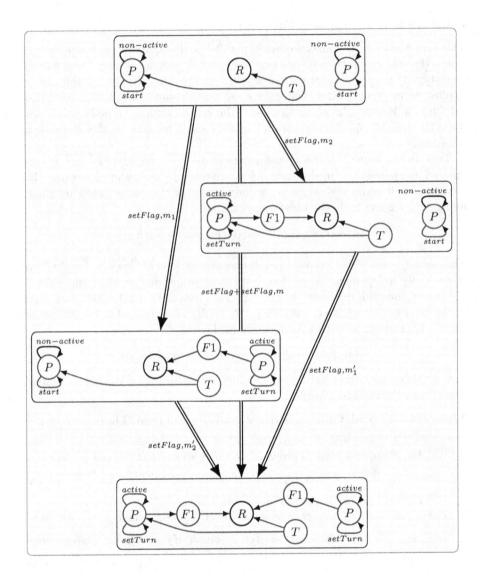

Fig. 10. Example Local Church-Rosser and Parallelism Theorem

that maps the process node in L to the other process node in S. These two direct typed graph transformations are parallel independent: in the intersection of the matches, there is only the resource node, which is a gluing point with respect to both transformations. Applying the Local Church–Rosser and Parallelism Theorem, we can apply *setFlag* again switching the matches leading to the same typed graph, as well as it is possible to apply *setFlag* + *setFlag* directly to S with the same result as shown in Fig. 10.   □

Concurrency Theorem

In contrast to the Local Church–Rosser Theorem, the Concurrency Theorem is concerned with the execution of (typed) graph transformations which may be sequentially dependent. This means that, in general, we cannot commute subsequent direct (typed) graph transformations, as done for independent transformations in the Local Church–Rosser Theorem, nor are we able to apply the corresponding productions in parallel, as done in the Parallelism Theorem. Nevertheless, it is possible to apply both transformations concurrently using a so-called E-concurrent (typed) graph production $p_1 *_E p_2$. Given an arbitrary sequence $G \stackrel{p_1,m_1}{\Longrightarrow} H \stackrel{p_2,m_2}{\Longrightarrow} G'$ of direct (typed) graph transformations, it is possible to construct an E-concurrent (typed) graph production $p_1 *_E p_2$. The "epimorphic overlap graph" E can be constructed as a subgraph of H from $E = n_1(R_1) \cup m_2(L_2)$, where n_1 and m_2 are the first comatch and the second match, and R_1 and L_2 are the right- and the left-hand side of p_1 and p_2, respectively. Note that the restrictions $e_1 : R_1 \rightarrow E$ of n_1 and $e_2 : L_2 \rightarrow E$ of m_2 are jointly surjective. The E-concurrent (typed) graph production $p_1 *_E p_2$ allows one to construct a direct (typed) graph transformation $G \stackrel{p_1*_Ep_2}{\Longrightarrow} G'$ from G to G' via $p_1 *_E p_2$. Vice versa, each direct (typed) graph transformation $G \stackrel{p_1*_Ep_2}{\Longrightarrow} G'$ via the E-concurrent (typed) graph production $p_1 *_E p_2$ can be sequentialized, leading to an E-related (typed) graph transformation sequence $G \stackrel{p_1,m_1}{\Longrightarrow} H \stackrel{p_2,m_2}{\Longrightarrow} G'$ of direct (typed) graph transformations via p_1 and p_2, where "E-related" means that n_1 and m_2 overlap in H as required by E.

Theorem 2 (Concurrency Theorem). *Given two (typed) graph productions p_1 and p_2, and an E-concurrent (typed) graph production $p_1 *_E p_2$, we have:*

- *Given an E-related (typed) graph transformation sequence $G \Rightarrow H \Rightarrow G'$ via p_1 and p_2, then there is a* synthesis construction *leading to a direct (typed) graph transformation $G \Rightarrow G'$ via $p_1 *_E p_2$.*

- *Given a direct (typed) graph transformation $G \Rightarrow G'$ via $p_1 *_E p_2$, then there is an* analysis construction *leading to an E-related (typed) graph transformation sequence $G \Rightarrow H \Rightarrow G'$ via p_1 and p_2.*

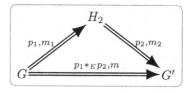

Example 6. The first two steps $S \Rightarrow G_1 \Rightarrow G_2$ of the typed graph transformations in Fig. 7 are sequentially dependent, because the *setTurn*-loop needed to apply the typed graph production *setTurn1* to G_1 is created by *setFlag*. The E-concurrent production for this transformation sequence is shown in the top row of Fig. 11, leading to the depicted E-related typed graph transformation. $\square$

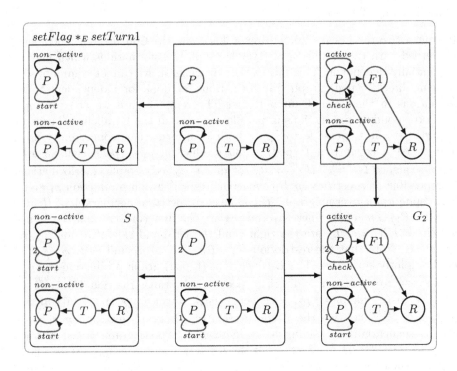

Fig. 11. Example Concurrency Theorem

Embedding and Extension Theorem

For the Embedding and Extension Theorem, we analyze under what conditions a (typed) graph transformation $t : G_0 \overset{*}{\Rightarrow} G_n$ can be extended to a (typed)

graph transformation $t' : G'_0 \overset{*}{\Rightarrow} G'_n$ via an extension morphism $k_0 : G_0 \to G'_0$. The idea is to obtain an extension diagram (1), where the same (typed) graph productions $p_1, \ldots, p_n$ are applied in the same order in t and t'.

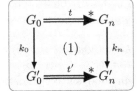

Unfortunately, this is not always possible, but we are able to give a necessary and sufficient consistency condition to allow such an extension. This result is important for all kinds of applications where we have a large (typed) graph G'_0, but only small subparts of G'_0 have to be changed by the (typed) graph productions $p_1, \ldots, p_n$. In this case we choose a suitably small subgraph G_0 of G'_0 and construct a (typed) graph transformation $t : G_0 \overset{*}{\Rightarrow} G_n$ via $p_1, \ldots, p_n$ first. In a second step, we extend $t : G_0 \overset{*}{\Rightarrow} G_n$ via the inclusion $k_0 : G_0 \to G'_0$ to a (typed) graph transformation $t' : G'_0 \overset{*}{\Rightarrow} G'_n$ via the same (typed) graph productions $p_1, \ldots, p_n$.

Now we are going to formulate the consistency condition which allows us to extend $t : G_0 \stackrel{*}{\Rightarrow} G_n$ to $t' : G'_0 \stackrel{*}{\Rightarrow} G'_n$ via $k_0 : G_0 \rightarrow G'_0$, leading to the extension diagram (1) above. The idea is to first construct a boundary graph B and a context graph C for $k_0 : G_0 \rightarrow G'_0$, such that G'_0 is the gluing of G_0 and C along B, i.e. $G'_0 = G_0 +_B C$. In fact, this boundary graph B is the smallest subgraph of G_0 which contains the identification points IP and the

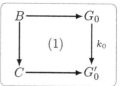

dangling points DP of $k_0 : G_0 \rightarrow G'_0$, considered as a match morphism. Now the (typed) graph morphism $k_0 : G_0 \rightarrow G'_0$ is said to be consistent with $t : G_0 \stackrel{*}{\Rightarrow} G_n$ if the boundary graph B is preserved by t. This means that none of the (typed) graph production $p_1, \ldots, p_n$ deletes any item of B.

Theorem 3 (Embedding and Extension Theorem). *Given a (typed) graph transformation* $t : G_0 \stackrel{*}{\Rightarrow} G_n$ *and a (typed) graph morphism* $k_0 : G_0 \rightarrow G'_0$ *which is consistent with respect to t, then there is an extension diagram over t and k_0.*

Given a (typed) graph transformation $t : G_0 \stackrel{*}{\Rightarrow} G_n$ *with an extension diagram (1), and the boundary B and the context graph C of $k_0 : G_0 \rightarrow G'_0$, then we have:*

1. *k_0 is consistent with respect to $t : G_0 \stackrel{*}{\Rightarrow} G_n$.*
2. *There is a (typed) graph production $der(t) = (G_0 \stackrel{d_0}{\leftarrow} D_n \stackrel{d_n}{\rightarrow} G_n)$, called the derived span of $t : G_0 \stackrel{*}{\Rightarrow} G_n$, leading to a direct (typed) graph transformation $G'_0 \Rightarrow G'_n$ via $der(t)$.*
3. *G'_n is the gluing of C and G_n along B, i.e. $G'_n = G_n +_B C$.*

Example 7. We embed the start graph S, with the typed graph morphism k_0, into a larger context graph H, where an additional resource is available that is also assigned to the first process. The boundary B and context graph C for k_0 are shown in the left-hand side of Fig. 12. Since, in the boundary graph, there is only the first process node, which is preserved by every step of the typed graph transformation $t : S \stackrel{*}{\Rightarrow} G$, we can extend t over k_0 to H and obtain a typed graph transformation $t' : H \stackrel{*}{\Rightarrow} H'$ shown in Fig. 12. Note that H' is the gluing of C and G along B. □

Critical Pairs and Local Confluence Theorem

A (typed) graph transformation system is called *confluent* if, for all (typed) graph transformations $G \stackrel{*}{\Rightarrow} H_1$ and $G \stackrel{*}{\Rightarrow} H_2$, there is a (typed) graph X together with

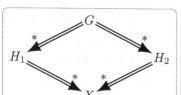
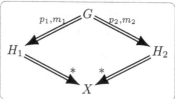

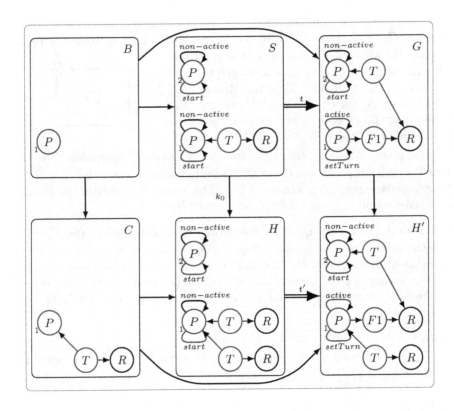

Fig. 12. Example Embedding and Extension Theorem

(typed) graph transformations $H_1 \stackrel{*}{\Rightarrow} X$ and $H_2 \stackrel{*}{\Rightarrow} X$. *Local confluence* means that this property holds for all pairs of direct (typed) graph transformations $G \Rightarrow H_1$ and $G \Rightarrow H_2$.

Confluence is an important property of a (typed) graph transformation system, because, in spite of local nondeterminism concerning the application of a (typed) graph production, we have global determinism for confluent (typed) graph transformation systems. *Global determinism* means that, for each pair of terminating (typed) graph transformations $G \stackrel{*}{\Rightarrow} H$ and $G \stackrel{*}{\Rightarrow} H'$ with the same source graph, the target graphs H and H' are equal or isomorphic. A (typed) graph transformation $G \stackrel{*}{\Rightarrow} H$ is called *terminating* if no (typed) graph production in the (typed) graph transformation system is applicable to H anymore.

The Local Church–Rosser Theorem shows that, for two parallel independent direct (typed) graph transformations $G \stackrel{p_1,m_1}{\Longrightarrow} H_1$ and $G \stackrel{p_2,m_2}{\Longrightarrow} H_2$, there is a (typed) graph G' together with direct (typed) graph transformations $H_1 \stackrel{p_2,m_2'}{\Longrightarrow} G'$ and $H_2 \stackrel{p_1,m_1'}{\Longrightarrow} G'$. This means that we can apply the (typed) graph productions p_1 and p_2 with given matches in an arbitrary order. If each pair of productions is parallel independent for all possible matches, then it can be shown that the corresponding (typed) graph transformation system is confluent.

In the following, we discuss local confluence for the general case in which $G \Rightarrow H_1$ and $G \Rightarrow H_2$ are not necessarily parallel independent. According to a general result for rewriting systems, it is sufficient to consider local confluence, provided that the (typed) graph transformation system is terminating.

The main idea is to study critical pairs. A pair $P_1 \overset{p_1,o_1}{\Longleftarrow} K \overset{p_2,o_2}{\Longrightarrow} P_2$ of direct (typed) graph transformations is called a critical pair if it is parallel dependent, and minimal in the sense that the pair (o_1, o_2) of matches $o_1 : L_1 \to K$ and $o_2 : L_2 \to K$ is jointly surjective. This means that each item in K has a preimage in L_1 or L_2. In other words, K can be considered as a suitable gluing of L_1 and L_2. It can be shown that every pair of parallel dependent direct (typed) graph transformations is an extension of a critical pair.

In order to show local confluence, it is sufficient to show strict confluence of all its critical pairs. As discussed above, confluence of a critical pair $P_1 \Leftarrow K \Rightarrow P_2$ means the existence of a (typed) graph K' together with (typed) graph transformations $P_1 \overset{*}{\Rightarrow} K'$ and $P_2 \overset{*}{\Rightarrow} K'$.

Strictness is a technical condition which means, intuitively, that the largest subgraph N of K which is preserved by the critical pair $P_1 \Leftarrow K \Rightarrow P_2$ is also preserved by $P_1 \overset{*}{\Rightarrow} K'$ and $P_2 \overset{*}{\Rightarrow} K'$. In [10], it has been shown that confluence of critical pairs without strictness is not sufficient to show local confluence.

Theorem 4 (Local Confluence Theorem). *A (typed) graph transformation system is locally confluent if all its critical pairs are strictly confluent.*

Example 8. We analyze our typed graph grammar $MutualExclusion$ and take a closer look at the typed graph productions $setFlag$ and $setTurn1$. For a typed graph K that may lead to a critical pair, we have to consider overlappings of the left-hand sides L_1 of $setFlag$ and L_2 of $setTurn1$. The typed graph transformations $K \overset{setFlag}{\Longrightarrow} P_1$ and $K \overset{setTurn1}{\Longrightarrow} P_2$ are parallel dependent if the loop in L_2 typed *non-active* is deleted by $setFlag$. This leads to the two critical overlappings K and K', and we have the critical pairs $P_1 \overset{setFlag}{\Longleftarrow} K \overset{setTurn1}{\Longrightarrow} P_2$ and $P_1' \overset{setFlag}{\Longleftarrow} K' \overset{setTurn1}{\Longrightarrow} P_2'$ shown in Fig. 13.

There are many more critical pairs for other pairs of typed graph transformations in our grammar. All these critical pairs are strictly confluent. Therefore the typed graph transformation system is locally confluent. However, as we can see in the derivation graph, the typed graph grammar is not terminating; nevertheless, it is confluent.   □

Graph Constraints and Application Conditions

(Typed) graph constraints allow us to formulate properties for (typed) graphs. In particular, we are able to formulate the condition that a (typed) graph G must (or must not) contain a certain subgraph G'. Beyond that, we can require that G contains C (conclusion) if it contains P (premise). Application conditions, similarly to the gluing condition, allow us to restrict the application of (typed) graph productions. Both concepts are important for increasing the expressive power of (typed) graph transformation systems.

Definition 6 (Graph constraint). *An* atomic (typed) graph constraint *is of the form* $PC(a)$, *where* $a : P \to C$ *is a (typed) graph morphism.*

A (typed) graph constraint *is a Boolean formula over atomic (typed) graph constraints. This means that true and every atomic (typed) graph constraint are (typed) graph constraints, and, for (typed) graph constraints* c *and* c_i *with* $i \in I$ *for some index set* I, $\neg c$, $\wedge_{i \in I} c_i$, *and* $\vee_{i \in I} c_i$ *are (typed) graph constraints.*

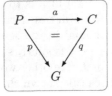

A (typed) graph G satisfies a (typed) graph constraint c, written $G \models c$, if

- $c = true$;
- $c = PC(a)$ *and, for every injective (typed) graph morphism* $p : P \to G$, *there exists an injective (typed) graph morphism* $q : C \to G$ *such that* $q \circ a = p$;
- $c = \neg c'$ *and* G *does not satisfy* c';
- $c = \wedge_{i \in I} c_i$ *and* G *satisfies all* c_i *with* $i \in I$;
- $c = \vee_{i \in I} c_i$ *and* G *satisfies some* c_i *with* $i \in I$.

Now we introduce application conditions for a match $m : L \to G$, where L is the left-hand side of a (typed) graph production p. The idea is that the (typed) graph production cannot be applied at m if m violates the application condition.

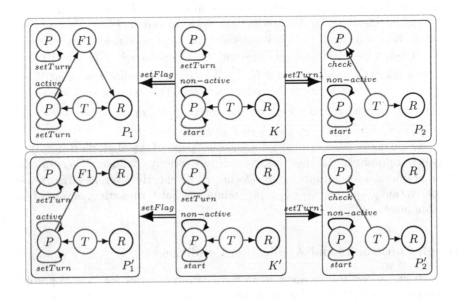

Fig. 13. Example critical pairs

Definition 7 (Application condition). *An atomic application condition over a (typed) graph L is of the form $P(x, \vee_{i \in I} x_i)$, where $x : L \to X$ and $x_i : X \to C_i$ with $i \in I$ for some index set I are (typed) graph morphisms.*

An application condition over L is a Boolean formula over atomic application conditions over L. This means that true and every atomic application condition are application conditions, and, for application conditions acc and acc_i with

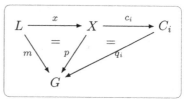

$i \in I$, $\neg acc$, $\wedge_{i \in I} acc_i$, *and* $\vee_{i \in I} acc_i$ *are application conditions.*

A (typed) graph morphism $m : L \to G$ satisfies an application condition acc, written $m \models acc$, if

- *acc = true;*
- *acc = $P(x, \vee_{i \in I} x_i)$ and, for all injective (typed) graph morphisms $p : X \to G$ with $p \circ x = m$, there exists an $i \in I$ and an injective (typed) graph morphism $q_i : C_i \to G$ with $q_i \circ x_i = p$;*
- *acc = $\neg acc'$ and m does not satisfy acc';*
- *acc = $\wedge_{i \in I} acc_i$ and m satisfies all acc_i with $i \in I$;*
- *acc = $\vee_{i \in I} acc_i$ and m satisfies some acc_i with $i \in I$.*

Given a (typed) graph production $p = (L \xleftarrow{l} K \xrightarrow{r} R)$, an *application condition* $A(p) = (A_L, A_R)$ for p consists of a left application condition A_L over L and a right application condition A_R over R. A direct (typed) graph transformation $G \overset{p,m}{\Rightarrow} H$ with a comatch $n : R \to H$ satisfies the application condition $A(p) = (A_L, A_R)$ if $m \models A_L$ and $n \models A_R$. Otherwise, p cannot be applied to G via m.

A widely used variant of application conditions are negative application conditions. A *negative application condition* is of the form $NAC(x)$, where $x : L \to X$ is a (typed) graph morphism. A (typed) graph morphism $m : L \to G$ satisfies $NAC(x)$ if there *does not* exist an injective (typed) graph morphism $p : X \to G$ with $p \circ x = m$. A negative application condition $NAC(x)$ is equivalent to an application condition of the form $P(x, \vee_{i \in I} x_i)$ with an empty index set I.

Example 9. We consider the typed graph constraint $PC(a : P \to C)$ in Fig. 14 for the typed graphs of the graph grammar *MutualExclusion*. A typed graph G satisfies this constraint if, for each resource node R, there is a turn variable that connects it to a process. The start graph S obviously satisfies this constraint – there is only one resource, which is connected to the first process node.

For an example of an application condition, we add a new production *addResource* to our typed graph grammar *MutualExclusion*, as shown in Fig. 15. This production inserts a new resource node and a new turn node, connected to a given process. For the application of this production, we define the left negative application condition $NAC(x)$ as depicted. With $NAC(x)$, we forbid the possibility that the process that the turn will be connected to is already active.   □

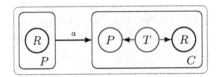

Fig. 14. Example typed graph constraint

It is possible to construct for each (typed) graph constraint an equivalent right application condition and for each right application condition an equivalent left application condition. This allows us to make sure that a derived (typed) graph H satisfies a given (typed) graph constraint $PC(a)$, provided that the match $m : L \rightarrow G$ of the direct (typed) graph transformation $G \overset{p,m}{\Longrightarrow} H$ satisfies the corresponding left application condition acc.

3   Transformations in Adhesive HLR Systems

In this section, we generalize the basic concepts of the algebraic approach from graphs in Section 2 to high-level structures. The concept of weak adhesive high-level replacement (HLR) categories is introduced as a suitable categorical framework for graph transformation in this more general sense.

In addition to pushouts we also need pullbacks. The intuitive idea of a pullback G_0 of injective morphisms $f_1 : G_1 \rightarrow G_3$ and $f_1 : G_2 \rightarrow G_3$ is that G_0 is the intersection of G_1 and G_2 with injective morphisms $g_1 : G_0 \rightarrow G_1$ and $g_2 : G_0 \rightarrow G_2$ leading to the commutative diagram (1).

$$\begin{array}{ccc} G_0 & \overset{g_1}{\longrightarrow} & G_1 \\ {\scriptstyle g_2}\downarrow & (1) & \downarrow{\scriptstyle f_1} \\ G_2 & \underset{f_2}{\longrightarrow} & G_3 \end{array}$$

If f_1 is an inclusion and f_2 an arbitrary morphism then G_0 can be considered as the preimage $f_2^{-1}(G_1)$.

The intuitive idea of weak adhesive HLR categories is that of categories with suitable pushouts and pullbacks which are compatible with each other. More precisely, the definition is based on van Kampen squares.

The idea of a van Kampen (VK) square is that of a pushout which is stable under pullbacks, and, vice versa, that pullbacks are stable under combined pushouts and pullbacks.

Definition 8 (Van Kampen square). *A pushout (1) is a van Kampen square if, for any commutative cube (2) with (1) in the bottom and where the back faces are pullbacks, the following statement holds: the top face is a pushout iff the front faces are pullbacks.*

It might be expected that, at least in the category **Sets**, every pushout is a van Kampen square. Unfortunately, this is not true. However, at least pushouts along injective functions or monomorphisms are VK squares in **Sets** and several other categories.

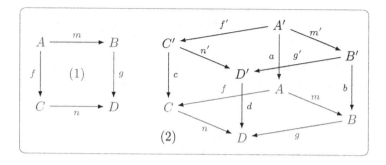

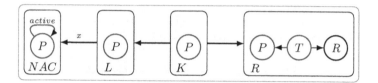

Fig. 15. Example negative application condition

Definition 9 (Weak adhesive HLR category). *A category* **C** *with a morphism class* $\mathcal{M}$ *is called a* weak adhesive HLR category *if:*

1. $\mathcal{M}$ *is a class of monomorphisms closed under isomorphisms, composition* $(f : A \to B \in \mathcal{M}, g : B \to C \in \mathcal{M} \Rightarrow g \circ f \in \mathcal{M})$, *and decomposition* $(g \circ f \in \mathcal{M}, g \in \mathcal{M} \Rightarrow f \in \mathcal{M})$.
2. **C** *has pushouts and pullbacks along* $\mathcal{M}$-*morphisms, and* $\mathcal{M}$-*morphisms are closed under pushouts and pullbacks.*
3. *Pushouts in* **C** *along* $\mathcal{M}$-*morphisms are weak VK squares, i.e. the VK square property holds for all commutative cubes with* $m \in \mathcal{M}$ *and* $(f \in \mathcal{M}$ *or* $b, c, d \in \mathcal{M})$.

For historical reasons, these categories are called weak adhesive HLR categories. In [11] and related work, adhesive categories are used as the categorical framework for deriving process congruences from reaction rules. The step from adhesive to adhesive HLR categories is justified by the fact that there are some important examples – such as algebraic specifications and typed attributed graphs – which are not adhesive categories. However, they are adhesive HLR categories for a suitable subclass $\mathcal{M}$ of all monomorphisms. Thus, the main difference between adhesive HLR categories and adhesive categories is that a distinguished class $\mathcal{M}$ of monomorphisms is considered instead of all monomorphisms, so that only pushouts along $\mathcal{M}$-morphisms have to be VK squares. Another important example – the category **PTNets** of place/transition nets with the class $\mathcal{M}$ of injective morphisms – fails to be an adhesive HLR category, but is a weak adhesive HLR category. This justifies the step to weak adhesive HLR categories.

Weak adhesive HLR categories are closed under product, slice, coslice, functor, and comma category constructions. This means that we can construct new weak adhesive HLR categories from given ones.

Theorem 5 (Construction Theorem). *If* $(\mathbf{C}, \mathcal{M}_1)$ *and* $(\mathbf{D}, \mathcal{M}_2)$ *are weak adhesive HLR categories, then the following categories are weak adhesive HLR categories:*

1. *the* product category $(\mathbf{C} \times \mathbf{D}, \mathcal{M}_1 \times \mathcal{M}_2)$,
2. *the* slice category $(\mathbf{C}\backslash X, \mathcal{M}_1 \cap \mathbf{C}\backslash X)$,
3. *the* coslice category $(X\backslash \mathbf{C}, \mathcal{M}_1 \cap X\backslash \mathbf{C})$,
4. *the* functor category $([\mathbf{X}, \mathbf{C}], \mathcal{M} - functor\ transformations)$,
5. *the* comma category $(ComCat(F, G; \mathcal{I}), (\mathcal{M}_1 \times \mathcal{M}_2) \cap Mor_{ComCat})$, *where* $F : \mathbf{C} \to \mathbf{X}$ *preserves pushouts along* $\mathcal{M}_1$-*morphisms and* $G : \mathbf{D} \to \mathbf{X}$ *preserves pullbacks (along* $\mathcal{M}_2$-*morphisms).*

Examples for weak adhesive HLR categories are the categories **Sets** of sets, **Graphs** of graphs, **Graphs**$_{\mathbf{TG}}$ of typed graphs, **Hypergraphs** of hypergraphs, **ElemNets** of elementary Petri nets, **PTNets** of place/transition nets and **AHLNets** of algebraic high-level nets, all together with the class $\mathcal{M}$ of injective morphisms, as well as the category **Spec** of algebraic specifications with the class $\mathcal{M}_{strict}$ of strict injective specification morphisms and the category **AGraphs**$_{\mathbf{ATG}}$ of typed attributed graph with the class $\mathcal{M}_{D-iso}$ of injective graph morphisms with isomorphic data part. After proving that **Sets** is a weak adhesive HLR category, the proofs for most of these categories can be done using the Construction Theorem.

Analogously to the (typed) graph case, we can define productions, transformations, and adhesive HLR systems and grammars, where we replace injective morphisms by $\mathcal{M}$-morphisms.

Definition 10 (Adhesive HLR system and grammar). *A production* $p = (L \xleftarrow{l} K \xrightarrow{r} R)$ *consists of objects* L, K, *and* R, *and two morphisms* l *and* r *with* $l, r \in \mathcal{M}$.

Given a production p, *an object* G, *and a morphism* $m : L \to G$, *called match, a* direct transformation $G \overset{p,m}{\Longrightarrow} H$ *from* G *to an object* H *is given by the pushouts (1) and (2).*

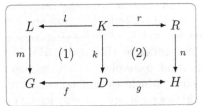

An adhesive HLR system $AS = (\mathbf{C}, \mathcal{M}, P)$ *consists of a weak adhesive HLR category* $(\mathbf{C}, \mathcal{M})$ *and a set of productions* P.

An adhesive HLR grammar $AG = (AS, S)$ *consists of an adhesive HLR system* AS *and a start object* S.

The language L *of an adhesive HLR grammar* AG *is defined by*

$$L = \{G \mid \exists\ transformation\ S \overset{*}{\Rightarrow} G\}.$$

Under a few additional conditions, it has been shown in [7] that all the results for (typed) graph transformations given in Subsection 2.2 are valid in adhesive HLR systems. Hence they can be applied to all the examples of weak adhesive HLR categories discussed above.

4   Conclusion

In this paper, we have given an overview of several concepts and results of algebraic graph transformation based on gluing constructions and the double pushout approach. Basic results concerning independence, parallelism, concurrency, embedding, critical pairs and confluence have been introduced and explained by examples.

As a generalization, we have defined the categorical framework of adhesive high-level replacement systems for unified constructions and proofs, which allows to instantiate the rich theory not only to graphs and typed graphs, but also to many different high-level structures. As a consequence we obtain a rigorous approach to various transformation systems providing as fundamental results the Local Church-Rosser and Parallelism, Concurrency, Embedding and Extension, and the Local Confluence Theorems.

For a detailed presentation of all the concepts, results and proofs we refer to our book [7].

References

[1] Rozenberg, G.: Handbook of Graph Grammars and Computing by Graph Transformation, Foundations, vol. 1. World Scientific, Singapore (1997)
[2] Ehrig, H., Engels, G., Kreowski, H.-J., Rozenberg, G.: Handbook of Graph Grammars and Computing by Graph Transformation, Applications, Languages and Tools, vol. 2. World Scientific, Singapore (1999)
[3] Ehrig, H., Kreowski, H.-J., Montanari, U., Rozenberg, G.: Handbook of Graph Grammars and Computing by Graph Transformation, Concurrency, Parallelism and Distribution, vol. 3. World Scientific, Singapore (1999)
[4] Ehrig, H., Habel, A., Kreowski, H.-J., Parisi-Presicce, F.: From Graph Grammars to High Level Replacement Systems. In: Ehrig, H., Kreowski, H.-J., Rozenberg, G. (eds.) Graph Grammars and Their Application to Computer Science. LNCS, vol. 532, pp. 269–291. Springer, Heidelberg (1991)
[5] Ehrig, H., Habel, A., Kreowski, H.-J., Parisi-Presicce, F.: Parallelism and Concurrency in High-Level Replacement Systems. MSCS 1(3), 361–404 (1991)
[6] Ehrig, H., Habel, A., Padberg, J., Prange, U.: Adhesive High-Level Replacement Categories and Systems. In: Ehrig, H., Engels, G., Parisi-Presicce, F., Rozenberg, G. (eds.) ICGT 2004. LNCS, vol. 3256, pp. 144–160. Springer, Heidelberg (2004)
[7] Ehrig, H., Ehrig, K., Prange, U., Taentzer, G.: Fundamentals of Algebraic Graph Transformation. Springer, Heidelberg (2006)
[8] Lack, S., Sobociński, P.: Adhesive Categories. In: Walukiewicz, I. (ed.) FOSSACS 2004. LNCS, vol. 2987, pp. 273–288. Springer, Heidelberg (2004)

[9] Lynch, N.: Distributed Algorithms. Morgan Kaufmann, San Mateo, CA (1996)

[10] Plump, D.: On Termination of Graph Rewriting. In: Nagl, M. (ed.) WG 1995. LNCS, vol. 1017, pp. 88–100. Springer, Heidelberg (1995)

[11] Sobociński, P.: Deriving Process Congruences from Reaction Rules. PhD thesis, BRICS (2004)

Deterministic Two-Dimensional Languages over One-Letter Alphabet[*]

Marcella Anselmo[1] and Maria Madonia[2]

[1] Dipartimento di Informatica ed Applicazioni, Università di Salerno I-84084 Fisciano (SA) Italy
anselmo@dia.unisa.it
[2] Dip. Matematica e Informatica, Università di Catania, Viale Andrea Doria 6/a, 95125 Catania, Italy
madonia@dmi.unict.it

Abstract. We study the family DREC(1) of deterministic tiling recognizable two-dimensional languages in the case of a one-letter alphabet. The family coincides with both the class of languages accepted by deterministic on-line tessellation acceptors ($\mathcal{L}(\text{DOTA})(1)$) and the one of languages recognized by 2-way alternating finite automata ($\mathcal{L}(2\text{AFA})(1)$). We show that DREC(1) is complex enough to contain languages that cannot be realized by classical operations, while other languages constructed using classical operations cannot be deterministically recognized. Furthermore we prove that there are unambiguously recognizable languages that cannot be deterministically recognized even in the case of one-letter alphabet. In particular $\mathcal{L}(\text{DOTA})(1)$ is different from $\mathcal{L}(\text{OTA})(1)$ (its non-deterministic counterpart).

Keywords: Automata and Formal Languages. Determinism. Two-dimensional languages.

1 Introduction

Two-dimensional languages, or picture languages, are a generalization of the classical string languages: they are sets of two-dimensional arrays of symbols over a finite alphabet. In the literature many different classes of picture languages have been considered and such classes are interesting as formal methods of image recognition as well as mathematical objects in their own right. In particular, the class of two-dimensional languages defined by finite state devices was deeply studied and in [9] the family $\text{REC}(\Sigma)$ of *recognizable picture languages* was introduced as a generalization of the class of recognizable string languages. This definition follows from a characterization of recognizable string languages in terms of local languages and projections (cf. [6]): the pair of a local picture language and a projection is called *tiling system*.

---

[*] This work was partially supported by PRIN project *Linguaggi Formali e Automi: aspetti matematici e applicativi*.

S. Bozapalidis and G. Rahonis (Eds.): CAI 2007, LNCS 4728, pp. 147–159, 2007.

It is also noteworthy that the definition of REC(Σ) is implicitly non-deterministic and it seems not possible to eliminate this non-determinism without a loss in power of recognition: any deterministic finite model for two-dimensional languages defines a class that is strictly included in REC(Σ) (see e.g. [5,11,21]). This result fits the fact that REC family is not closed under complementation whereas any deterministic family must have this closure property.

On the other hand deterministic languages play an important role in the recognition process of pictures. Indeed the parsing problem for two-dimensional languages in REC(Σ) is a NP-complete problem [16]. In order to decide whether a given picture p belongs to the language recognized by a given tiling system we have to scan p in order to find a picture p', in the local language, whose projection is equal to p. In general, such recognition process is non-deterministic: at each step, one can have a backtracking on all already scanned positions. So it is important to have tiling systems that lead to computations with no backtracking. After this, deterministic tiling systems were recently defined in [1], and the given definition generalizes the one-dimensional (string) case. Indeed a tiling system corresponds, in one dimension, to a set of undirected transitions. To consider determinism we have to fix a computation direction. Remark that determinism, also in the one dimensional case, is a notion related to a fixed direction (usually understood): we have determinism, along the left-to-right direction, and co-determinism, along the right-to-left direction. In two dimensions this reasoning leads to define determinism along four main directions, each one starting in one of the four corners. Deterministic languages are defined as languages that can be recognized by a tiling system that is deterministic according to some corner-to-corner direction. The class of deterministic languages over an alphabet Σ is denoted DREC(Σ). Once again, the generalization from one to two dimensions, results in a more complex notion.

In this paper we study deterministic languages over a one-letter alphabet, whose class is denoted DREC(1). Note that the investigation on one-letter alphabet is a necessary step in studying recognizability: if a language belongs to REC(Σ) then necessarily its projection over a one-letter alphabet must belong to REC(1). Considering a one-letter alphabet is equivalent to study the shapes of pictures before their contents.

The tiling recognizability of unary languages has been considered in [7,9]. More precisely there are considered languages of pictures where the number of columns is a function of the number of rows (or vice-versa) and it is shown that such functions cannot grow quicker than an exponential function or slower than a logarithmic one. Regular expressions for languages over one-letter alphabet are studied in [19,2]. Some comparisons between the different kind of automata accepting two-dimensional languages, in the special case of a one-letter alphabet, are contained in [11,12,18,17]. In general, the same separation results hold in the one-letter case as in the several-letters case. Very recently the authors of [4] investigated the complexity of unary tiling-recognizable picture languages.

This paper mainly focuses on DREC(1) family, but it concerns other families of unary languages too. First it is shown (Proposition 2) that DREC(1) coincides with both $\mathcal{L}(\text{DOTA})(1)$, the family of languages accepted by deterministic on-line tessellation acceptors, and $\mathcal{L}(2\text{AFA})(1)$, the class of languages accepted by 2-way alternating finite automata: when the cardinality of the alphabet is 1, these three approaches are equivalent. Hence any result stated for one of these families immediately holds for the other two ones too. Then we prove a necessary condition for languages in DREC(1) expressing some periodicity in the local language corresponding to a deterministic tiling system (Proposition 3). As application, we provide an example of a language L_{mult} not in DREC(1) (Proposition 4), carefully analyzing its local pictures.

Proposition 4 has several applications. It allows us to show that in the *one-letter case*: 1) there are unambiguously but not deterministically recognizable languages (Proposition 5); 2) there exist languages accepted by on-line tessellation acceptors (OTA), but not by deterministic on-line tesselation acceptors (DOTA) (Corollary 3); 3) DREC(1) (and hence $\mathcal{L}(\text{DOTA})(1)$, and $\mathcal{L}(2\text{AFA})(1)$ too) is not closed under row and column star operations (Proposition 6).

In Section 5 we compare DREC(1) with some other families of one-letter languages defined using boolean operations, row-, column-, diagonal-concatenations and stars. We show that the structure of DREC(1) cannot be captured by such operations: there are languages in DREC(1) that cannot be expressed using union, concatenations and stars and there are languages constructed with these operations that are not in DREC(1). Recall that this is also the case for the whole REC(Σ): the only known characterization needs also some alphabetic projection [9]. Finally in Section 6 we state some open problems.

2   Preliminaries

We introduce some definitions about two-dimensional languages. The notations used, some examples and results and more details can be mainly found in [9].

A *two-dimensional string* (or a *picture*) over a finite alphabet Σ is a two-dimensional rectangular array of elements of Σ. The set of all pictures over Σ is denoted by Σ^{**} and a *two-dimensional language* over Σ is a subset of Σ^{**}. Given a picture $p \in \Sigma^{**}$, let $\ell_1(p) = m$, the number of rows and $\ell_2(p) = n$ the number of columns; the pair (m, n) is the *size* of p. Note that when a one-letter alphabet is concerned, a picture p is totally defined by its size (m, n), and we will write $p = (m, n)$. Unlike the one-dimensional case, we can define an infinite number of empty pictures, namely all the pictures of size $(m, 0)$ and of size $(0, n)$, for all $m, n \geq 0$, that we denote by $\lambda_{m,0}$ and $\lambda_{0,n}$ respectively. For any picture p of size (m, n), we consider the *bordered picture* $\hat{p}$ of size $(m + 2, n + 2)$ obtained by surrounding p with a special *boundary symbol* $\# \notin \Sigma$.

A *tile* is a picture of size $(2, 2)$ and $B_{2,2}(p)$ is the set of all sub-pictures of size $(2, 2)$ of a picture p. Given an alphabet Γ, a two-dimensional language $L \subseteq \Gamma^{**}$ is *local* if there exists a finite set Θ of tiles over $\Gamma \cup \{\#\}$ such that $L = \{p \in \Gamma^{**} | B_{2,2}(\hat{p}) \subseteq \Theta\}$ and we will write $L = L(\Theta)$.

A *tiling system* is a quadruple $(\Sigma, \Gamma, \Theta, \pi)$ where Σ and Γ are finite alphabets, Θ is a finite set of tiles over $\Gamma \cup \{\#\}$ and $\pi : \Gamma \to \Sigma$ is a projection.

A two-dimensional language $L \subseteq \Sigma^{**}$ is *tiling recognizable* if there exists a tiling system $(\Sigma, \Gamma, \Theta, \pi)$ such that $L = \pi(L(\Theta))$ (extending π in the usual way). We denote by $\mathrm{REC}(\Sigma)$ the family of all *tiling recognizable* picture languages over Σ. Note that when a unary alphabet is considered a tiling system can be specified by giving only the local alphabet and the set of tiles; we will write (Γ, Θ).

Furthermore, in this paper, for any family of languages $\mathcal{F}(\Sigma)$ over an alphabet Σ, we denote by $\mathcal{F}(1)$ the corresponding family over a one-letter alphabet.

Example 1. Consider the language $L_{m,m} = \{(m, m) \mid m \geq 0\}$ of square pictures over a one-letter alphabet, say $\Sigma = \{a\}$, that is pictures with same number of rows and columns. $L_{m,m}$ belongs to $\mathrm{REC}(1)$. Indeed it can be obtained as projection of the language of squares over the alphabet $\{0, 1\}$ in which all the symbols in the main diagonal are 1, whereas the remaining positions carry symbol 0.

Let p and q be two pictures over an alphabet Σ. The *column concatenation* of p and q (denoted by $p \oplus q$) and the *row concatenation* of p and q (denoted by $p \ominus q$) are partial operations, defined only if $\ell_1(p) = \ell_1(q)$ and if $\ell_2(p) = \ell_2(q)$, respectively and are given by:

$$p \oplus q = \boxed{\begin{array}{c|c} p & q \end{array}} \qquad\qquad p \ominus q = \boxed{\begin{array}{c} p \\ \hline q \end{array}}.$$

Only in the case of one-letter alphabet, one can also define the diagonal concatenation [2]. The *diagonal concatenation* of $p = (m, n)$ and $q = (m', n')$ is the picture $p \oslash q = (m + m', n + n')$. It can be represented by

$$p \oslash q = \boxed{\begin{array}{c|c} p & \\ \hline & q \end{array}}$$

The definitions of picture concatenations can be extended to languages. By iterating these operations, one can define the row-, column- and diagonal- transitive closures of languages, denoted by $*\ominus$, $*\oplus$, and $*\oslash$, respectively. $\mathrm{REC}(\Sigma)$ family is closed under row and column concatenation and their closures, under union, intersection and rotation; on the contrary it is *not* closed under complementation (see [9]) even in the case of one-letter alphabet [20]. Further $\mathrm{REC}(1)$ is closed under diagonal concatenation and its closure.

Example 2. Let $L_{2m,2n}$ be the language of pictures over a one-letter alphabet with even dimensions, that is $L_{2m,2n} = \{p \mid \ell_1(p) = 2m, \ell_2(p) = 2n, \; m, n \geq 0\}$. We have that $L_{2m,2n} = \{\{(2, 2)\}^{*\ominus}\}^{*\oplus}$, and also $L_{2m,2n} = \{\lambda_{0,2}\}^{*\oplus} \oslash \{\lambda_{2,0}\}^{*\oslash}$.

Example 3. Let $L_{fc=lc}$ be the language of pictures over $\Sigma = \{a, b\}$ whose first column is equal to the last one. Language $L_{fc=lc} \in \mathrm{REC}(\Sigma)$. Informally we can define a local language where information about first column symbols of a picture p is brought along horizontal direction, by means of subscripts, to match the last column of p (see [3]).

Consider also the language $L_{fc=c'}$ of pictures where the first column is equal to some i-th column, $i \neq 1$. Note that $L_{fc=c'} = L_{fc=lc} \oplus \Sigma^{**}$ and thus $L_{fc=c'} \in \mathrm{REC}(\Sigma)$. Similarly $L_{c'=lc} = \Sigma^{**} \oplus L_{fc=lc}$ and $L_{fc=c'} \cap L_{c'=lc}$ are all in $\mathrm{REC}(\Sigma)$.

Two-dimensional on-line tessellation acceptors (OTA) were introduced in [11]. A run of a OTA on a picture consists in associating a state to each position of the picture. Such state for some position (i, j) is given by the transition function and depends on the symbol in that position and on the states already associated to positions $(i, j-1)$, $(i-1, j-1)$ and $(i-1, j)$ (note that an equivalent definition is possible with the state not depending on the state in the top-left corner, $(i-1, j-1)$ [9]). A deterministic version of this model is obtained, as usual, requiring that the state associated by the transition function to any position is unique; a deterministic OTA is referred to as DOTA. We have that the family of two-dimensional languages recognized by a DOTA ($\mathcal{L}(\text{DOTA})(\Sigma)$) is strictly included in the family of two-dimensional languages recognized by a OTA ($\mathcal{L}(\text{OTA})(\Sigma)$). Moreover, despite this kind of automaton is quite difficult to manage, this is actually the machine counterpart of a tiling system: $\text{REC}(\Sigma) = \mathcal{L}(\text{OTA})(\Sigma)$ [14].

Another model of automaton recognizing two-dimensional languages is the 4-way automaton (4NFA or 4DFA for the deterministic version): a four-way automaton is defined as an extension of the two-way automaton that recognizes strings (cf. [5]) by allowing it to move in four directions: *Left, Right, Up, Down*. A 2NFA is a 4NFA that can move right and down only.

An alternating finite automaton (AFA) [15] is a generalization of a finite automaton where a state can be either existential or universal. A computation that meets a universal (existential, resp.) state accepts if every (at least one, resp.) path from that state is accepting. A two-way two-dimensional alternating automaton (here denoted 2AFA) is an AFA that can move right and down only.

3   Deterministic Recognizable Languages

Very recently the definition of deterministic tiling systems and deterministic languages was introduced and discussed [1].

Deterministic recognizable languages are defined according to one of the four corner-to-corner directions: from top-left corner towards the bottom-right one (*tl2br* for short), and all the others *corner-to-corner directions* in the set $C2C = \{tl2br, tr2bl, bl2tr, br2tl\}$. A tiling system $(\Sigma, \Gamma, \Theta, \pi)$ is *tl2br-deterministic* if for any $\gamma_1, \gamma_2, \gamma_3 \in \Gamma \cup \{\#\}$ and $\sigma \in \Sigma$ there is at most one tile $\begin{array}{|c|c|} \hline \gamma_1 & \gamma_2 \\ \hline \gamma_3 & \gamma_4 \\ \hline \end{array} \in \Theta$, with $\pi(\gamma_4) = \sigma$. Similarly d-deterministic tiling systems are defined, for any $d \in C2C$. A recognizable two-dimensional language L is *deterministic*, if it is recognized by a d-deterministic tiling system for some corner-to-corner direction d. Moreover, $DREC(\Sigma)$ denotes the class of deterministic recognizable two-dimensional languages over the alphabet Σ. According to our notation, $DREC(1)$ will denote the class of deterministic languages over a one-letter alphabet.

Example 4. The tiling system sketched in Example 1 for the language $L_{m,m}$ of square pictures is d-deterministic for $d = tl2br, br2tl$, but not for the other directions. Anyway $L_{m,m} \in DREC(1)$.

Example 5. Let $L_{fc=lc}$ be the language of pictures over $\Sigma = \{a, b\}$ whose first column is equal to the last one, as defined in Example 3. The tiling system described is d-deterministic for any $d \in C2C$ and hence $L_{fc=lc} \in \mathrm{DREC}(\Sigma)$.

Now we study closure properties of $\mathrm{DREC}(\Sigma)$ under the boolean operations, in order to compare the general alphabet case with the one-letter case.

Proposition 1. *Let Σ be a finite alphabet. Then $\mathrm{DREC}(\Sigma)$ is closed under complementation, but it is not closed under union and intersection.*

Proof. The closure under complementation is in [1]. Let $L_{fc=c'}$ and $L_{c'=lc}$ as in Example 3. These languages are both in $\mathrm{DREC}(\Sigma)$, but their intersection is not [1]. Hence $\mathrm{DREC}(\Sigma)$ is not closed under union (otherwise the closure under union and complementation would yield the one under intersection). □

4   Properties of DREC(1)

In this section we study some properties of the family $\mathrm{DREC}(1)$ of deterministic recognizable two-dimensional languages over a one-letter alphabet. We state a necessary condition for languages in $\mathrm{DREC}(1)$; it will provide an example of a recognizable language that cannot be deterministically recognized. Such example will bring to several important consequences.

In [1], $\mathrm{DREC}(\Sigma)$ is characterized as the closure by rotation of $\mathcal{L}(\mathrm{DOTA})(\Sigma)$. When $|\Sigma| = 1$, the characterization can be strengthened as follows.

First remark that if $L \in \mathrm{DREC}(1)$ then, for any $d \in C2C$, there exists a d-deterministic tiling system recognizing L, as in the following example.

Example 6. The tiling system for language $L_{m,m}$ of square pictures shown in Example 1 is not tr2bl-deterministic, but we can obtain a tr2bl-deterministic tiling system recognizing $L_{m,m}$, by replacing its tiles by their mirror images. This way, the local image of a square is a square with symbol 1 on the counter-diagonal and 0 elsewhere. The same holds for the other corner-to-corner directions.

Proposition 2. *$\mathrm{DREC}(1) = \mathcal{L}(\mathrm{DOTA})(1) = \mathcal{L}(2AFA)(1)$.*

Proof. $L \in \mathcal{L}(\mathrm{DOTA})(1)$ iff L is recognized by a tl2br-deterministic tiling system; thus $\mathcal{L}(\mathrm{DOTA})(1) \subseteq \mathrm{DREC}(1)$. Moreover, if L is recognized by a d-deterministic tiling system for some $d \in C2C$ then it is also recognized by a tl2br-deterministic tiling system, constructed making some mirror image of the tiles, as in Example 6; thus $\mathrm{DREC}(1) = \mathcal{L}(\mathrm{DOTA})(1)$. In [13] the authors show that a language is in $\mathcal{L}(\mathrm{DOTA})(\Sigma)$ iff its 180° rotation is in $\mathcal{L}(2AFA)(\Sigma)$. But when $|\Sigma| = 1$ then any language coincides with its 180° rotation. □

Let us now state a necessary condition on the local language associated to a language in $\mathrm{DREC}(1)$. It says that the (local) pictures with a same number m of rows satisfy some "periodicity" condition. Note that, in the general case of a language $L \in \mathrm{REC}(\Sigma)$, the Horizontal Iteration Lemma holds (cf. [9]): any sufficiently long (local) picture with m rows has a factor that can be arbitrarily repeated still remaining in L. In the case of $L \in \mathrm{DREC}(1)$ the result is much stronger: because of determinism, all (local) pictures with m rows in L can be obtained (column) concatenating a fixed picture (x_m) with some repetitions of another picture (y_m) (any local picture is the prefix of the longer ones). The result is also strongly based on the cardinality 1 of the alphabet and it does not hold in general (for example the local language for $L_{fc=lc}$ on a two-letter alphabet as in Example 3 does not satisfy this condition, even if it is tl2br-deterministic).

Let us write $y' \prec y$ if $y = y' \oplus y''$ for some $y'' \in \Sigma^{**}$, say that y' is a *prefix* of y, and denote by $Pref(L) = \{y' \mid y' \prec y \text{ for some } y \in L\}$. Further we introduce for any picture p of size (m, n), the *half-bordered picture* $\tilde{p}$ of size $(m+1, n+1)$ obtained by surrounding p with the boundary symbol only on its top and left borders. We will denote by $\tilde{L}(\Theta) = \{p \in \Gamma^{**} \mid B_{2,2}(\tilde{p}) \subseteq \Theta\}$ and $\tilde{L}_m(\Theta)$ the set of pictures in $\tilde{L}(\Theta)$ with m rows.

Proposition 3. *Let $L \in DREC(1)$ and (Γ, Θ) be a tl2br-deterministic tiling system for L with $|\Gamma| = \gamma$.*

For any $m > 0$, there exist $x_m, y_m \in \Gamma^{**}$, with $\ell_2(x_m), \ell_2(y_m) < \gamma^m$, such that $\forall p \in \tilde{L}_m(\Theta)$, with $\ell_2(p) > \gamma^m$, we have $p \in Pref(x_m \oplus (y_m^{*\oplus}))$.

Moreover, if $\forall m > 0$, $\bar{y}_m$ denotes some y_m constructed as above with minimal number of columns, then $\ell_2(\bar{y}_{m+1}) = c \cdot \ell_2(\bar{y}_m)$ for some $c \in \{1, 2, \ldots, \gamma\}$.

Proof. Let $m > 0$: every picture in $\tilde{L}_m(\Theta)$ can have at most γ^m distinct columns. So, consider the picture $p_0 \in \tilde{L}_m(\Theta)$ with $\ell_2(p_0) = \gamma^m + 1$: in p_0 there exist two columns, say the i-th and the j-th ones, with $i < j$, that are equal. Clearly $1 \le i \le \gamma^m$ (such considerations are similar to the ones in the proof of the Horizontal Iteration Lemma [9]). Set x_m the picture of size $(m, i-1)$ such that $x_m \prec p_0$, and y_m the picture of size $(m, j-i)$ such that $x_m \oplus y_m \prec p_0$. Since (Γ, Θ) is tl2br-deterministic, then for any picture $p \in \tilde{L}_m(\Theta)$ with $\ell_2(p) > \gamma^m$, $p_0 \prec p$. So the i-th column of p is equal to its j-th one. Furthermore determinism implies that also the $(i+1)$-th column of p is equal to the $(j+1)$-th one, and, in general, the n-th column of p is equal to the $(n + \ell_2(y_m))$-th one, for any $n > i$. Therefore we have $p \in Pref(x_m \oplus (y_m^{*\oplus}))$.

Moreover, if we choose in p_0 the indexes i and j such that $(j - i)$ is minimal, then in any $p \in \tilde{L}_m(\Theta)$ with $\ell_2(p) > \gamma^m$, there cannot exist two equal columns at a distance less than $j - i = \ell_2(y_m)$ (apply again the determinism).

Now, for any $m > 0$, let us choose y_m and y_{m+1} with minimal number of columns and denote them by $\bar{y}_m$ and $\bar{y}_{m+1}$: we show that $\ell_2(\bar{y}_{m+1}) = c \cdot \ell_2(\bar{y}_m)$ for some $c \in \{1, 2, \ldots, \gamma\}$. Indeed, any $q \in \tilde{L}_{m+1}(\Theta)$, with $\ell_2(q) > \gamma^{m+1}$, is in $Pref(x_{m+1} \oplus (\bar{y}_{m+1}^{*\oplus}))$. By erasing the last row of q we obtain a picture $p \in \tilde{L}_m(\Theta)$, that is in $Pref(x_v \oplus (y_v^{*\oplus}))$, with $\ell_2(\bar{y}_{m+1}) = \ell_2(y_v)$. On the other hand, we have $p \in Pref(x_m \oplus (\bar{y}_m^{*\oplus}))$. In such situation we have that necessarily $\ell_2(\bar{y}_{m+1}) = \ell_2(y_v)$ is a multiple of $\ell_2(\bar{y}_m)$ by some factor $c \in \{1, 2, \ldots, \gamma\}$. Clearly, we cannot have $\ell_2(\bar{y}_{m+1}) < \ell_2(\bar{y}_m)$. Moreover it cannot be $\ell_2(\bar{y}_{m+1}) \equiv h \bmod \ell_2(\bar{y}_m)$, $h \ne 0$. Otherwise, $y_v = y_0 \oplus \ldots \oplus y_0 \oplus y_0'$ with $y_0 = y_0' \oplus y_0''$ and $\ell_2(y_0') = h$. This would imply that in y_0 the first column and the $(h + 1)$-th one are equal, against the minimality of $\ell_2(\bar{y}_m)$. At last, it cannot be $\ell_2(\bar{y}_{m+1}) = k \cdot \ell_2(\bar{y}_m)$ with $k > \gamma$, otherwise $y_v = x_0 \oplus \ldots \oplus x_0$, k times, for some x_0 with $\ell_2(x_0) = \ell_2(\bar{y}_m)$. But, since below the first column of x_0, in p, can occur at most γ different symbols, this is against the minimality of $\ell_2(\bar{y}_{m+1}) = k \cdot \ell_2(x_0)$.   □

The necessary condition just stated for the local language associated to a language in DREC(1) has a weaker consequence on the language itself.

Corollary 1. *Let $L \in DREC(1)$. Then there exists a constant $\gamma > 0$ such that for any $m > 0$, there exist positive integers $c_m, p_m < \gamma^m$ and for any $n > c_m$ we have $(m, n) \in L$ iff $(m, n + p_m) \in L$.*

Moreover, if $\forall m > 0$, $\bar{p}_m$ denotes the minimal p_m constructed as above, then $\bar{p}_{m+1} = c\bar{p}_m$ for some $c \in \{1, 2, \ldots, \gamma\}$.

Example 7. Let $L = \{(m, m+2k) \mid m, k \geq 0\}$. A tl2br-deterministic tiling system recognizing L is the one associating, for any $m, k \geq 0$, to the m-th row of length $m + 2k$, the local row $0^{m-1}1(ab)^k$, and to the m-th row of length $m + 2k + 1$, the local row $0^{m-1}1(ab)^k a$. In this case the minimal x_m is a picture of size (m, m) (with 1 on the diagonal, 0 under the diagonal and a proper prefix of $(ab)^*$ in each row up the diagonal) and the minimal y_m is a picture of size $(m, 2)$ (where rows ab and ba properly alternate).

Example 8. Let $L = \{(m, 2^m) \mid m \geq 0\}$. The tiling system for L that can be constructed from the DOTA given in [11] (following the canonical construction, cf. [9]) provides a more involved example of a tl2br-deterministic tiling system, where the number of columns of both pictures x_m and y_m grows as m grows.

Given a tl2br-deterministic tiling system (Γ, Θ), let us denote for any $m > 0$ by $k_m^{(\Gamma, \Theta)}$ (or simply k_m when the tiling system is clearly stated) the number of columns of $\bar{y}_m$ constructed as in Proposition 3.

Corollary 2. *Let $L \in DREC(1)$, let (Γ, Θ) be a tl2br-deterministic tiling system for L and let $|\Gamma| = \gamma$. Then $\forall m > 0$, we have $k_m = 1^{h_1} 2^{h_2} 3^{h_3} \ldots \gamma^{h_\gamma}$ for some $h_i \geq 0$, $i = 1, \ldots, \gamma$.*

Proof. The proof is by induction on m. Since $L \in DREC(1)$, k_1 is at most γ. For the inductive step note that, from Proposition 3, we have $k_{m+1} = ck_m$ for some $c \in \{1, 2, \ldots, \gamma\}$ and that for k_m the inductive hypothesis holds.   □

We now apply the necessary condition for DREC(1) stated in Proposition 3 in order to show that the language $L_{mult} = \{(m, km) \mid m \geq 0, k \geq 0\}$ does not belong to DREC(1). Note that we cannot use Corollary 1 for this goal. For this we need to deeply analyze the local pictures in L_{mult}, from a computational point of view, looking at what local columns must "represent" and from a more analytical point of view, looking at the periodicity of the divisors (less than or equal to a given threshold) in a sequence of consecutive integers. First, for any $m, n > 0$, let us denote by $D_m(n)$ the set of all the divisors of n that are less than or equal to m and by $lcm(1, 2, \ldots, m)$ the lowest common multiple of $1, 2, \ldots, m$.

Proposition 4. *The language L_{mult} does not belong to DREC(1).*

Proof. The proof is by contradiction and it consists in showing that if $L_{mult} \in$ DREC(1) then Corollary 2 does not hold. Suppose that $L_{mult} \in$ DREC(1) and let (Γ, Θ) be a tl2br-deterministic tiling system recognizing it with $|\Gamma| = \gamma$. From Proposition 3, for any $m > 0$ there exist $x_m, \bar{y}_m \in \Gamma^{**}$, such that for any $p \in \tilde{L}_m(\Theta)$, with $\ell_2(p) > \gamma^m$, we have $p \in Pref(x_m \oplus (\bar{y}_m^{*\oplus}))$ with $\ell_2(\bar{y}_m) = k_m$. We are going to show that, under such hypothesis, for any $m > 0$, k_m is a multiple of $lcm(1, 2, \ldots, m)$. This contradicts Corollary 2, since for any prime integer z such that $z > \gamma$, k_z would be a multiple of z, against the fact that the prime divisors of k_z must be less than or equal to γ.

Let us show that for any $m > 0$, k_m is a multiple of $lcm(1, 2, \ldots, m)$.

First we show that, if $p \in \tilde{L}_m(\Theta)$ is such that its i-th column is equal to its j-th one and $j > i > \ell_2(x_m) + \ell_2(\bar{y}_m)$, then $D_m(i) = D_m(j)$. Let us denote for any $d \leq m$, $i \leq \ell_2(p)$ by $p_{d,i}$ the subpicture of p consisting of its d rows and its first i columns. Note that the periodicity of p (i.e. $p \in Pref(x_m \oplus (\bar{y}_m^{*\oplus}))$) implies a similar periodicity for any subpicture $p_{d,i}$. Furthermore, since (Γ, Θ) is a tl2br-deterministic tiling system, if $(d, i) \in L_{mult}$ then its (unique) counter-image in $L(\Theta)$ is $p_{d,i}$. Now suppose that let $d \in D_m(j)$; then $(d, j) \in L_{mult}$ (note that d is a divisor of n iff the picture $(d, n) \in L_{mult}$). The first d symbols in the j-th column of p match the $\#$ symbols (by mean of allowed tiles in Θ). Then so is for the first d symbols in the i-th column of p, $p_{d,i} \in L(\Theta)$ and, hence, $(d, i) \in L_{mult}$ i.e. $d \in D_m(i)$. Conversely, if $d \in D_m(i)$ then $(d, i) \in L_{mult}$, that is $p_{d,i}$ is in $L(\Theta)$. We show that also $p_{d,j}$ is in $L(\Theta)$. Indeed the top and the left borders match $\#$ symbols since $p \in \tilde{L}_m(\Theta)$; the right border matches $\#$ symbols since it is equal to the right border of $p_{d,i}$; and finally the bottom border of $p_{d,j}$ matches $\#$ symbols because the bottom border of $p_{d,i}$ does and the remaining bottom tiles are a repetition of some previous ones (recall that $j > i > \ell_2(x_m) + \ell_2(\bar{y}_m)$). This concludes the proof that, if $p \in \tilde{L}_m(\Theta)$ is such that its i-th column is equal to its j-th one and $j > i > \ell_2(x_m) + \ell_2(\bar{y}_m)$, then $D_m(i) = D_m(j)$. In particular we have that for any $n > 2\gamma^m$, since the n-th column of p is equal to the $(n + k_m)$-th one, then $D_m(n) = D_m(n + k_m)$.

Now, using this fact, we are able to show that, for any $m > 0$, k_m is a multiple of $lcm(1, 2, \ldots, m)$. It suffices to show that for any $h \in \{1, 2, \ldots, m\}$, k_m is a multiple of h, i.e. $k_m \equiv 0 \bmod h$. Suppose that there exists $\bar{h} \in \{1, 2, \ldots, m\}$ such that k_m is not a multiple of $\bar{h}$. Then there exists l, $1 \leq l < \bar{h}$, such that $k_m \equiv l \bmod \bar{h}$. Take n such that $n > 2\gamma^m$ and $n \equiv (\bar{h} - l) \bmod \bar{h}$. Then $n + k_m \equiv 0 \bmod \bar{h}$ and this is against $D_m(n) = D_m(n + k_m)$.   □

This result has some immediate, but very meaningful consequences. Firstly, consider the notion of unambiguity in tiling system recognizability [8]. A tiling system for $L \subseteq \Sigma^{**}$ is *unambiguous* when any picture in L has a unique local counter-image and L is *unambiguous* if it is recognized by an unambiguous tiling system. UREC(Σ) denotes the family of all unambiguous two-dimensional languages over Σ. As one may expect, determinism implies unambiguity. Furthermore in [1] the proper inclusion of DREC(Σ) in UREC(Σ) is shown. We are now able to show that it holds even for a one-letter alphabet.

Proposition 5. *DREC(1) is strictly contained in UREC(1).*

Proof. Consider the language L_{mult}. Proposition 4 shows that it does not belong to DREC(1). On the other hand we can construct an unambiguous tiling system recognizing L_{mult}. Starting from a tiling system $\mathcal{T}$ recognizing the language of square pictures, as sketched in Example 1, we can yield a tiling system for L_{mult}, following the construction of a tiling system for the column star of a language in [9]. We make two disjoint copies of $\mathcal{T}$ and we force it to alternate starting with the first copy. The resulting tiling system is unambiguous, since for any picture (m, km) the value k is unique. □

It is well known that OTA are more powerful than DOTA [11] but the examples given in the literature are on alphabets with cardinality greater than one. The language L_{mult} gives an example of a language that is in $\mathcal{L}(OTA)(1)$ but not in $\mathcal{L}(DOTA)(1)$.

Corollary 3. *$\mathcal{L}(DOTA)(1)$ is strictly contained in $\mathcal{L}(OTA)(1)$.*

Remark 1. Note that a different proof of this result could be obtained observing that $\mathcal{L}(OTA)(1)$=REC(1) and REC(1) is not closed under complementation [20], while $\mathcal{L}(DOTA)(1)$ is closed under complementation [11].

Another consequence of Proposition 4 regards closure properties of DREC(1). Also note that, because of Proposition 2, the same non-closure properties hold for $\mathcal{L}(DOTA)(1)$, and $\mathcal{L}(2AFA)(1)$ (as far as we know, the properties are not stated even in those frameworks).

Proposition 6. *DREC(1), $\mathcal{L}(DOTA)(1)$, and $\mathcal{L}(2AFA)(1)$ are not closed neither under $*①$ nor under $*\ominus$.*

Proof. Consider the language L_{mult}. In Example 4 we have shown that $L_{m,m} \in$ DREC(1). On the other hand Proposition 4 shows that $L_{mult} = L_{m,m}^{*①}$ does not belong to DREC(1). In a similar way, the 90° rotation of L_{mult} is an example of non-closure under $*\ominus$. □

5   DREC(1) and Some Regular Families

In this section we will compare DREC(1) with some families REG(1), $\mathcal{L}(D)$, $\mathcal{L}(CRD)$, of languages over a one-letter alphabet, that can be constructed using union, row-, column- and diagonal-concatenations and their closures. REG(Σ) is defined in [19], while $\mathcal{L}(D)$, $\mathcal{L}(CRD)$ are defined in [2].

- REG(1) is the smallest family containing the singleton languages and closed under union, row- and column-concatenations and stars.
- $\mathcal{L}(D)$ is the smallest family containing the empty set, $\lambda_{0,0}$, $\lambda_{0,1}$, $\lambda_{1,0}$ and closed under union, diagonal-concatenation and star.
- $\mathcal{L}(CRD)$ is the smallest family containing the empty set, $\lambda_{0,0}$, $\lambda_{0,1}$, $\lambda_{1,0}$ and closed under union, row-, column-, and diagonal-concatenations and stars.

Now, let us show some properties and characterizations of these classes that will be useful later, to yield some comparison results. First we state the closure of DREC(1) under the boolean operations. The result follows from the closure of $\mathcal{L}$(DOTA)(Σ) [11] and Proposition 2. It holds for the unary alphabet and not in general (Proposition 1) since in general mixing tiling systems that are deterministic from different corner-to-corner directions does not yield any determinism.

Corollary 4. *DREC(1) is closed under union, intersection and complementation.*

Consider now the family REG(1). In [19] the author showed that also REG(1) is closed under the boolean operations and that the following characterization holds. Roughly it says that languages in REG(1) contain pictures whose number of rows is somehow independent from the number of columns.

Proposition 7. *([19]) L is in REG(1) if and only if it is a finite union of Cartesian product of ultimately periodic sets.*

Example 9. Let $L_{m,m}$ be the language of square pictures (see Example 1). We have $L_{m,m} = \{(1,1)\}^{*\oslash} = \{\lambda_{0,1} \oslash \lambda_{1,0}\}^{*\oslash} \in \mathcal{L}(D)$, while $L_{m,m} \notin$ REG(1). In fact in $L_{m,m}$ there are an infinite number of pairs of pictures (n,n) and (n',n') with $n \neq n'$ while $(n,n') \notin L_{m,m}$ (use Proposition 7).

On the other hand in [2] it is shown that $L \in \mathcal{L}(D)$ iff the set of sizes of pictures in L is a rational relation and iff L is accepted by a 2NFA. We give here another characterization, more similar to the one in Proposition 7 for REG(1).

Proposition 8. $L \in \mathcal{L}(D)$ *if and only if it is a finite union of languages of the form* $c^{*\oslash} \oslash P^{*\oslash}$, *where c is a single picture and P is a finite set of pictures. Furthermore $\mathcal{L}(D)$ is closed under union, intersection and complementation.*

Proof. We use the characterization of rational relations over the unary alphabet in terms of semilinear sets (or Presburger sets) of $\mathbb{N}^2$ given in [10]. □

We are now able to show the following relations among DREC(1), REG(1), $\mathcal{L}$(D), and $\mathcal{L}$(CRD).

Proposition 9. *DREC(1) is incomparable with $\mathcal{L}$(CRD).*

Proof. The language L_{mult} is in $\mathcal{L}$(CRD) since it is $L_{m,m}^{*\oslash}$ with $L_{m,m} \in \mathcal{L}$(D) (Example 9) and $\mathcal{L}(D) \subseteq \mathcal{L}$(CRD) by definition. On the other hand $L_{mult} \notin$ DREC(1) (see Proposition 4).

Consider now $L' = \{(m, 2^m) \mid m \geq 0\}$. We have $L' \in \mathcal{L}$(DOTA)(1) (see [11]) and hence $L' \in$ DREC(1) (see Proposition 2). On the other hand $L' \notin \mathcal{L}$(CRD) (cf. [2]). □

Proposition 10. *REG(1)$\subset \mathcal{L}$(D)$\subseteq$ DREC(1)$\cap \mathcal{L}$(CRD).*

Proof. REG(1)$\subset \mathcal{L}$(D). Indeed from Proposition 7 any language in REG(1) can be accepted by a 2NFA that first verifies the number of columns moving right on the first row (simulating a classical finite automaton on strings), and then verifies the number of rows moving down on the last column. The inclusion is strict: for example $L_{m,m} \in \mathcal{L}$(D)\REG(1) (see Example 9).

$\mathcal{L}$(D)$\subseteq$DREC(1) since $\mathcal{L}$(D)= $\mathcal{L}$(2NFA)$\subset \mathcal{L}$(2AFA)=DREC(1).

$\mathcal{L}$(D)$\subseteq \mathcal{L}$(CRD) follows from definition.   □

Remark 2. Previous propositions say that the structure of DREC(1) cannot be captured by classical operations: DREC(1) does not coincide neither with $\mathcal{L}$(D) nor with $\mathcal{L}$(CRD). And the same result holds if we also consider other operations, such as intersection and/or complementation, since $\mathcal{L}$(D) is closed under intersection and complementation, while considering also intersection and/or complementation of languages in $\mathcal{L}$(CRD) would result in a class equal or bigger than $\mathcal{L}$(CRD), but never equal to DREC(1).

6   Conclusions and Open Questions

We have shown some properties of DREC(1) that point out the richness and complexity of deterministic two-dimensional languages, even for a one-letter alphabet. The investigation is surely not complete.

First we do not know whether DREC(1) is closed under row or column concatenation. Recall that this would also solve the analogous problem for the classes $\mathcal{L}$(DOTA)(1) or $\mathcal{L}$(2AFA)(1).

It is also an open question whether $\mathcal{L}$(D)=DREC(1)$\cap\mathcal{L}$(CRD) or not. Following the characterizations given in this paper, this equivalence would also say that $\mathcal{L}$(2NFA)(1)= $\mathcal{L}$(2AFA)(1)$\cap\mathcal{L}$(CRD), that is a language is accepted by a 2AFA and it is in $\mathcal{L}$(CRD) iff it can be accepted with only existential states.

Further, we have shown that DREC(1) is strictly contained in UREC(1). It would be interesting to understand whether UREC(1) is strictly contained in REC(1) or the two classes collapse.

References

1. Anselmo, M., Giammarresi, D., Madonia, M.: From determinism to nondeterminism in recognizable two-dimensional languages. In: Harju, T., Karhumäki, J., Lepistö, A. (eds.) DLT 2007. LNCS, vol. 4588, pp. 36–47. Springer, Heidelberg (2007)
2. Anselmo, M., Giammarresi, D., Madonia, M.: New Operators and Regular Expressions for two-dimensional languages over one-letter alphabet. Theoretical Computer Science 340(2), 408–431 (2005)
3. Anselmo, M., Giammarresi, D., Madonia, M., Restivo, A.: Unambiguous Recognizable two-dimensional languages. RAIRO: Theoretical Informatics and Applications 40(2), 227–294 (2006)
4. Bertoni, A., Goldwurm, M., Lonati, V.: On the complexity of unary tiling-recognizable picture languages. In: Thomas, W., Weil, P. (eds.) STACS 2007. LNCS, vol. 4393, pp. 381–392. Springer, Heidelberg (2007)

5. Blum, M., Hewitt, C.: Automata on a two-dimensional tape. In: IEEE Symposium on Switching and Automata Theory, pp. 155–160 (1967)
6. Eilenberg, S.: Automata, Languages and Machines, vol. A. Academic Press, London (1974)
7. Giammarresi, D.: Two-dimensional languages and recognizable functions. In: Rozenberg, G., Salomaa, A. (eds.) Procs. in Dev. on Language Theory 1993, pp. 290–301. World Scientific Publishing Co., Singapore (1994)
8. Giammarresi, D., Restivo, A.: Recognizable picture languages. Int. Journal Pattern Recognition and Artificial Intelligence 6(2&3), 241–256 (1992)
9. Giammarresi, D., Restivo, A.: Two-dimensional languages. In: Rozenberg, G., et al. (eds.) Handbook of Formal Languages, vol. III, pp. 215–268. Springer, Heidelberg (1997)
10. Ginsburg, S., Spanier, E.: Semigroups, Presburger formulas, and languages. Pacific Journal of Mathematics 16, 285–296 (1966)
11. Inoue, K., Nakamura, A.: Some properties of two-dimensional on-line tessellation acceptors. Information Sciences 13, 95–121 (1977)
12. Inoue, K., Nakamura, A.: Two-dimensional finite automata and unacceptable functions. Intern. J. Comput. Math. A7, 207–213 (1979)
13. Ito, A., Inoue, K., Takanami, I.: Deterministic two-dimensional On-line tesselation Acceptors are equivalent to two-way two-dimensional alternating finite automata through 180°-rotation. Theor. Comp. Sc. 66, 273–287 (1989)
14. Inoue, K., Takanami, I.: A characterization of recognizable picture languages. In: Nakamura, A., Saoudi, A., Inoue, K., Wang, P.S.P., Nivat, M. (eds.) ICPIA 1992. LNCS, vol. 654, Springer, Heidelberg (1992)
15. Inoue, K., Takanami, I., Taniguchi, H.: Two-dimensional alternating Turing machines. Theor. Comp. Sc. 27, 61–83 (1983)
16. Lindgren, K., Moore, C., Nordahl, M.: Complexity of two-dimensional patterns. Journal of Statistical Physics 91(5-6), 909–951 (1998)
17. Kari, J., Moore, C.: Rectangles and squares recognized by two-dimensional automata. In: Karhumäki, J., Maurer, H., Păun, G., Rozenberg, G. (eds.) Theory Is Forever. LNCS, vol. 3113, pp. 134–144. Springer, Heidelberg (2004),
    http://www.santafe.edu/$\sim$moore/pubs/picture.html
18. Kinber, E.B.: Three-way Automata on Rectangular Tapes over a One-Letter Alphabet. Information Sciences 35, 61–77 (1985)
19. Matz, O.: Regular expressions and Context-free Grammars for picture languages. In: Reischuk, R., Morvan, M. (eds.) STACS 97. LNCS, vol. 1200, pp. 283–294. Springer, Heidelberg (1997)
20. Matz, O.: Dot-depth, monadic quantifier alternation, and first-order closure over grids and pictures. Theoretical Computer Science 270(1-2), 1–70 (2002)
21. Potthoff, A., Seibert, S., Thomas, W.: Nondeterminism versus determinism of finite automata over directed acyclic graphs. Bull. Belgian Math. Soc. 1, 285–298 (1994)

Recognizable Picture Languages and Polyominoes

Giusi Castiglione and Roberto Vaglica

Università di Palermo, Dipartimento di Matematica e Applicazioni,
via Archirafi, 34 - 90123 Palermo, Italy
{giusi, vaglica}@math.unipa.it

Abstract. We consider the problem of recognizability of some classes of polyominoes in the theory of picture languages. In particular we focus our attention on the problem posed by Matz of finding a non-recognizable picture language for which his technique for proving the non-recognizability of picture languages fails. We face the problem by studying the family of L-convex polyominoes and some closed families that are similar to the recognizable family of all polyominoes but result to be non-recognizable. Furthermore we prove that the family of L-convex polyominoes satisfies the necessary condition given by Matz for the recognizability and we conjecture that the family of L-convex polyominoes is non-recognizable.

1 Introduction

The main ingredients of the paper are polyominoes and the theory of picture languages (or two-dimensional languages).

First, we recall that a *discrete set* is a finite subset of the Cartesian plane $\mathbb{Z} \times \mathbb{Z}$ defined up to translations. A *polyomino* is a discrete set whose points are represented by cells (unitary squares) and whose interior is connected. Polyominoes are very famous objects first studied by Golomb in 1954 (cf.[16],[15]) and popularized by Gardner in 1957 (cf.[12]). They are related to many problems such as tiling (cf.[14],[17]), enumeration (cf.[2]) and discrete tomography (cf.[1],[19]) and find applications in the study of lattice models in physics (cf.[11],[18]). Cause the difficulty of the general problems, in the different areas, several subclasses were defined by using the geometrical notion of convexity. Thus, many classes of polyominoes were born such as, for example, the class of *h-convex* polyominoes (whose rows are connected), *v-convex* polyominoes (whose columns are connected) and convex polyominoes (both h-convex and v-convex).

Recently, it has been introduced a classification of convex polyominoes that has in the first level the class of *L-convex* polyominoes. This class has been considered by several points of view with nice results. In [4] it is shown that the family of L-convex polyominoes is a well ordering according to the subpicture order. In [5] and [6] combinatorial aspects of L-convex polyominoes are analyzed by giving the result of the enumerations according to the semiperimeter and the area. Furthermore, L-convex polyominoes have been characterized with

S. Bozapalidis and G. Rahonis (Eds.): CAI 2007, LNCS 4728, pp. 160–171, 2007.

respect to uniqueness from the tomographical point of view (see [7]). Discrete tomography is based on the concept of projections, in our case horizontal and vertical projections. The horizontal (resp. vertical) projections of a discrete set is an integer vector whose components represent the number of cells in each row (resp. column) ordinately. A discrete set is said to be *unique* if another discrete set with the same projections does not exist. As regards L-convex polyominoes we have that a discrete set is an L-convex polyomino if and only if it is convex and unique.

Another field of research in which polyominoes recently took relevance is that one of two-dimensional languages. In this field, in recent years, new formal approaches have been proposed in order to extend results and techniques used on words to pictures. In [13] the authors give the basic definition of *tiling system* and *tiling recognizability* of picture languages and in [21] Matz gives a technique for showing that a picture language is non-recognizable. More precisely, Matz gives a necessary condition satisfied by tiling recognizable picture language and poses the problem of finding a non-recognizable picture language for which his technique fails to prove the non-recognizability.

As regards languages of polyominoes, in [9] the authors give the idea to investigate whether the main classes of polyominoes are represented by recognizable picture languages. They give the tiling systems for languages representing the classes of h-convex, v-convex and convex polyominoes. Reinhardt, in a very technical paper, proves that the picture language that represents the class of the polyominoes is tiling recognizable (see [22]).

Our aim is investigating about recognizability of L-convex polyominoes. Our first motivation is given by the before mentioned problem posed by Matz. Indeed, in this paper, we prove that the family of L-convex polyominoes satisfies the condition given by Matz for tiling recognizable languages and we conjecture that L-convex polyominoes are non-recognizable. Also in [9] the authors study the language of L-convex polyominoes. They try to prove its recognizability but they get the result only by limiting the number of maximal rectangles contained in each polyomino. Here we use another kind of approach. In particular we use the study of some classes strictly related to L-convexes and the tomographical characterization of L-convex polyominoes. We analyze polyominoes that are unique and h-convex (resp. v-convex) and we prove that they are non-recognizable. As we will see these polyominoes are very closed both L-convex polyominoes and polyominoes. In this way we obtain the class of L-convex polyominoes as the intersection of two non-recognizable languages, the unique h-convex and the unique v-convex polyominoes. At the same time, by proving that unique discrete set are non-recognizable, we have that the class of L-convex polyominoes is the intersection of a recognizable and a non-recognizable picture language i.e. the convex polyominoes and the unique discrete set respectively. The mentioned results do not allow us to prove that L-convex polyominoes are non-recognizable but lead us to conjecture the non-recognizability of L-convex polyominoes. If our conjecture is true it gives a counterexample to the inversion of Matz's lemma and, furthermore, to prove it would mean to introduce a new method to prove the non-recognizability. In any case, the results of this paper show that the language of L-convex polyominoes is on the borderline between recognizability and non-recognizability, and it is an interesting object of study in the theory of picture languages.

2 Polyominoes

In this section we give some basic definitions and notations about polyominoes. For a complete background we refer to [15,16,17]. A *discrete set* is a finite subset of the Cartesian plane $\mathbb{Z} \times \mathbb{Z}$ defined up to translations. A discrete set can be represented as a binary matrix in $\{0,1\}$ or by a set of cells (unitary squares) as depicted in Fig. 1. If we consider the minimum rectangle r containing the discrete set s we can number its rows and columns from the top to the bottom and from the left to the right respectively. In the sequel we will refer to both the two mentioned representations and we will denote by $s(i,j)$ the entry of the matrix of the i-th row and the j-th column. An entry 1 (resp. 0) corresponds to the presence (resp. absence) of a cell.

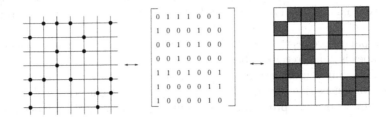

Fig. 1. A discrete set and its representations in terms of a binary matrix and a set of cells

A very famous class of discrete sets is the class of *polyominoes*. A *Polyomino* is a discrete set in which every pair of its cells is connected. A polyomino is said to be *v-convex* (resp. *h-convex*) if each of its column (resp. row) is connected. A polyomino is said to be *convex* if it is both v-convex and h-convex (see Fig. 2). We denote by $\mathcal{V}$ (resp. $\mathcal{H}$) the set of v-convex (resp. h-convex) polyominoes and by $\mathcal{C}$ the set of convex polyominoes.

In [3] it was introduced a new property of convexity that involves the kind of paths connecting pairwise the cells. The authors call *step* a pair of adjacent cells, distinguishing four types of steps: *north* $N = (0,1)$, *south* $S = (0,-1)$, *east* $E = (1,0)$, and *west* $W = (-1,0)$. A *path* is a sequence of steps. A self-avoiding path (i.e. made of distinct cells) is said to be *monotone* if it consists of steps of, at most, two types. In the mentioned paper, it was proved that a polyomino is convex if and only if every pair of its cells is joined by a monotone path. Hence they gave a classification of convex polyominoes by taking into account the minimum number of changes of direction in their monotone paths. Let $k \in \mathbb{N}$,

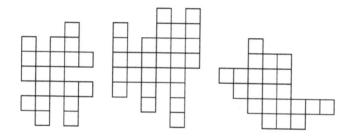

Fig. 2. A polyomino, a v-convex polyomino and a convex polyomino

a convex polyomino is said to be k-convex if every pair of its cells can be joined by a monotone path with at most k changes of direction.

Definition 1. *An L-convex polyomino p is a convex polyomino in which every pair of cells is connected by a path, of cells of p, with at most one change of direction. Such a path is called (cause its shape) L-path.*

The class of such polyominoes is here denoted by $\mathcal{L}$. In Figure 3 it is depicted an L-convex polyomino where the shaded cells form an *L-paths.*

L-convex polyominoes has been studied with regards to reconstruction from partial information. A very famous approach in this field is *Discrete Tomography.* It considers the problem to reconstruct a discrete set from measurements, called *projections*, of the number of cells in the set that lie on lines with fixed direction (see [19] and also [1], [20], for a survey). In the special case of a convex polyomino p, one considers horizontal and vertical projections, i.e. the two vectors H and V whose components give the the number of cells of p in each row and in each column, respectively. See Fig. 4 for an example.

Given two vectors H and V the class of discrete sets having H and V as projections is denoted by $\mathcal{U}(H, V)$. It is well known that in the general case the knowledge of the horizontal and vertical projections of a discrete set are not sufficient to univocally determine it, see for example in Fig. 4 two polyominoes with the same projections (cf. [10]). Many papers have been devoted to the study of cardinality of this set each time by considering different convexity hypothesis

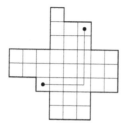

Fig. 3. An L-convex polyomino s and an L-path included in it

to reduce the ambiguity. A classical problem of discrete tomography is the finding of unique sets with respect to H and V.

Definition 2. *A discrete sets p is said to be* unique *with respect to (H, V) if* $\mathcal{U}(H, V) = \{p\}$.

A classical approach (cf. [19]) suggests to test uniqueness of discrete set by finding some *switching components* in it.

Definition 3. *A switching component of a discrete set p is a set* $\{p(i_1, j_1),$ $p(i_1, j_2),\ p(i_2, j_1),\ p(i_2, j_2)\}$ *such that:*

$$p_{i_1 j_1} = p_{i_2 j_2} = 1 - p_{i_1 j_2} = 1 - p_{i_2 j_1}.$$

In particular the following basic theorem holds:

Theorem 1. *(Ryser's Theorem) A discrete set is nonunique (with respect to its horizontal and vertical projections) if and only if it has a switching component.*

In Fig. 4, we can see an example of two different discrete sets with the same projections where we highlight a switching component.

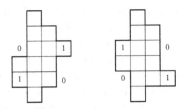

Fig. 4. Two polyominoes with the same projections $H = (1, 2, 3, 3, 3, 1)$ and $V = (2, 6, 4, 1)$. The two polyominoes differ from one another in a switching component.

As regards the L-convexity, we can easily observe that L-convex polyominoes do not have any switching components. In [7] the following lemma is proved.

Lemma 1. *An L-convex polyomino p is unique with respect to its horizontal and vertical projections.*

Finally, recall that an integer vector $X = (x_1, \ldots, x_k)$ is *unimodal*, if there exists $0 \le i \le k$, such that $x_1 \le x_2 \le \cdots \le x_i$ and $x_i \ge x_{i+1} \ge \cdots \ge x_k$. The element x_i is here called *mode*. Hence, we have the theorem (cf. [7]):

Theorem 2. *Let H, V be two unimodal integer vectors and p a discrete set of $\mathcal{U}(H, V)$.*

$$\mathcal{U}(H, V) = \{p\} \Leftrightarrow p \text{ is an L-convex polyomino}$$

As a simple consequence, we have the following remark that will be referred in the next sections because it will be the core of some new definitions.

Remark 1. Let p be a discrete set.

$$p \text{ is convex and unique} \Leftrightarrow p \text{ is a L-convex polyomino.}$$

3   Recognizable Picture Languages and Polyominoes

Here we briefly recall some definitions regarding the theory of two-dimensional languages and tiling systems. For general background we refer the reader to [13]. A *picture* is a two dimensional rectangular array of elements in a finite alphabet. Given a finite alphabet Σ, let $\Sigma^{*,*}$ denote the set of all the pictures over Σ and $\Sigma^{m,n}$ the set of pictures with m columns and n rows, i.e. who's size is (m, n). A *picture language* (or a two-dimensional language) is a set of pictures. A subpicture of a picture p is a subarray of p. A $(2, 2)$ subpicture is called a *tile*. Moreover, we denote by $B_{2,2}(p)$ the set of all the tiles of p.

As usual, we identify the boundaries of a picture by surrounding it with a special symbol. More precisely, given a picture $p \in \Sigma^{m,n}$, we define $\hat{p}$ as the picture of size $(m+2, n+2)$ obtained by surrounding p with the special boundary symbol $\sharp \notin \Sigma$. If $p, q \in \Sigma^{*,*}$ are two pictures with the same number of rows, with pq we will denote the column-concatenation of p and q (see [13] for details).

Definition 4. *A picture language L over Σ is called* local *if there exists a finite set Θ of tiles over $\Sigma \cup \{\sharp\}$ such that $L = \{p | p \in \Sigma^{*,*}$ and $B_{2,2}(\hat{p}) \subseteq \Theta\}$.*

If Θ is the set of tiles that identifies the local picture language L, we will write $L = L(\Theta)$.

Definition 5. *A Tiling system (TS) is a 4-tuple $\tau = (\Sigma, \Gamma, \Theta, \pi)$, where Σ and Γ are two finite alphabets, Θ is a finite set of tiles over the alphabet $\Gamma \cup \{\sharp\}$ and $\pi : \Gamma \to \Sigma$ is a projection.*

The tiling system τ defines the language $L = \pi(L(\Theta))$, given by the projection π on the local language $L(\Theta)$. We will briefly summarize it by writing $L = L(\tau)$. We say that a picture language $L \subseteq \Sigma^{*,*}$ is *recognizable by tiling system* (or equivalently *tiling recognizable*) if there exists a tiling system $\tau = (\Sigma, \Gamma, \Theta, \pi)$ such that $L = L(\tau)$.

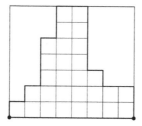

Fig. 5. A stack polyomino

The family of discrete sets and, in particular, sets of polyominoes, are trivially picture languages, over the binary alphabet, that can be or not be tiling recognizable. In the following we give an example of a language of polyominoes that is tiling recognizable.

Example 1. A *stack* polyomino (see Fig. 5) is a convex polyomino containing two adjacent vertices of its minimal bounding rectangle (cf. [2]). If we call $\mathcal{S}$ the family of stack polyominoes we have that $\mathcal{S}$ is tiling recognizable (cf. [9]). Let $p \in \mathcal{S}$, let us observe that the set of entries 0 is composed of two convex distinct parts A and B located, respectively, at the two vertices, not belonging to p, of the rectangle. To each stack p a picture can be associated by representing with a 1 the cells of p, and with a and b the cells of A and B respectively (see Fig. 6). Let $\mathcal{L}_\mathcal{S}$ the language of such pictures over the alphabet $\{1, a, b\}$.

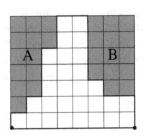

#	#	#	#	#	#	#	#	#	#
#	a	a	a	1	1	b	b	b	#
#	a	a	a	1	1	b	b	b	#
#	a	a	1	1	1	b	b	b	#
#	a	a	1	1	1	b	b	b	#
#	a	a	1	1	1	1	b	b	#
#	a	1	1	1	1	1	1	1	#
#	1	1	1	1	1	1	1	1	#
#	#	#	#	#	#	#	#	#	#

Fig. 6. A stack polyomino p and its representation in $\mathcal{L}_\mathcal{S}$

Let us consider the following sets of tiles

$$\Theta_R = \left\{ \begin{array}{cc} \# & \# \\ \# & \# \end{array}, \begin{array}{cc} \# & \# \\ \# & 1 \end{array}, \begin{array}{cc} \# & \# \\ 1 & \# \end{array}, \begin{array}{cc} \# & 1 \\ \# & \# \end{array}, \begin{array}{cc} 1 & \# \\ \# & \# \end{array}, \begin{array}{cc} \# & \# \\ 1 & 1 \end{array}, \begin{array}{cc} 1 & 1 \\ \# & \# \end{array}, \begin{array}{cc} \# & 1 \\ \# & 1 \end{array}, \begin{array}{cc} 1 & \# \\ 1 & \# \end{array}, \begin{array}{cc} 1 & 1 \\ 1 & 1 \end{array} \right\},$$

$$\Theta_A = \left\{ \begin{array}{cc} a & a \\ a & a \end{array}, \begin{array}{cc} \# & \# \\ \# & a \end{array}, \begin{array}{cc} \# & \# \\ a & a \end{array}, \begin{array}{cc} \# & a \\ \# & a \end{array}, \begin{array}{cc} \# & \# \\ a & 1 \end{array}, \begin{array}{cc} \# & a \\ \# & 1 \end{array}, \begin{array}{cc} a & a \\ a & 1 \end{array}, \begin{array}{cc} a & 1 \\ a & 1 \end{array}, \begin{array}{cc} a & a \\ 1 & 1 \end{array}, \begin{array}{cc} a & 1 \\ 1 & 1 \end{array} \right\},$$

$$\Theta_B = \left\{ \begin{array}{cc} b & b \\ b & b \end{array}, \begin{array}{cc} \# & \# \\ b & \# \end{array}, \begin{array}{cc} \# & \# \\ b & b \end{array}, \begin{array}{cc} b & \# \\ b & \# \end{array}, \begin{array}{cc} \# & \# \\ 1 & b \end{array}, \begin{array}{cc} b & \# \\ 1 & \# \end{array}, \begin{array}{cc} b & b \\ 1 & b \end{array}, \begin{array}{cc} 1 & b \\ 1 & b \end{array}, \begin{array}{cc} b & b \\ 1 & 1 \end{array}, \begin{array}{cc} 1 & b \\ 1 & 1 \end{array} \right\}.$$

$\mathcal{L}_\mathcal{S}$ is local over the alphabet $\Sigma_\mathcal{S} = \{1, a, b\}$ and $\mathcal{L}_\mathcal{S} = L(\theta_R \cup \theta_A \cup \theta_B)$. If we consider the projection $\pi_\mathcal{S} : \Sigma_\mathcal{S} \rightarrow \{0, 1\}$, such that $\pi_\mathcal{S}(a) = \pi_\mathcal{S}(b) = 0$ and $\pi_\mathcal{S}(1) = 1$ we have $\pi_\mathcal{S}(\mathcal{L}_\mathcal{S}) = \mathcal{S}$ that is $\mathcal{S}$ is tiling recognizable.

In [21], O. Matz gives in a lemma a necessary condition of tiling recognizability whose statement is the following.

Lemma 2. *Let $L \subseteq \Sigma^{*,*}$ be tiling recognizable. Let $\{\mathcal{M}_n\}_{n \in \mathbb{N}}$ be a sequence of sets $\mathcal{M}_n \subseteq \Sigma^{n,+} \times \Sigma^{n,+}$ such that $\forall n$ following relations hold:*

$$\forall (p, q) \in \mathcal{M}_n \text{ we have } pq \in L \qquad (1)$$

$$\forall (p,q) \neq (p',q') \in \mathcal{M}_n \text{ we have } \{pq', p'q\} \nsubseteq L. \qquad (2)$$

Then $|\mathcal{M}_n|$ is $2^{\mathcal{O}(n)}$ [21].

Furthermore, in the same paper, he raises the question of finding a picture language for which this lemma cannot be applied to prove it is not recognizable. He conjectures the language of square pictures over $\{a,b\}$ with number of a's equal to number of b's as a possible solution. In fact, as it is easy to see, for this language one can not constructs a sequence $\{\mathcal{M}_n\}$ of sets satisfying the two conditions of lemma but with order definitively greater than $2^{\mathcal{O}(n)}$. However, this conjecture has been shown to be false. Indeed, K.Reinhardt shows the following more general result.

Theorem 3. *The language of picture over $\{a,b\}$, where the number of a's is equal to the number of b's and having a size (n,m) with $m < 2^n$ and $n < 2^m$, is recognizable (cf. [23]).*

With the motivations explicated in the introduction we want to continue the study of languages of polyominoes taking into consideration L-convexity.

Theorem 4. *Let $\{\mathcal{M}_n\}_{n \in \mathbb{N}}$ be a sequence of sets $\mathcal{M}_n \subseteq \Sigma^{n,+} \times \Sigma^{n,+}$, with $\Sigma = \{0,1\}$. For all $n \in \mathbb{N}$, if $\mathcal{M}_n$ satisfies relations (1) and (2) with respect to the language $\mathcal{L}$ of L-convex polyominoes then $|\mathcal{M}_n|$ is $2^{\mathcal{O}(n)}$.*

Proof. Firstly, let us observe that for any $(p,q) \in \Sigma^{n,+} \times \Sigma^{n,+}$, if $pq \in \mathcal{L}$ then $p, q \in \mathcal{L}$. Hence, we have that for any $n \in \mathbb{N}$, a set $\mathcal{M}_n$ satisfying first condition is a subset of $\mathcal{L} \times \mathcal{L}$.

Let p an L-convex polyomino having vertical projections $(c_1, c_2, ..., c_k)$. We know that the vector of vertical projections is unimodal. Let $1 \leq m \leq k$ such that c_m is the mode and $c_{m+1} \neq c_m$ if it exists. To p we can associate, by a suitable deletion of columns, the L-convex polyomino $\tilde{p}$ with vertical projections $(a_1, ..., a_h, c_m, b_1, ..., b_l)$, with $a_i \neq a_j$, for $1 < j < h$, $b_i \neq b_j$, for $1 < j < l$ and such that the sets

$$\{c_1, ..., c_m\} = \{a_1, ..., a_h, c_m\} \text{ and } \{c_{m+1}, ..., c_k\} = \{b_1, ..., b_l\}.$$

Let us observe that $\tilde{p} \in \mathcal{L}$. So, we can define in $\mathcal{L}$ an equivalence relation $\sim$ as follows

$$p \sim q \text{ iff } \tilde{p} = \tilde{q}.$$

If we denote by $[p]$ the equivalence class represented by $p \in \mathcal{L}$ and $\tilde{\mathcal{L}}$ the closure of $\mathcal{L}$ with respect to $\sim$, by definition, we have that:

  $-$ $\forall p \in \mathcal{L}, \tilde{p} \in [p]$;
  $-$ $\forall (p,q) \neq (p',q') \in [p] \times [q]$, such that $pq, p'q' \in \mathcal{L}$ we have $pq', p'q \in \mathcal{L}$;

Our aim is to know $|\mathcal{M}_n|$. For this reason we compute $|\tilde{\mathcal{L}} \cap \Sigma^{n,+}|$ i.e. we enumerate $\tilde{\mathcal{L}}$ according to the number of rows. In [6] and [8], in order to enumerate L-convex polyominoes by rows and columns, the authors represent L-convex polyominoes in terms of words of a regular language. By using the same coding we obtain the regular expression for the language whose words represent the elements of $\tilde{\mathcal{L}}$. That is

$$a \left(b + c^+a + c^+d + c^+da + bda + bd + ba\right)^* b . \tag{3}$$

Let $\tilde{l}_{i,j}$ be the number of polyominoes of $\tilde{\mathcal{L}}$ with $i+1$ rows and $j+1$ columns. From (3), removing the first and the last letter, we can obtain the generating function for these numbers, as described in [24], after setting $a = d = y$ and $b = c = x$:

$$\tilde{L}(x,y) = \sum_{i,j \geq 0} \tilde{l}_{i,j}\, x^i y^j = \cfrac{1}{1 - x - 2xy - xy^2 - \cfrac{2xy}{1-x} - \cfrac{xy^2}{1-x}}$$

by setting $y = 1$, i.e. considering the generating function with respect to the number of rows we have

$$\tilde{L}(x) = \sum_{n \in \mathbb{N}} \tilde{l}_n x^n = \frac{1-x}{4x^2 - 8x + 1}$$

and closed formula for $\tilde{l}_n$, i.e. the number of L-convex polyominoes of $\tilde{\mathcal{L}}$ having n rows, is

$$\tilde{l}_n = \frac{1}{8}\left[\left(\frac{2}{2+\sqrt{3}}\right)^n + \left(\frac{2}{2-\sqrt{3}}\right)^n\right] = |\tilde{\mathcal{L}} \cap \Sigma^{n,+}|.$$

Hence, the thesis follows.

In next section we give some results to introduce our conjecture about the language of L-convex polyominoes.

4 L_h-convex and L_v-convex polyominoes

Lemma 1 states that all the convex polyominoes that are unique with respect to horizontal and vertical projections are L-convex and viceversa. Now we define a more general family of polyominoes by considering those polyominoes that are unique and horizontally (resp. vertically) convex.

Definition 6. *A polyomino p is called $L_h - convex$ (resp. $L_v - convex$) if it is h-convex (resp. v-convex) and it has no switching component.*

We call these polyominoes L_h-convex (resp. $L_v - convex$) and denote the family by $\mathcal{L}_h$ (resp. $\mathcal{L}_v$). Actually, this concept is not only a set-theoretic generalization but introduces a more general convexity property than the L-convexity. More precisely, it is easy to see that an L_h (resp. L_v) polyomino p is such that each pair

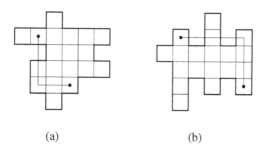

(a)   (b)

Fig. 7. (a) An $L_h - convex$ polyomino and an L-path whose vertical arm is partially included in the polyomino; (b) an $L_v - convex$ polyomino and an L-path with horizontal arm partially included in the polyomino

of its cells can be joined by an L-path whose horizontal (resp. vertical) arm is entirely contained in p and whose vertical (resp. horizontal) arm can result to be partially contained in p (see Fig. 7). Note that, in both the two cases of convexity, the cell in which the L-path changes direction must be a cell of p. Hence, we have, trivially, that cells of a unique discrete set p have the property that they pairwise can be joined by an L-path whose both the arms can be partially contained in p. So, we can resume with the following set-theoretic relations

$$\mathcal{L} \subset \mathcal{L}_h \subset \mathcal{U} \, , \; \mathcal{L} \subset \mathcal{L}_v \subset \mathcal{U},$$

$$\mathcal{L} = \mathcal{L}_h \cap \mathcal{V} = \mathcal{L}_v \cap \mathcal{H} = \mathcal{C} \cap \mathcal{U}$$

and

$$\mathcal{L} = \mathcal{L}_h \cap \mathcal{L}_v \, .$$

Theorem 5. $\mathcal{L}_h$ *(resp.* $\mathcal{L}_v$*) is not tiling recognizable.*

Proof. The proof is based on Matz's lemma. Roughly speaking, what we are doing is to construct a sequence of pair of special picture sets whose elements satisfy relations (1) and (2) of the lemma. However, as we will see, the cardinality of an element of the sequence is definitively greater than $2^{\mathcal{O}(n)}$. Let $\Sigma = \{0, 1\}$ and $\sigma \in S_n$ (i.e. a permutation of the symmetric group of degree n). We define p_σ as the picture of size (n, n), whose i-th row contains 0 in its first $n - \sigma(i)$ positions and 1 in the remainders. With p_σ^s we denote the mirror image of p_σ with respect to the vertical line at the right of its last column. It is easy to see that the picture $p_\sigma p_\sigma^s$ ($\forall \sigma \in S_n$) has not any switching component (see Fig. 8). Vice versa we note that the picture $p_\sigma p_\gamma^s$, with $\sigma \neq \gamma$, has got, at least, one switching component. Indeed, let I and J the two indexes such that

$$\sigma(I) = \gamma(J) = Max\{\sigma(i) : \sigma(i) = \gamma(j) \text{ and } i \neq j\}_{i,j=1,\dots,n}.$$

As it is easy to check, a switching component is given by the intersection of the two rows with indexes I and J with the columns with indexes $n - \sigma(I) + 1$ and $n + \gamma(J)$ (see Fig. 8). We can resume by saying that for all partitions σ, γ

the picture $p_\sigma p_\gamma^s$ has no switching component if and only if $\sigma = \gamma$. This implies that for all $n \geq 1$ the set $M_n = \{(p_\sigma, p_\sigma^s) \mid \sigma \in S_n\}$ satisfies the conditions of Lemma 2. However, since $\mid M_n \mid = \mid S_n \mid = n!$, from Lemma 2 we have that $\mathcal{L}_h$ is not tiling recognizable.

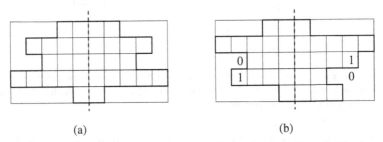

(a)   (b)

Fig. 8. The picture $p_\sigma p_\gamma^s$ with: (a) $\sigma = \gamma = \{2, 4, 3, 5, 1\}$; (b) $\sigma = \{2, 5, 3, 4, 1\}$ and $\gamma = \{1, 5, 4, 2, 3\}$

By considering the same sequence of the previous proof we have that the following theorem holds.

Theorem 6. $\mathcal{U}$ *is not tiling recognizable.*

Hence, we can resume that the family of L-convex polyominoes can be written either as the intersection of a tiling recognizable picture language ($\mathcal{H}$) and a not tiling recognizable picture languages ($\mathcal{L}_v$), or as the intersection of two not tiling recognizable picture languages ($\mathcal{L}_v$ and $\mathcal{L}_h$). However, although we are able to isolate the main properties that lead to the L-convexity (as unicity and convexity) and to give an answer about the tiling recognizability for any of this subclass, both the previous two relations do not bring us to resolve the open problem but led us to advance the following conjecture.

Conjecture 1. $\mathcal{L}$ *is not tiling recognizable.*

References

1. Barcucci, E., Del Lungo, A., Nivat, M., Pinzani, R.: Reconstructing convex polyominoes from horizontal and vertical projections. Theoret. Comput. Sci. 155, 321–347 (1996)
2. Bousquet-Mèlou, M.: A method for the enumeration of various classes of column-convex polygons. Dis. Math. 154, 1–25 (1996)
3. Castiglione, G., Restivo, A.: Reconstruction of L-convex Polyominoes. Electronic Notes in Discrete Mathematics 12 (2003)
4. Castiglione, G., Restivo, A.: Ordering and Convex Polyominoes. In: Margenstern, M. (ed.) MCU 2004. LNCS, vol. 3354, pp. 128–139. Springer, Heidelberg (2005)
5. Castiglione, G., Frosini, A., Restivo, A., Rinaldi, S.: Enumeration of L-convex Polyominoes. Theoret. Comput. Sci. 347, 336–352 (2005)

6. Castiglione, G., Frosini, A., Munarini, E., Restivo, A., Rinaldi, S.: Enumeration of L-convex Polyominoes, II. Bijection and area. In: FPSAC 2005, Taormina (June 20–25, 2005)
7. Castiglione, G., Frosini, A., Restivo, A., Rinaldi, S.: A Tomographical Characterization of L-convex Polyominoes. In: Andrès, É., Damiand, G., Lienhardt, P. (eds.) DGCI 2005. LNCS, vol. 3429, pp. 115–125. Springer, Heidelberg (2005)
8. Castiglione, G., Frosini, A., Munarini, E., Restivo, A., Rinaldi, S.: Combinatorial aspects of L-convex polyominoes. In: European Journal of Combinatorics (In Press)
9. De Carli, F., Frosini, A., Rinaldi, S., Vuillon, L.: On the Tiling System Recognizability of Various Classes of Convex Polyominoes. In: Annals of combinatorics (To appear)
10. Del Lungo, A., Nivat, M., Pinzani, R.: The number of convex polyominoes reconstructible from their orthogonal projections. Discrete Math. 157, 65–78 (1996)
11. Dhar, D.: Equivalence of two-dimensional directed animal problem to a onedimensional path problem. Adv. in Appl. Math. 9, 959–962 (1988)
12. Gardner, M.: Mathematical Games. Scientific American 196, 126–134 (1957)
13. Giammarresi, D., Restivo, A.: Two-dimensional languages. In: Salomaa, A., Rozemberg, G. (eds.) Handbook of Formal Languages, vol. 3, pp. 215–267. Springer, Heidelberg (1997)
14. Girault-Beauquier, D., Nivat, M.: Tiling the plane with one tile, Proceedings of the sixth annual symposium on Computational geometry, Berkley, California, United States, June 07-09, 1990 (1990)
15. Golomb, S.W.: Checker boards and polyominoes. Amer. Math. Monthly 61, 675–682 (1954)
16. Golomb, S.W.: Polyominoes. Scribner, New York (1965)
17. Golomb, S.W.: Polyominoes: Puzzles, Patterns, Problems and Packing. Princeton Academic Press, London (1996)
18. Hakim, V., Nadal, J.P.: Exact result for 2D directed lattice animals on a strip of finite width. J. Phys. A: Math. 16, L213–L218 (1983)
19. Herman, G.T., Kuba, A.: Discrete Tomography: Foundations, Algorithms and Applications, Birkhauser Boston, Cambridge, MA on of convex 2D discrete sets in polynomial time. Theoret. Comput. Sci. 283, 223–242 (2002)
20. Kuba, A., Balogh, E.: Reconstruction of convex 2D discrete sets in polynomial time. Theoret. Comput. Sci. 283, 223–242 (2002)
21. Matz, O.: On piecewise testable, starfree, and recognizable picture languages. In: Nivat, M. (ed.) ETAPS 1998 and FOSSACS 1998. LNCS, vol. 1378, pp. 203–210. Springer, Heidelberg (1998)
22. Reinhardt, K.: On some recognizable picture-languages. In: Brim, L., Gruska, J., Zlatuška, J. (eds.) MFCS 1998. LNCS, vol. 1450, pp. 760–770. Springer, Heidelberg (1998)
23. Reinhardt, K.: The #a = #b Pictures are Recognizable. In: Ferreira, A., Reichel, H. (eds.) STACS 2001. LNCS, vol. 2010, pp. 527–538. Springer, Heidelberg (2001)
24. Salomaa, A., Soittola, M.: Automata-theoretic aspects of formal power series. Springer, New York (1978)

An Algebra for Tree-Based Music Generation

Frank Drewes and Johanna Högberg

Department of Computing Science, Umeå University
S–90187 Umeå, Sweden
{drewes,johanna}@cs.umu.se

Abstract. We present an algebra whose operations act on musical pieces, and show how this algebra can be used to generate music in a tree-based fashion. Starting from input which is either generated by a regular tree grammar or provided by the user via a digital keyboard, a sequence of tree transducers is applied to generate a tree over the operations provided by the music algebra. The evaluation of this tree yields the musical piece generated.

1 Introduction

The purpose of this paper is to show that certain musical structures can be generated in a grammatical manner by using a general method known as tree-based generation. Known from the areas of graph and picture generation [Eng97, Dre06], a tree-based generator consists of a tree generator and a Σ-algebra. The tree generator is any formal device (e.g., a grammar), that generates a set of trees over Σ. The algebra is then used to evaluate these trees.

Thus, the type of objects being generated depends on the domain of the algebra. Here, we exploit this fact for the generation of music. To be precise, we use the term music as a shorthand for "sound structures that adhere to certain basic rules of composition". Rather than trying to imitate human composers, we are interested in finding out how and in how far the formal structures found in music can be captured using the devices of formal language theory and, in particular, tree-based generators. A precursor of the approach presented here is the system Willow [Hög05], which consists of a regular tree grammar g and a sequence $td_1, \dots, td_n$ of top-down tree transducers (td transducers, see, e.g. [GS97, FV98]). Intuitively, g generates a (tree representing a) coarse rhythmical structure. This tree is then passed through $td_1, \dots, td_n$. Each of them enriches the tree to add a certain musical attribute, e.g. tempo, chord progression, or melodic arc. Finally, the output tree of td_n represents the musical piece generated.

In Willow, the generated trees are interpreted in an ad hoc way rather than using a formally defined algebra. In this paper, we present such an algebra and show how it can be used to generate music in a fully tree-based manner. Another extension of the previous system is that some musical attributes are realised by macro tree transducers (mt transducers) rather than td transducers, as this

S. Bozapalidis and G. Rahonis (Eds.): CAI 2007, LNCS 4728, pp. 172–188, 2007.

device is particularly suited to model recurring variations on a pattern. Moreover, we now allow the user to input themes by means of a digital keyboard.

Formally, if user input is used, it is translated into a tree that becomes the input of (macro or top-down) tree transducers $tt_1, \ldots, tt_n$, thus replacing the initial tree generated by the regular tree grammar g. However, whereas g generates trees over a finite output signature, user input gives rise to trees containing symbols which are unknown at the time the tree transducers $tt_1, \ldots, tt_n$ are designed. Such a situation cannot be handled by ordinary tree transducers; tt_1 would simply reject any tree containing an unknown symbol. In contrast, the appropriate behaviour would be to "tolerate" them, i.e. copy them to the output without doing anything. Therefore, we propose a proper generalisation of the mt transducer called tolerant mt transducer (tmt transducer). A tmt transducer has, as a subset of its set of states, a set of so-called tolerant states. Whenever an unknown input symbol is encountered in a tolerant state, this symbol is simply copied to the output and the computation continues on its subtrees in the same state. In all other situations, a tmt transducer behaves like an ordinary mt transducer. As mt transducers in their turn generalise td transducers, we also get a tolerant version of the td transducer (ttd transducer).

To the best of our knowledge, no tree-based approach (in the sense described above) to the generation of music has been proposed earlier (except for the first attempt made in [Hög05]). However, there are some publications, often with a somewhat different focus, that propose other approaches to music generation using techniques from formal language theory.

In [Jur06], Jurish gives a characterisation of generic musical structure within the framework of formal language theory. Tojo and Oka [TON06] present an analysis system for chord progressions based on head-driven phrase structure grammars. In [Pru86], Prusinkiewicz explores the generation of musical scores by means of L-systems and the so-called turtle geometry. This approach is further developed by Worth and Stepney, who describe their search for simultaneously 'pleasing' graphical and musical renderings of languages generated by L-systems in [WS05]. Another device that has been used for the generation of music is the cellular automaton. A survey of this type of work is found in [BELM04].

Related approaches can also be found in the enclosing field of algorithmic composition. Markov models appear frequently in music theory, where they are used both as a compositional device [Ame89, Vis04], and for attribute classification [SXK04, CV01]. Genetic algorithms have also become increasingly popular [HG91, Jac95], as have approaches to deriving music from fractals and chaotic systems [Cha03]. Synthesis and analysis based on neural networks and learning systems are described in [Bha88, BDPV94, Tod91] and [Moz94]. However, most established is probably the linguistic approach. Two representatives of this line of research are Baroni [Bar83] and Moorer [Moo72].

The structure of this paper is as follows. In the next section, we compile the terminology around trees and tree generation required, including the new concept of tolerant mt transducers. In Section 3, the music algebra is introduced. An example is discussed in Section 4, and Section 5 concludes the paper.

An implementation of the music algebra has been added to TREEBAG [Dre06], a system that allows to define and execute tree grammars and tree transducers, and interpret the resulting trees by means of some algebra. In this way, the example presented in Section 4 has been realized. Both the implementation and the example can be downloaded from http://www.cs.umu.se/~johanna/algebra.

2   Tree Generation

Throughout this paper, we denote the natural numbers (including 0) by $\mathbb{N}$, and $\mathbb{Z}$ and $\mathbb{R}^+$ denote the integers and the positive reals, resp. The set $\mathbb{R}^+ \cup \{0\}$ of nonnegative reals is denoted by $\mathbb{R}_0^+$. For $n \in \mathbb{N}$, $[n]$ denotes the set $\{1, \ldots, n\}$. We denote the power set of a set S by $\wp(S)$ and the set of all finite strings over S by S^*; the empty string is denoted by λ. The transitive and reflexive closure of a binary relation $\rightarrow$ is denoted by $\rightarrow^*$.

Let B be the set consisting of the three special symbols '[', ']', and ','. The set $\mathcal{T}$ of all *trees* is the smallest set of strings such that, for every symbol $f \notin B$ and all trees $t_1, \ldots, t_k \in \mathcal{T}$ $(k \in \mathbb{N})$, the string $f[t_1, \ldots, t_k]$ is in $\mathcal{T}$ as well. As usual, the set of all subtrees of $f[t_1, \ldots, t_k]$ contains the tree itself and all subtrees of its *direct subtrees* $t_1, \ldots, t_k$. As a notational simplification, we identify the tree $f[]$ with f. In this sense, every single symbol is a tree. As another notational simplification, a tree of the form $g[t_1]$ (i.e., having only one direct subtree) may be denoted by $g\,t_1$.

For a tree $t = f[t_1, \ldots, t_k]$, the set $\mathrm{nod}(t)$ of *nodes of* t is the subset of $\mathbb{N}^*$ given by $\mathrm{nod}(t) = \{\lambda\} \cup \{iv \mid i \in [k] \text{ and } v \in \mathrm{nod}(t_i)\}$. For a node $u \in \mathrm{nod}(t)$, t/u denotes the subtree of t rooted at u, and $t[\![u \leftarrow s]\!]$ denotes the tree obtained from t by replacing t/u with $s \in \mathcal{T}$. Thus,

$$t/u = \begin{cases} t & \text{if } u = \lambda \\ t_i/v & \text{if } u = iv,\ i \in [k] \end{cases}$$

and

$$t[\![u \leftarrow s]\!] = \begin{cases} s & \text{if } u = \lambda \\ f[t_1, \ldots, t_{i-1}, t_i[\![v \leftarrow s]\!], t_{i+1}, \ldots, t_k] & \text{if } u = iv,\ i \in [k]. \end{cases}$$

A (not necessarily finite) *alphabet* is a set Σ of symbols such that $B \cap \Sigma = \emptyset$. Given an alphabet Σ and a set T of trees, $\Sigma(T)$ denotes the set of all trees $f[t_1, \ldots, t_k]$ such that $f \in \Sigma$ and $t_1, \ldots, t_k \in T$ for some $k \in \mathbb{N}$. The set $\mathcal{T}_\Sigma(T)$ of all trees over Σ with subtrees in T is the smallest set of trees such that $T \cup \Sigma(\mathcal{T}_\Sigma(T)) \subseteq \mathcal{T}_\Sigma(T)$. The set $\mathcal{T}_\Sigma(\emptyset)$ of trees over Σ is briefly denoted by $\mathcal{T}_\Sigma$.

A *ranked alphabet* is an alphabet Σ which is given as a (not necessarily disjoint) union $\Sigma = \bigcup_{k \in \mathbb{N}} \Sigma^{(k)}$. A symbol in $\Sigma^{(k)}$ is said to be of rank k and may be denoted by $f^{(k)}$ to indicate its rank. A ranked alphabet is called finite if only finitely many of the $\Sigma^{(k)}$ are nonempty, and each of them is finite. If symbols from ranked alphabets are used to build trees, we apply the additional restriction that each occurrence of $f^{(k)}$ must have exactly k subtrees. Thus, in this case,

$\Sigma(T)$ is the set of all trees $f[t_1, \ldots, t_k]$ such that $f^{(k)} \in \Sigma$ for some $k \in \mathbb{N}$, and $t_1, \ldots, t_k \in T$.

Throughout this paper, let $\mathcal{U}$ be a universe of symbols, where $\mathcal{U} \cap B = \emptyset$. The intuition behind $\mathcal{U}$ is that it is our supply of "ordinary" symbols. Symbols that do not belong to $\mathcal{U}$ have an auxiliary character. Variables provide an example of the latter: we let X be a countably infinite alphabet of variables, where $X \cap \mathcal{U} = \emptyset$. Thus, as usual in the theories of term rewriting, tree languages, and tree transformation, variables are just a special sort of symbols, rather than being variables in the mathematical sense. In trees, variables will only appear as leaves, i.e., they will be considered as symbols of rank 0. A mapping $\sigma \colon X' \to T$, where $X' \subseteq X$, is a *substitution*. For a tree t, $t\sigma$ denotes the tree obtained from t by simultaneously replacing every occurrence of $x \in X'$ with $\sigma(x)$. As a recursive definition, $x\sigma = \sigma(x)$ for all $x \in X'$, and if $t = f[t_1, \ldots, t_l]$ with $l > 0$ or $f \notin X'$, then $t\sigma = f[t_1\sigma, \ldots, t_l\sigma]$.

A *term rewrite system* is a set R of rules of the form $l \to r$, where $l, r \in T$ are such that all variables in r occur in l as well. For trees s, t, there is a *rewrite step* $s \to_R t$ if there are a rule $l \to r$ in R, a node $v \in \mathrm{nod}(s) \cap \mathrm{nod}(t)$, and a substitution σ, such that $s/v = l\sigma$ and $t = s[\![v \leftarrow r\sigma]\!]$.

Regular tree grammars are defined as usual. Thus, such a grammar is a system $g = (N, \Sigma, R, S)$ consisting of a ranked alphabet $N = N^{(0)}$ of nonterminals, where $N \cap \mathcal{U} = \emptyset$, a ranked alphabet $\Sigma \subseteq \mathcal{U}$ of terminals, a set R of rules, and an initial nonterminal S. The sets N, Σ, and R are required to be finite. Every rule in R is of the form $A \to t$, where $A \in N$ and $t \in T_\Sigma(N)$. The language generated by g, called a regular tree language, is given by $L(g) = \{t \in T_\Sigma \mid S \to_R^* t\}$.

As mentioned in the introduction, our general approach for generating music is adopted from [Hög05]: a tree denoting a musical piece is generated by starting with a very simple tree and then applying to it a sequence of macro and top-down tree transducers, each of which is responsible for adding a specific musical aspect. However, ordinary macro and top-down tree transducers are not ideal for this purpose, because they have a finite input alphabet and cannot process input trees that contain other symbols. Instead, we would like the transformation to "step over" such symbols, at least if they are encountered in certain states. For these reasons, we now introduce the so-called tolerant macro tree transducer.

Definition 1 (tolerant mt transducer). *A tolerant macro tree transducer (tmt transducer, for short) is a system $mt^t = (\Sigma, \Sigma', Q, Q^t, R, q_0)$, where*

- *$\Sigma, \Sigma' \subseteq \mathcal{U}$ are finite ranked alphabets, the input and output alphabets,*
- *Q with $Q^{(0)} = \emptyset$ and $Q \cap \mathcal{U} = \emptyset$ is a ranked alphabet of states,*
- *$Q^t \subseteq Q$ is the set of tolerant states,*
- *R is a finite set of rules of the form*

$$q[f[x_1, \ldots, x_k], y_1, \ldots, y_m] \to t$$

with $q \in Q^{(m+1)}$, $f^{(k)} \in \Sigma$ for some $k \in \mathbb{N}$, $x_1, \ldots, x_k, y_1, \ldots, y_m \in X$ being pairwise distinct, and $t \in T$, where the set T is recursively defined as
   - *$y_i \in T$, for all $i \in [m]$,*

- $g[t_1, \ldots, t_l] \in T$, for every $g \in \Sigma'$ of rank l and $t_1, \ldots, t_l \in T$, and
- $q'[x_i, t_1, \ldots, t_l] \in T$ for every $q' \in Q^{(l+1)}$, $i \in [k]$, and $t_1, \ldots, t_l \in T$.
- $q_0 \in Q^{(1)}$ is the initial state.

From the above definition, we obtain the *tolerant top-down tree transducer* (abbreviated ttd transducer) by considering the special case when $Q = Q^{(1)}$. Furthermore, mt transducers and td transducers are a special case of tmt transducers and ttd transducers, resp., by taking $Q^t = \emptyset$.

Given an input tree $t \in \mathcal{T}_\mathcal{U}$, a computation of mt^t starts with $q_0 t$ and applies the term rewrite rules in R until a tree in $\mathcal{T}_\mathcal{U}$ is reached. Whenever a symbol not in Σ is reached in a tolerant state, this symbol is simply copied to the output. Formally, this reads as follows.

Definition 2 (computed tree transduction). *In what follows, let $mt^t = (\Sigma, \Sigma', Q, Q^t, R, q_0)$ be a tolerant mt transducer. For trees $s, t \in \mathcal{T}_{\mathcal{U} \cup Q}$, there is a computation step $s \rightarrow_{mt^t} t$ if $s \rightarrow_{R \cup R'} t$, where*

$$R' = \{q[f[x_1, \ldots, x_k], y_1, \ldots, y_m] \rightarrow \\ f[q[x_1, y_1, \ldots, y_m], \ldots, q[x_k, y_1, \ldots, y_m]] \mid q^{(m+1)} \in Q^t, f \in \mathcal{U} \setminus \Sigma^{(k)}\} \ .$$

A tree $t \in \mathcal{T}_{\mathcal{U} \cup Q}$ is a sentential tree (with respect to mt^t) if $q_0 s \rightarrow^*_{mt^t} t$ for some $s \in \mathcal{T}_\mathcal{U}$. The tree transduction computed by mt^t, called a tolerant mt transduction, is the mapping $mt^t \colon \mathcal{T}_\mathcal{U} \rightarrow \wp(\mathcal{T}_\mathcal{U})$ such that

$$mt^t(s) = \{t \in \mathcal{T}_\mathcal{U} \mid q_0 s \rightarrow^*_{mt^t} t\}$$

for all $s \in \mathcal{T}_\mathcal{U}$. For a set $T \subseteq \mathcal{T}_\mathcal{U}$, we let $mt^t(T) = \bigcup\{mt^t(t) \mid t \in T\}$.

As usual, the tree transduction computed by mt^t is considered to be a partial function $mt^t \colon \mathcal{T}_\mathcal{U} \rightarrow \mathcal{T}_\mathcal{U}$ if $|mt^t(t)| \leq 1$ for all $t \in \mathcal{T}_\mathcal{U}$.

Let us return for a moment to Definition 1. The way the rules of an mt transducer are defined allows for sentential trees with nested states. If the mt transducer is deterministic[1], then the way in which computation steps are made does not matter: the output tree (if it exists at all) is completely determined by the input tree. In the nondeterministic case, however, the input tree may yield different sets of output trees, depending on the order in which the nested states are processed. In [EV85], the following strategies are discussed.

- *Innermost-Outermost* (IO, for short) applies the rewriting rules bottom-up. A state can only be rewritten when its direct subtrees, except the first, consist exclusively of symbols in $\mathcal{U}$.
- *Outermost-Innermost* (OI, when abbreviated) applies the rewriting rules top-down. In other words, a state can only be rewritten when the path going from the root to that state is labelled with symbols in $\mathcal{U}$ only.
- *Unrestricted* allows the rewriting rules to be applied in any order.

---

[1] There do not exist two distinct rules with left-hand sides $q[f[x_1, \ldots, x_k], y_1, \ldots, y_m]$ and $q[f[x'_1, \ldots, x'_k], y'_1, \ldots, y'_k]$, resp., i.e., the left-hand sides differ only in the naming of variables.

As shown in [EV85], the translations realized by OI and the unrestricted mode coincides. Let us now compare the IO and OI strategies at an intuitive level. In an OI transduction, subtrees may be copied (by nonlinear rules) *before* they are processed. Continuing the derivation, the copies can thus be turned into several non-isomorphic subtrees in the output. In an IO transduction, a subtree is only copied *after* it has been processed, and may thus yield a number of identical subtrees in the output. Both the IO and the OI behaviour can be beneficial in the generation of musical pieces. Whereas IO allows for regularity and a clear structure, it is more convenient to use OI when we want to endow the generated piece with a more spontaneous quality.

3   An Algebra for Music

In this section, we introduce the main contribution of the paper, the so-called music algebra. The operations of this algebra can be used to assemble a musical piece in a stepwise manner. Let us first summarise a few basics regarding music that are needed for a better understanding of the elements of the music algebra.

We identify a musical piece with a sequence of notes. In general, a note is characterised by its *tone, length, accent,* and *timbre*. The tone of a note is the ratio between its frequency and that of a fixed reference note. Similarly, the length of a note is measured relative to the length of the reference note. A note whose duration is equal to that of the reference note, is a *whole note*, a note whose duration is but half of that of the reference note is a *half note*, and so on. Even accent is a relative property; a note is accented if it is played, for example, louder than any surrounding note. The timbre of a note is the subjective quality which lets us distinguish between instruments. For the sake of simplicity, we disregard the accent and timbre of a note.

A *scale* is a set of tones, and just as there are an infinite number of tones, so are there an infinite number of conceivable scales. However, only some of them are used in practice. The most important of these is probably the chromatic scale, which is constructed as follows: First a reference tone is ordained, which would normally be the so-called *a above middle c*, an alias for 440 Hz (for convenience, we henceforth restrict ourselves to chromatic scales built around this tone). By doubling this pitch, the tone one *octave* higher is found, and by dividing it by two, the tone one octave lower. The whole of the audible interval is split into octaves, and every octave is divided into 12 tones, in such a way that the ratio between two consecutive tones amounts to $2^{1/12}$. The tones can thus be referred to by integers: 0 refers to the reference tone, and for every tone t, $t-1$ and $t+1$ refer to the next lower and higher tone, resp., in the chromatic scale. In other words, in our setting, tone t has a frequency of $2^{t/12} \cdot 440$ Hz.

The music algebra, which we denote by $\mathbf{M}$, is a many-sorted algebra whose data domains are the sets $\mathbb{Z}$, $\mathbb{R}^+$, and $\mathcal{P}$. While $\mathcal{P}$, to be defined below, denotes the set of all musical pieces, elements of $\mathbb{Z}$ are always interpreted as tones, and elements of $\mathbb{R}^+$ are always in one way or another related to time.

Definition 3 (note and musical piece). *A note is a pair $n = (t, l) \in \mathbb{Z} \times \mathbb{R}^+$ consisting of the tone t of length l. The set of all notes is denoted by $\mathcal{N}$.*

The set $\mathcal{P}$ of musical pieces *is the set of all pairs $P = (N, L)$ such that*

- *$N \subseteq \mathcal{N} \times \mathbb{R}_0^+$ is a finite set of* played notes, *where, for a played note $(n, s) \in N$, s is the point in time where n starts to be played;*
- *$L \in \mathbb{R}^+$ is the length of the entire piece, with $L \geq s + l$ for all $((t, l), s) \in N$.*

Note that there are no pieces whose length is 0, and that a piece cannot end before its last note has been played (owing to the requirement placed on L). In the following, we may denote the components of a note n by t_n and l_n, resp. Similarly, the components of a piece P may be denoted by N_P and L_P, resp.

We now define a number of operations on musical pieces. For this, an auxiliary notation turns out to be useful. Given a set N of played notes and a mapping f on played notes, we let $N[\![(n, s) \mapsto f(n, s)]\!] = \{f(n, s) \mid (n, s) \in N\}$. Similarly, if f is a mapping on notes, we let $N[\![n \mapsto f(n)]\!] = \{(f(n), s) \mid (n, s) \in N\}$.

Now, consider pieces $P = (N, L)$ and $P' = (N', L')$.

- LENGTH$(P) = L$ returns the length of P.
- If $N \neq \emptyset$, let $q = \max\{s + l_n \mid (n, s) \in N\}$, i.e., q is the point in time where the last notes in P end. The highest tone at the end of P is returned by

$$\text{HIGHEST}(P) = \begin{cases} \max\{t_n \mid (n, s) \in N \text{ and } s + l_n = q\} & \text{if } N \neq \emptyset \\ 0 & \text{otherwise.} \end{cases}$$

(Using the operations defined below, this gives also access to the lowest tone at the end of P and the highest and lowest tones at the beginning of P.)
- For every factor $a \in \mathbb{R}^+$,

$$\text{SCALE}(P, a) = (N[\![(n, s) \mapsto ((t_n, a \cdot l_n), a \cdot s)]\!], a \cdot L)$$

scales P by the factor a.
- Inversion of all tones of notes in P is obtained by

$$\text{INV}(P) = (N[\![n \mapsto (-t_n, l_n)]\!], L) \ .$$

- For every tone $t \in \mathbb{Z}$,

$$\text{RAISE}(P, t) = (N[\![n \mapsto (t_n + t, l_n)]\!], L)$$

raises every tone of notes in P by t.
- P is played backwards by

$$\text{BACK}(P) = (N[\![(n, s) \mapsto (n, L - s - l_n)]\!], L) \ .$$

- MUTE$(P) = (\emptyset, L)$ returns a silent piece having the same length as P.

- OVERLAY(P, P') yields the overlay of P and P', given by

$$\text{OVERLAY}(P, P') = (N \cup N', \max(L, L')) \ .$$

For the sake of convenience, we extend OVERLAY to any number of arguments, i.e. for pieces $P_1, \ldots, P_k \ (k \geq 1)$,

$$\text{OVERLAY}(P_1, \ldots, P_k) = \text{OVERLAY}(P_1, \cdots \text{OVERLAY}(P_{k-1}, P_k) \cdots)$$
$$= (\textstyle\bigcup_{i \in [k]} N_{P_i}, \max_{i \in [k]} L_{P_i}) \ .$$

- The concatenation of P and P' is given by

$$\text{CONCAT}(P, P') = (N \cup N' [\![(n, s) \mapsto (n, L + s)]\!], L + L').$$

Similarly to OVERLAY, CONCAT is extended to any positive number of arguments $P_1, \ldots, P_k$, i.e.

$$\text{CONCAT}(P_1, \ldots, P_k) = \text{CONCAT}(P_1, \cdots \text{CONCAT}(P_{k-1}, P_k) \cdots) \ .$$

- Finally, let $S \subseteq \mathbb{Z}$ be a finite nonempty set of tones. Then SNAP$_S(P)$ adjusts all tones of notes in P to the nearest tone in S. Formally, for $t \in \mathbb{Z}$, let $\Delta(t) = \min_{s \in S} |t - s|$, and let $\alpha(t) \in S$ be given by

$$\alpha(t) = \begin{cases} t + \Delta(t) \text{ if } t + \Delta(t) \in S \\ t - \Delta(t) \text{ otherwise.} \end{cases}$$

Thus, $\alpha(t)$ is the adjusted value of t, where we select the higher tone if both $t + \Delta(t)$ and $t - \Delta(t)$ belong to S. Now, we let

$$\text{SNAP}_S(P) = (N [\![n \mapsto (\alpha(t_n), l_n)]\!], L).$$

The music algebra $\mathbf{M}$ contains all of the operations defined above. In addition, it contains the binary operations $+$ and $-$ (addition and subtraction, resp.) on $\mathbb{Z}$, as well as the binary operations $+$, $\cdot$, max, and min (addition, multiplication, maximum, and minimum, resp.) on $\mathbb{R}^+$. For the sake of better readability, whenever the binary operations $+$, $-$, and $\cdot$ occur in trees, we use the customary infix notation. For example, a tree of the form $+[t_1, t_2]$ is written as $t_1 + t_2$.

Let $\Sigma_{\mathbf{M}}$ denote the ranked alphabet consisting of

- the operations of $\mathbf{M}$, viewed as symbols of appropriate ranks, and
- all elements of $\mathbb{Z} \cup \mathbb{R}^+ \cup \mathcal{N}$, viewed as constants, i.e. symbols of rank 0. Here, every note $n \in \mathcal{N}$ is identified with the corresponding one-note piece $(\{(n, 0)\}, l_n)$.

For a well-typed tree $t \in \mathcal{T}_\Sigma$ (where well-typedness is defined in the obvious way), we let val$_{\mathbf{M}}(t)$ denote its value, obtained by recursively evaluating subtrees and applying the operation in the root of the tree to the results.

For the sake of convenience, our implementation extends $\mathbf{M}$ in such a way that every symbol $f^{(k)} \notin \Sigma_{\mathbf{M}} \ (k \geq 1)$ is interpreted as CONCAT$^{(k)}$, i.e. as

k-ary concatenation of musical pieces. In particular, for $k = 1$, f is interpreted as the identity, which is very convenient as it allows us to use such symbols as markers in the generated trees, providing information for subsequent tree transducers, without interfering with the evaluation process. Note, however, that this property of $\mathbf{M}$ is by no means essential for the power of the approach as it is straightforward to add a tree transducer that removes such symbols.

Given any tree generator Γ, the pair $G = (\Gamma, \mathbf{M})$ is called a tree-based music generator. Here, a *tree generator* is any device Γ that defines a tree language $L(\Gamma) \subseteq \mathcal{T}_\mathcal{U}$; the set of musical pieces generated by G is then

$$L(G) = \{\mathrm{val}_\mathbf{M}(t) \mid t \in L(\Gamma)\}.$$

In the example discussed in the next section, the tree generator Γ is composed of a regular tree grammar g and tolerant top-down and macro tree transducers $tt_1, \ldots, tt_k$. In this case, we define $L(\Gamma) = tt_k(\cdots tt_1(L(g)) \cdots)$.

4   Variations and Canons

This section describes how simple *variations*[2] and *canons* can be generated. In a variation, the subject is introduced together with an answer, an imitation of the subject, and possibly a countersubject – a substantive figure that is meant to sound well when played parallel to the subject. If there are several voices available, then the subject appears at some point in all of them. The piece concludes after the subject (or answer) has appeared in the last voice. The subject can be explored and developed by performing it in *inversion* (upside-down), *retrograde* (back-to-front), *diminution* (with shorter note values) or *augmentation* (with longer note values). The subject can also appear in *stretto*, meaning that it is played as a canon, or as a *false entry*, in that it is fractioned or incomplete.

In our implementation, the subject is either generated by the regular tree grammar SUBJECT, or derived from a midi file. A suitable subject is a short sequence of notes that is easy to recognise, contained within an octave, and has a relatively simple rhythm. We do not make any assumptions about the time measure, that is to say, the input completely determines the rhythm of the generated piece, and is not modified to fit a standard meter. Since, in our algebra, the length of a note can be any positive real number, this is not a problem. However, as a consequence (mentioned in the introduction), our tree transducers must be able to tolerate input symbols that are not known a priori.

Before explaining how the actual generation process works, let us discuss a detail that illustrates the usefulness of ttd transducers. Suppose we want to use a subject provided by the user. In a first step, we turn the corresponding midi file into a tree t_{init} that evaluates to the subject. The symbols in this tree are CONCAT[(2)], OVERLAY[(2)], and MUTE[(1)] as internal nodes, and an unknown set of notes as leaves. Some of the subsequent tree transducers work by descending down the input tree until a note is reached, which is then modified by applying some operation. However, there is a problem with this. Once

---

[2] We use this term to denote a toy fugue.

a computation has reached such a note, the only thing that can be done is to tolerate the symbol, because it is unknown (unless the whole computation is aborted). Therefore, we need a preprocessing step that replaces every note n in t_{init} by NOTE$[n]$. Thus, NOTE is a marker signifying that a leaf will be reached in the next step. Interestingly, the preprocessing can be implemented by a ttd transducer. To see this, let PRE $= (\Sigma, \Sigma', \{q, q'\}, \{q'\}, R, q)$, where $\Sigma = \{\text{CONCAT}^{(2)}, \text{OVERLAY}^{(2)}, \text{MUTE}^{(1)}\}$, $\Sigma' = \Sigma \cup \{\text{NOTE}^{(1)}\}$, and

$$R = \{qf[x_1, \ldots, x_k] \to f[t_1, \ldots, t_k] \mid f^{(k)} \in \Sigma \text{ and }$$
$$t_i \in \{q[x_i], \text{NOTE}[q'[x_i]]\} \text{ for all } i \in [k]\}.$$

If we disregard the case where t_{init} is a single note, there is exactly one successful computation for each input tree t_{init}. It descends down the tree in state q, guessing nondeterministically when it has reached a note. At that point, it adds the required occurrence of NOTE and continues in state q'. Since there are no rules at all for q', an incorrect guess means that the computation will fail. This also happens if a note is reached in state q, because q is not tolerant.

If we want to derive a variation on a subject, then we proceed as follows. Regardless of whether the subject is grammatically derived or provided by the user, we generate a template *exposition*. An exposition is basically a way in which to organise a set of themes in time and over a number of voices, usually ranging from Bass to Soprano. An example exposition is shown in Figure 1. The rtg SUBJECT & EXPOSITION produces an output tree $t_{sub\&tmp}$ of the form

$$t_{sub\&tmp} = \text{CONCAT}[\text{SUBJECT}[t_{sub}], \text{EXPOSITION}[t_{tmp}]] \;,$$

where t_{sub} and t_{tmp} are the tree representations of the subject and the exposition template, respectively. The leaves of the latter are the symbols *sub*, *ans*, *cnt*, and *acc*, which act as placeholders: the tree $t_{sub\&tmp}$ is passed to the tmt transducer ARRANGE, which derives from the subtree t_{sub} an answer, a countersubject, and an accompaniment and substitutes these for the placeholders. The answer is the subject played in retrograde and raised by 7 tones, so t_{ans} is given by RAISE[BACK[t_{sub}], 7]. The countersubject is the an inverted version of the subject, i.e. t_{cnt} equals INV[t_{sub}]. Furthermore, in both the answer and the countersubject some notes are lengthened at the others' expense (see below). This is also done to in the accompaniment, which is basically a simplified version of the subject.

SOPRANO				subject	countersubject
ALTO		answer		countersubject	
TENOR	subject	countersubject			
BARITONE					answer
BASS		accompaniment		accompaniment	accompaniment

Fig. 1. One possibility to weave a musical piece around a subject

As mentioned in the introduction, the example has been implemented in the system TREEBAG. The declaration of ARRANGE in TREEBAG is listed in Figure 2.

generators.tmtTransducer("Arrange"):
(
    {CONCAT: 2, SUB: 0, ANS: 0, CNT: 0, ACC: 0},
    {#include(../signature)},
    {INI: 1, ARR: 5, SUB: 1, ANS: 1, CNT: 1, ACC: 1, LUT: 1, EXT: 1},
    {ARR: 5, LUT: 1, EXT: 1},
    {

$$\text{INI}[\text{CONCAT}[x_1, x_2]] \quad \rightarrow \quad \text{ARR}[x_2, \text{SUB}[x_1], \text{ANS}[x_1], \text{CNT}[x_1], \text{ACC}[x_1]],$$

$$\text{SUB}[\text{SUBJECT}[x_1]] \quad \rightarrow \quad \text{LUT}[x_1],$$
$$\text{ANS}[\text{SUBJECT}[x_1]] \quad \rightarrow \quad \text{RAISE}[\text{BACK}[\text{EXT}[x_1]], 7],$$
$$\text{CNT}[\text{SUBJECT}[x_1]] \quad \rightarrow \quad \text{INV}[\text{EXT}[x_1]],$$
$$\text{ACC}[\text{SUBJECT}[x_1]] \quad \rightarrow \quad \text{EXT}[x_1],$$

$$\text{ARR}[\text{SUB}, y_1, y_2, y_3, y_4] \quad \rightarrow \quad y_1,$$
$$\text{ARR}[\text{ANS}, y_1, y_2, y_3, y_4] \quad \rightarrow \quad y_2,$$
$$\text{ARR}[\text{CNT}, y_1, y_2, y_3, y_4] \quad \rightarrow \quad y_3,$$
$$\text{ARR}[\text{ACC}, y_1, y_2, y_3, y_4] \quad \rightarrow \quad y_4,$$
$$\text{ARR}[\text{CONCAT}[x_1, x_2], y_1, y_2, y_3, y_4] \rightarrow$$
$$\text{CONCAT}[\text{ARR}[x_1, y_1, y_2, y_3, y_4], \text{ARR}[x_2, y_1, y_2, y_3, y_4]],$$

$$\text{LUT}[\text{CONCAT}[x_1, x_2]] \quad \rightarrow \quad \text{CONCAT}[\text{LUT}[x_1], \text{LUT}[x_2]],$$

$$\text{EXT}[\text{CONCAT}[x_1, x_2]] \quad \rightarrow \quad \text{CONCAT}[\text{EXT}[x_1], \text{EXT}[x_2]],$$
$$\text{EXT}[\text{CONCAT}[x_1, x_2]] \quad \rightarrow \quad \text{SCALE}[$$
$$\text{LUT}[x_1],$$
$$(\text{LENGTH}[\text{LUT}[x_1]] + \text{LENGTH}[\text{LUT}[x_2]])/\text{LENGTH}[\text{LUT}[x_1]]$$
$$]$$

    }, INI)

Fig. 2. The declaration of the tmt transducer ARRANGE

The first four components are: the input signature, the output signature, the set of states, and the set of tolerant states. The state names INI, ARR, SUB, ANS, CNT, ACC, LUT, and EXT abbreviate *initial state, arrange subjects, subject, answer, countersubject, accompaniment, leave untouched,* and *extend note value,* respectively. Out of these states, only ARR, EXT, and LUT are tolerant.

The fifth component is the set of rewrite rules. The first of these rules is illustrated in Figure 3. Its purpose is to initiate the rewriting of the second subtree, i.e. the exposition template, while simultaneously turning the subject into three themes and an accompaniment. As mentioned earlier, we want to lengthen some of the notes in the accompaniment, but not all of them. When, during the generation of the accompaniment, a node of rank two labelled CONCAT is come across, we find that there are two applicable rules: one which leaves the local configuration untouched and simply proceeds downwards, and one which lengthens the notes found in the first subtree and discards those in the second. Using a feature of TREEBAG that allows to add weights to the rules, the first of these can be made twice as likely for application as the second. This decreases the risk of ending up with an accompaniment that is but a single long note. The

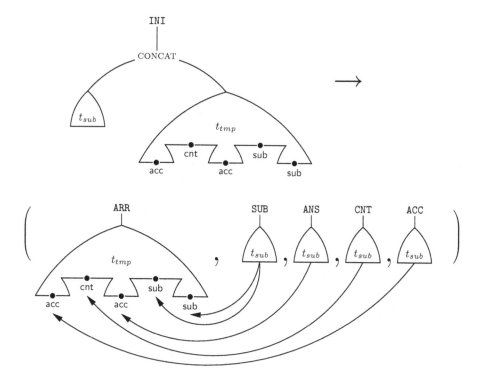

Fig. 3. A pictorial representation of the first rule in the tmt transducer ARRANGE (see Fig. 2)

sixth and last component is the initial state, in this case INI. The output from ARRANGE is the tree t_{exp}.

If t_{exp} is interpreted as a piece of music, i.e. $\text{val}_M(t_{\text{exp}})$ is played, then it is very likely to contain dissonances, as the theme is played against itself both in retrograde and inversion. To clear these, and to add a sense of movement, we wish to label the tree with chords in such a way that when the assigned chords are read left-to-right, they appear in accordance with some common chord progression. If this progression can be expressed as a directed graph G, in which the individual chords are the nodes, then the labelling can be done by a ttd transducer PROGRESSION that operates along the following principle.

Its states are tuples of the form $\langle c, c' \rangle$, where c and c' are chords. We choose $\langle c_s, c_e \rangle$ as the initial state if we wish the progression to start with c_s and end with c_e. The rules of PROGRESSION can be divided into two types, which develop and settle the progression, respectively. A rule

$$\langle c, c' \rangle[\text{CONCAT}[x_1, x_2]] \rightarrow \text{CONCAT}[\langle c, c'' \rangle[x_1], \langle c'', c' \rangle[x_2]]$$

is included if there is a path from c to c' in G that passes through c'', and

$$\langle c, c' \rangle[\text{CONCAT}[x_1, x_2]] \rightarrow \text{SNAP}_{\hat{c}}[\text{CONCAT}[\text{P}[x_1], \text{P}[x_2]]]$$

is included if the distance from c to c' in G is less or equal to one. Here, P is an auxiliary state that simply copies the subtree below it to the output, and the only state that is tolerant with respect to OVERLAY, CONCAT and NOTE. Since the refinement of the progression cannot proceed below these symbols, we know that each chord is represented by at least one note, and that no two chords are played in parallel. If every state was tolerant, then this could not be attained.

In the second rule above, $\hat{c}$ is the closure of the notes in c under transposition by an octave.[3] This assures that the local notes belong to $\hat{c}$, and that the complete note sequence respects the chosen chord progression. For a more detailed discussion of how chord progressions are modelled, see [Hög05].

When we generate a canon, we begin as we did for variations: a subject is either derived by a regular tree grammar, or extracted from midi data. Copies of this subject are then arranged over four voices by the ttd transducer CANON, and this is done in such a way that there are frequent overlaps and many false entries. Because of the overlaps, it is now easier to generate one voice at a time, and then combine them using OVERLAY, rather than generating one measure at a time, and then concatenating the results. This can be done using an extended version of the rule

$$S[\text{SUBJECT}[x_1]] \rightarrow$$
$$\text{OVERLAY}[$$
$$\text{RAISE}[$$
$$\text{CONCAT}[\text{MUTE}[\text{H}[x_1]], \text{P}[x_1], \text{MUTE}[\text{H}[x_1]], \text{MUTE}[\text{T}[x_1]]],$$
$$12],$$
$$\text{RAISE}[$$
$$\text{CONCAT}[\text{P}[x_1], \text{H}[x_1], \text{P}[x_1]],$$
$$24],$$
$$\text{RAISE}[$$
$$\text{CONCAT}[\text{MUTE}[\text{H}[x_1]], \text{MUTE}[\text{H}[x_1]], \text{P}[x_1], \text{MUTE}[\text{T}[x_1]]],$$
$$36]].$$

The two states H and T select the first and the second half, respectively, of the subtree below them. For this approach to yield a nice result, the tree t_{subj} must not be comb-like.

To make the generated piece more interesting to listen to, we end the generation process by adding various ornaments. An *ornament* is a musical embellishment that is not part of the overall melody, but rather an added decoration. An example is the mordent: a single rapid alteration between a note of the melodic line, and the note immediately above it. The ornaments are added by the ttd transducer ORNAMENT, which forms a mordent using the rule

$$S[\text{NOTE}[x_1]] \rightarrow \text{SCALE}[$$
$$\text{CONCAT}[S[x_1], \text{RAISE}[S[x_1], 1], S[x_1]],$$
$$\tfrac{1}{3} \cdot \text{LENGTH}[S[x_1]]].$$

---

[3] In fact, strictly speaking, we cannot use $\hat{c}$ because it is infinite. Therefore, it is replaced with its restriction to the audible range. This has the additional advantage that no tones outside this range will be generated.

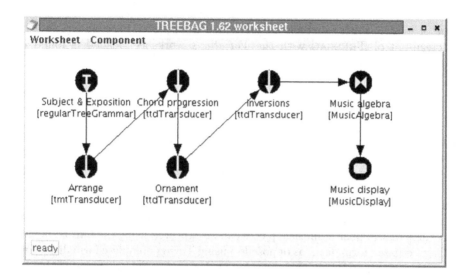

Fig. 4. A screenshot of TREEBAG with the *Variations* worksheet loaded

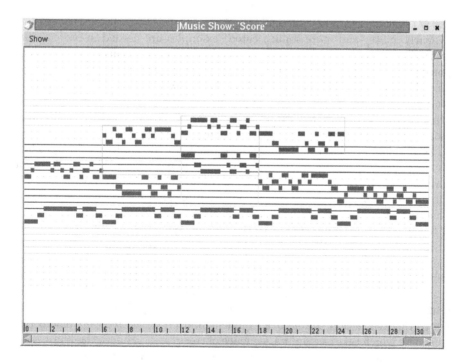

Fig. 5. A variation generated in above worksheet, here shown as a time/frequency diagram

The generated tree is interpreted by the algebra described in Section 3 as a piece of music, which can then be performed using the jMusic library [SB07]. A screenshot of TREEBAG with the *Variations* worksheet loaded is found in Figure 4, and a musical piece generated in this worksheet is shown in Figure 5.

5   Conclusion

In this paper, we have continued the work that was started in [Hög05]. In particular, we have presented an algebra for the tree-based generation of music. Moreover, we have shown that the generation process can make use of (tolerant) macro tree transducers in a natural manner, thus providing a greater flexibility and generative power than what can be achieved by using top-down tree transducers only. The motivation behind this work is to investigate how far typical structures that appear in musical pieces can be captured using the limited means of the tree-based formalism, as opposed to using Turing-complete formalisms for, e.g., imitating a particular human composer. Naturally, the discussion of the example in the previous section could not reveal much detail without becoming lengthy and repeating much of what has been said in [Hög05]. Readers who want to explore the details are invited to download the system and the example from http://www.cs.umu.se/~johanna/algebra.

Clearly, more (and more sophisticated) examples are needed in order to understand whether the operations of the algebra proposed in this paper are really appropriate. In fact, it will probably turn out that different types of music require different algebras and maybe also different types of tree generators. This situation is well known in the area of picture generation, where each choice consisting of a class of tree generators and a class of picture algebras results in a specific type of picture generator (see [Dre06]).

The problem of whether the concepts presented here can be used to produce "nice" or "interesting" music remains open. We do not expect this to be possible in a fully automatic manner, because we do not believe "nice music" to be a formally definable concept. However, our example shows that there are certain structural rules that formal grammatical systems can take care of. Thus, it is conceivable that a system similar to the one implemented in TREEBAG, though considerably more sophisticated, could become an interesting interactive tool for a human composer.

The current implementation is restricted in the sense that user-provided subjects can only be combined with a fixed (or finite number of) exposition templates rather than with an infinite set of grammatically derived templates. To remove this restriction, we need an implementation of the tolerant macro tree transducer which takes as input a sequence of trees $t_1, \ldots, t_k$ and, therefore, computes a mapping $mt^t : T_{\mathcal{U}}^k \to \wp(T_{\mathcal{U}})$. This can be achieved by replacing the initial state in such a tree transducer by an axiom of the form $q_1[t_1, q_2[t_2], \cdots, q_k[t_k]]$, where $q_1, \ldots, q_k$ are states.

Let us finally point out that the concept of tolerant macro tree transducers may be of independent interest, as it may be useful in other applications in which unknown symbols can occur. Thus, it could be worthwhile to study the theoretical properties of this type of tree transducer by, for example, comparing it with ordinary macro tree transducers.

Acknowledgement. We thank Albert Gräf and Carl Rehnberg for providing us with references regarding computer-generated music. The implementation of mt transducers in TREEBAG, from which the one of tmt transducers was derived by a few simple changes, has been made by Karl Azab (see also [Aza05]).

References

[Ame89]   Ames, C.: The Markov process as a compositional model: a survey and tutorial. Leonardo 22(2), 175–187 (1989)

[Aza05]   Azab, K.: Macro tree transducers in Treebag. Master thesis, Department of Computing Science, Umeå University (2005), http://www.cs.umu.se/education/examina/Rapporter/KarlAzab.pdf

[Bar83]   Baroni, M.: The concept of musical grammar. Music Analysis 2(2), 175–208 (1983)

[BDPV94]   Bresin, R., Poli, G.D., Vidolin, A.: A neural networks based system for automatic performance of musical scores. In: Proc. 1993 Stockholm Music Acoustic Conference, Royal Swedish Academy of Music, Stockholm, pp. 74–78 (1994)

[BELM04]   Burraston, D., Edmonds, E., Livingstone, D., Miranda, E.R.: Cellular automata in MIDI-based computer music (2004)

[Bha88]   Bharucha, J.: Neural net modeling of music. In: Proc. First Workshop on Artificial Intelligence and Music, American Association for AI, pp. 173–182 (1988)

[Cha03]   Chapel, R.H.: Realtime algorithmic music systems from fractals and chaotic functions: toward an active musical instrument. PhD thesis, Univ. Pompeu Fabra, Barcelona (2003)

[CV01]   Chai, W., Vercoe, B.: Folk music classification using hidden Markov models. In: Proc. Int. Conference on Artificial Intelligence (2001)

[Dre06]   Drewes, F.: Grammatical Picture Generation – A Tree-Based Approach. Texts in Theoretical Computer Science. An EATCS Series. Springer, Heidelberg (2006)

[Eng97]   Engelfriet, J.: Context-free graph grammars. In: Handbook of Formal Languages. Beyond Words, vol. 3, pp. 125–213. Springer, New York (1997)

[EV85]   Engelfriet, J., Vogler, H.: Macro tree transducers. Journal of Computer and System Sciences 31(1), 71–146 (1985)

[FV98]   Fülöp, Z., Vogler, H.: Syntax-Directed Semantics: Formal Models Based on Tree Transducers. Monographs in Theoretical Computer Science. An EATCS Series. Springer, Heidelberg (1998)

[GS97]   Gécseg, F., Steinby, M.: Tree languages. In: Rozenberg, G., Salomaa, A. (eds.) Handbook of Formal Languages. Beyond Words, ch. 1, vol. 3, pp. 1–68. Springer, Heidelberg (1997)

[HG91]   Horner, A., Goldberg, D.E.: Genetic algorithms and computer-assisted music composition. In: Proc. Fourth Int. Conference on Genetic Algorithms, San Diego, CA, pp. 437–441 (1991)

[Hög05]   Högberg, J.: Wind in the willows – generating music by means of tree trans-
           ducers. In: Farré, J., Litovsky, I., Schmitz, S. (eds.) CIAA 2005. LNCS,
           vol. 3845, pp. 153–162. Springer, Heidelberg (2006)

[Jac95]   Jacob, B.: Composing with genetic algorithms. In: Proc. ICMC, pp. 452–
           455 (1995)

[Jur06]   Jurish, B.: Music as a formal language. In: Zimmer, F. (ed.) Bang | Pure
           data, Wolke Verlag (2006)

[Moo72]   Moorer, J.A.: Music and computer composition. Commun. ACM 15(2),
           104–113 (1972)

[Moz94]   Mozer, M.: Neural network music composition by prediction: exploring the
           benefits of psychoacoustic constraints .... Connection-Science 6(2), 247–
           280 (1994)

[Pru86]   Prusinkiewicz, P.: Score generation with L-systems. In: Berg, P. (ed.) Proc.
           ICMC, Royal Conservatory, The Hague, Netherlands, vol. 1, pp. 455–457
           (1986)

[SB07]   Sorensen, A., Brown, A.: Introduction to jMusic. Internet resource. (Ac-
           cessed 27 Feb 2007), available at http://jmusic.ci.qut.edu.au/

[SXK04]   Shao, X., Xu, C., Kankanhalli, M.S.: Unsupervised classification of music
           genre using hidden Markov model. In: Proc. ICME, pp. 2023–2026 (2004)

[Tod91]   Todd, P.M: A connectionist approach to algorithmic composition. Com-
           puter Music Journal 13(4), 27–43 (1991)

[TON06]   Tojo, S., Oka, Y., Nishida, M.: Analysis of chord progression by HPSG.
           In: AIA'06: Proc. 24th IASTED International Conference on Artificial In-
           telligence and Applications, pp. 305–310. ACTA Press (2006)

[Vis04]   Visell, Y.: Spontaneous organisation, pattern models, and music. Organ-
           ised Sound (2004)

[WS05]   Worth, P., Stepney, S.: Growing music: musical interpretations of L-
           systems. In: Rothlauf, F., Branke, J., Cagnoni, S., Corne, D.W., Drech-
           sler, R., Jin, Y., Machado, P., Marchiori, E., Romero, J., Smith, G.D.,
           Squillero, G. (eds.) EvoWorkshops 2005. LNCS, vol. 3449, pp. 545–550.
           Springer, Heidelberg (2005)

Aperiodicity in Tree Automata[*]

Zoltán Ésik[1,2] and Szabolcs Iván[1]

[1] Dept. of Computer Sci. Univ. Szeged, Hungary
[2] GRLMC, Rovira i Virgili University, Tarragona, Spain

Abstract. We define and compare several different notions of aperiodicity in tree automata. We also relate these notions to the cascade product and logical definability of tree languages.

1 Introduction

By the well-known theorem of McNaughton and Papert [12], a word language is definable in the first order logic of linear order if and only if its syntactic semigroup is aperiodic, i.e., a finite semigroup containing no nontrivial group. The notion of syntactic semigroup was generalized to trees by Thomas in [16] giving rise to a notion of aperiodicity for tree automata, called context aperiodicity below. It is known that the minimal automaton of any tree language definable in the first order logic of finite trees with both the successor relations and the usual order relation is context aperiodic, cf. [16]. On the other hand, there exist regular tree languages with context aperiodic minimal automata which are not definable in first order logic, cf. [11,16].

In this paper we define a hierarchy of aperiodicity notions for tree automata by considering n-tuples of nontrivial trees (i.e., trees or terms different from the variables) in n variables and the semigroup of vector-valued term functions induced by them. We say that a (finite) tree automaton is n-aperiodic for some integer $n > 0$ if the semigroup of all such term functions is aperiodic. Moreover, we say that a tree automaton is strongly aperiodic if it is n-aperiodic for each n. We show that n-aperiodic tree automata form a proper hierarchy, and the class of 1-aperiodic tree automata is properly contained in the class of context aperiodic tree automata. We also show that the class of strongly aperiodic tree automata properly contains the class of definite tree automata. Moreover, we establish that each of these classes is a generalized cascade variety, i.e., it is closed under the generalized cascade product (wreath product) and taking subautomata and homomorphic images. We also study the complexity of deciding whether a tree automaton belongs to these classes. In particular, we establish a P-time algorithm for testing strong aperiodicity. We also provide an extension of the above aperiodicity notions which is motivated by the Krohn-Rhodes decomposition theory [4] and study a modified version of aperiodicity obtained by

---

[*] Research supported by the AUTOMATHA project of ESF and the National Foundation for Scientific Research of Hungary, grant T466886.

taking polynomial functions instead of term functions. Finally, we relate aperiodicity to logic.

Notation. For each $n \geq 0$, we let $[n]$ denote the set $\{1, \ldots, n\}$. When A is a set, $P(A)$ denotes its power set.

2   Aperiodicity in Tree Automata

In this section we define several notions of aperiodicity for tree automata and discuss the basic relationship between them. We start with some preliminary definitions. For a more detailed exposition, see [8].

A *rank type* R is a nonempty subset of the nonnegative integers. A *ranked alphabet* of rank type R is a finite set Σ which is the disjoint union of sets Σ_n, $n \geq 0$, such that Σ_n is not empty if and only if $n \in R$. The elements of Σ_n are called *function symbols* or *operation symbols* of rank n. Symbols of rank 0 are also called *constant symbols*. We will consider *finite* Σ-algebras $\mathbb{A} = (A, (\sigma^{\mathbb{A}})_{\sigma \in \Sigma})$, called Σ-*tree automata*. Here, for each $\sigma \in \Sigma_n$, $\sigma^{\mathbb{A}}$ is a function $A^n \to A$, the *interpretation* of σ in $\mathbb{A}$. The elements of A will be called *states*. Throughout the paper, the rank type R is fixed, but the ranked alphabet Σ may vary. For ease of notation, unless otherwise specified, the underlying set of a tree automaton denoted by a boldface letter will be denoted by the same letter in italics.

Let $X = \{x_1, x_2, \ldots\}$ denote a fixed countably infinite set of *variables*, and for each n, let $X_n = \{x_1, \ldots, x_n\}$. The set of Σ-*terms* (or Σ-*trees*, see below) over X_n will be denoted $T_\Sigma(X_n)$, or just T_Σ, when $n = 0$. Terms different from the variables are called *nontrivial*. A Σ-*context* (or just context, if Σ is understood) is a term $t \in T_\Sigma(X_1)$ containing exactly one occurrence of x_1. A nontrivial context is a context which is a nontrivial term. Note that there is a nontrivial context if and only if $1 \in R$, or $0 \in R$ and R contains an integer > 1. Sometimes it will be convenient to think of a term $t \in T_\Sigma$ as a finite directed ordered and rooted tree whose vertices are labeled in Σ such that the out-degree of a vertex is n if and only if its label is in Σ_n. Thus, the vertices labeled in Σ_0 are leaves. We will say that vertex v is the ith *successor* of vertex u if the ith out-edge of u connects u to v. Moreover, we say that v is a *descendant* of u, denoted $u < v$, if there is a nonempty path from u to v. The *subtree* of a tree rooted at a vertex v is determined by v and its descendants.

When $\mathbb{A}$ is a Σ-tree automaton, each $t \in T_\Sigma(X_n)$ induces a *term function* $A^n \to A$. Term functions induced by nontrivial trees will be called *proper*. By extension, each m-tuple $\underline{t} = (t_1, \ldots, t_m)$ of trees $t_i \in T_\Sigma(X_n)$ induces a vector-valued term function $\underline{t}^{\mathbb{A}} = \langle t_1^{\mathbb{A}}, \ldots, t_m^{\mathbb{A}} \rangle : A^n \to A^m$, which is the *target tupling* of the m functions $t_i^{\mathbb{A}} : A^n \to A$, $i = 1, \ldots, m$. When each $t_i^{\mathbb{A}}$ is proper, $\underline{t}^{\mathbb{A}}$ is also called proper. It is clear that for each n, the proper term functions $A^n \to A^n$ form a finite semigroup, denoted $S_n(\mathbb{A})$. In this semigroup, product is function composition. The subsemigroup of $S_1(\mathbb{A})$ consisting of the term functions induced by the nontrivial contexts will be denoted $\mathbb{C}(\mathbb{A})$.

For all unexplained notions from universal algebra and tree automata we refer to [8,9]. Recall from [4] that S *divides* T for finite semigroups S and T when S is a homomorphic image of a subsemigroup of T.

Proposition 1. *When $n \leq m$, $S_n(\mathbb{A})$ divides $S_m(\mathbb{A})$.*

Proof. First note that since $n \leq m$, we have $T_\Sigma(X_n) \subseteq T_\Sigma(X_m)$. Consider the functions $f = \langle t_1^\mathbb{A}, \ldots, t_m^\mathbb{A} \rangle \in S_m(\mathbb{A})$ such that for each $i \in [n]$, t_i is a nontrivial term in $T_\Sigma(X_m)$ not containing any occurrence of a variable x_j with $j > n$. These functions form a subsemigroup of $S_m(\mathbb{A})$. Moreover, $S_n(\mathbb{A})$ is a homomorphic image of T, one homomorphism being the map that takes any f of the above form to $\langle t_1^\mathbb{A}, \ldots, t_n^\mathbb{A} \rangle$, where now each t_i is considered as a tree in $T_\Sigma(X_n)$ inducing a function $A^n \to A$. ∎

The following fact is clear:

Proposition 2. *$C(\mathbb{A})$ is a subsemigroup of $S_1(\mathbb{A})$. When $R = \{1\}$, $C(\mathbb{A}) = S_1(\mathbb{A})$.*

Recall from [4] that a finite semigroup S is called aperiodic if it contains no nontrivial group, or equivalently, when there exists an integer $k \geq 1$ such that $s^k = s^{k+1}$, for all $s \in S$.

Definition 1. *We call a Σ-tree automaton $\mathbb{A}$ n-aperiodic for some $n \geq 1$ if the semigroup $S_n(\mathbb{A})$ is aperiodic. By extension, we call $\mathbb{A}$ strongly aperiodic if it is n-aperiodic for each $n \geq 1$.*

Definition 2. *We call a Σ-tree automaton $\mathbb{A}$ context aperiodic if $C(\mathbb{A})$ is aperiodic.*

The notion of context aperiodicity was introduced by Thomas in [16] under the name aperiodicity. In the "classical case" $R = \{1\}$, context aperiodicity and 1-aperiodicity coincide with aperiodicity, or counter-freeness, see below.

Example 1. In case of semigroups, the ranked alphabet contains a single binary symbol. Since there are no proper contexts, every finite semigroup is context aperiodic. We show that a finite semigroup S is 1-aperiodic if and only if its exponent is 1 or 2. (Recall that the exponent of S is the least positive integer d such that $s^k = s^{k+d}$ for all $s \in S$.)

Suppose first that S is a finite semigroup which is 1-aperiodic. For each integer $k > 1$, consider the term $x_1^k = x_1 \cdot (x_1 \cdot \ldots \cdot (x_1 \cdot x_1) \ldots)$ in the variable x_1. Since S is 1-aperiodic, there exists some integer m such that the equation $x_1^{k^{m+1}} = x_1^{k^m}$ holds in S. This implies that d is a divisor of $k^{m+1} - k^m = k^m(k - 1)$. Thus, each prime appearing in the prime decomposition of d must divide $k(k - 1)$. Since this holds for all $k > 1$, we conclude that $d = 1$ or d is a power of 2. But when $k = 3$, then d must divide $3^m \cdot 2$ for some m, so that the exponent is 1 or 2.

When the exponent is 1, S is aperiodic. It is then clear that S is 1-aperiodic. Assume that the exponent is 2. Then there is an integer n such that $s^{n+2} = s^n$

for all $s \in S$. Consider any term x_1^k, where $k \geq 2$. When $k^m \geq n$, we have that $s^{k^{m+1}} = s^{k^m}$ for all $s \in S$. This means that $x_1^{k^m}$ and $x_1^{k^{m+1}}$ induce equal term functions. It follows that S is 1-aperiodic.

Example 2. Recall that a semigroup S is *nilpotent* if there exist $0 \in S$ and an integer $n > 0$ such that $S^n = 0$. It is clear that any finite nilpotent semigroup is strongly aperiodic. We show that every 2-aperiodic finite semigroup S is nilpotent. To this end, first note that the exponent of S is 1 or 2 by the above argument. But if the exponent is 2, then there exist some $s \in S$ and $n > 0$ such that $s^n = s^{n+2}$ but $s^n \neq s^{n+1}$. Now both s^n and s^{n+1} are fixed points of the term function induced by x_1^3. Thus, by Proposition 5 below, S is not 2-aperiodic. We conclude that the exponent of S is 1, i.e., S is aperiodic, i.e., there exists $n > 0$ such that $x^{n+1} = x^n$ holds in S. Consider now the term vector (x_2^2, x_1^2). Since S is 2-aperiodic, there exists some k such that $2^k \geq n$ and $x_1^{2^k} = x_2^{2^{k+1}}$ holds in S. But since $2^k \geq n$, $x_1^{2^k} = x_1^n$ and $x_2^{2^{k+1}} = x_2^n$, so that $x_1^n = x_2^n$ also holds. Now this implies that S has a single idempotent e and $e = s^n$ for each $s \in S$. But then since $xy^n = xx^n = x^n$ and $y^n x = x^n$ also hold, $Se = eS = e$. Since all long enough products over S have a factorization which contains an idempotent, it follows that S is nilpotent.

Since any divisor of an aperiodic semigroup is aperiodic, from Propositions 1 and 2 we have:

Proposition 3. *Every 1-aperiodic tree automaton is context aperiodic, and when $n \leq m$, every m-aperiodic tree automaton is n-aperiodic.*

When A is a finite set and S is a semigroup of functions $A \to A$, then the pair (A, S) is called a *transformation semigroup*, cf. [4]. The *(left) action* of S on A is defined by $sa := s(a)$, for each $a \in A$ and $s \in S$. A transformation semigroup (A, S) is called *counter-free* when there exists no $s \in S$ which induces a nontrivial (cyclic) permutation of a subset of A. It is well-known that S is aperiodic if and only if (A, S) is counter-free. Note that for each n, $(A^n, S_n(\mathbb{A}))$ is a transformation semigroup as is $(A, C(\mathbb{A}))$.

Proposition 4. *A Σ-tree automaton $\mathbb{A}$ is n-aperiodic if and only if $(A^n, S_n(\mathbb{A}))$ is counter-free. Moreover, it is context aperiodic if and only if $(A, C(\mathbb{A}))$ is counter-free.*

For later use we prove:

Proposition 5. *For any integer $n > 0$, $2n$-aperiodic tree automaton $\mathbb{A}$ and proper term function $f : A^n \to A^n$ it holds that f has at most one fixed point.*

Proof. Let $n > 0$ be an integer, $\mathbb{A}$ a $2n$-aperiodic tree automaton, $t_1, \ldots, t_n \in T_\Sigma(X_n)$ nontrivial trees and $(a_1, \ldots, a_n), (b_1, \ldots, b_n) \in A^n$ two fixed points of $f = \langle t_1^{\mathbb{A}}, \ldots, t_n^{\mathbb{A}} \rangle$. We have to show that $a_i = b_i$ for all $i \in [n]$.

For any $i \in [n]$ let s_i denote the tree we get from t_i by replacing each occurrence of x_j by x_{n+j}, for each $j \in [n]$ (so we add n to the index of each variable). Now the proper term function

$$f' = \langle s_1^{\mathbb{A}}, \ldots, s_n^{\mathbb{A}}, t_1^{\mathbb{A}}, \ldots, t_n^{\mathbb{A}} \rangle : A^{2n} \to A^{2n}$$

satisfies both

$$f'(a_1, \ldots, a_n, b_1, \ldots, b_n) = (b_1, \ldots, b_n, a_1, \ldots, a_n)$$

and

$$f'(b_1, \ldots, b_n, a_1, \ldots, a_n) = (a_1, \ldots, a_n, b_1, \ldots, b_n).$$

Since $\mathbb{A}$ is $2n$-aperiodic, $(a_1, \ldots, a_n, b_1, \ldots, b_n) = (b_1, \ldots, b_n, a_1, \ldots, a_n)$, hence $a_i = b_i$ for all $i \in [n]$.   ∎

Proposition 6. *Suppose $\mathbb{A}$ is a finite tree automaton which is not n-aperiodic for the integer $n > 0$. Then there exists a proper term function $f : A^n \to A^n$ that has at least two different fixed points.*

Proof. If $\mathbb{A}$ is not n-aperiodic, then there exist a proper term function $f : A^n \to A^n$ and a subset $B \subseteq A^n$ with $k > 1$ elements such that the restriction of f to B is a cyclic permutation of B. But then each element of B is a fixed point of the proper term function f^k.   ∎

Corollary 1. *A tree automaton $\mathbb{A}$ is strongly aperiodic if and only if for each $n > 0$, no proper term function $A^n \to A^n$ has two or more fixed points.*

Remark 1. We can show that $\mathbb{A}$ is strongly aperiodic iff no proper term function $A^n \to A^n$ has two or more fixed points when $n = |A|^2$.

3   Aperiodicity and the Cascade Product

Let $\mathbb{A}$ be a Σ-tree automaton, $\mathbb{B}$ a Δ-tree automaton and $\alpha = \{\alpha_n : n \in R\}$ a collection of functions, where for each $n \in R$, α_n maps $A^n \times \Sigma$ to the set of all *nontrivial* trees in $T_\Delta(X_n)$. Then the *generalized cascade product* $\mathbb{A} \times_\alpha \mathbb{B}$ is defined as the Σ-algebra on the set $A \times B$ such that for any $\sigma \in \Sigma_n$ and $(a_1, b_1), \ldots, (a_n, b_n) \in A \times B$,

$$\sigma^{\mathbb{A} \times_\alpha \mathbb{B}}((a_1, b_1), \ldots, (a_n, b_n)) = (\sigma^{\mathbb{A}}(a_1, \ldots, a_n), t^{\mathbb{B}}(b_1, \ldots, b_n))$$

where $t = \alpha_n(a_1, \ldots, a_n, \sigma)$. When the range of each α_n only contains trees of the form $\delta(x_1, \ldots, x_n)$, where $\delta \in \Delta_n$, then we call the product a *cascade product*.

Remark 2. Note that the direct product is a special case of the cascade product. When $\mathbb{A} \times_\alpha \mathbb{B}$ is a cascade product, we may view each α_n as a function $A^n \times \Sigma_n \to \Delta_n$.

Remark 3. The generalized cascade product can be defined in terms of the cascade product. We say that a Δ-tree automaton $\mathbb{B}'$ is a *derived automaton* of a Δ-tree automaton $\mathbb{B}$ if $\mathbb{B}$ and $\mathbb{B}'$ have the same set of states and each basic operation $\delta^{\mathbb{B}'}$ is a proper term function of $\mathbb{B}$. Any generalized cascade product $\mathbb{A} \times_\alpha \mathbb{B}$ is a cascade product $\mathbb{A} \times_\beta \mathbb{B}'$ for some derived automaton $\mathbb{B}'$ of $\mathbb{B}$ and for some β.

Definition 3. *We say that a nonempty class of tree automata is a* (generalized) *cascade variety if it is closed under taking subautomata, homomorphic images, and the* (generalized) *cascade product.*

Remark 4. A cascade variety is a generalized cascade variety if and only if it is closed under derived automata.

In the rest of this section we show that for each $n > 0$, the class of all n-aperiodic tree automata is a generalized cascade variety. Hence, the class of all aperiodic tree automata is also a generalized cascade variety.

The following fact is clear.

Lemma 1. *Let $n > 0$ be an integer and $\mathbb{A}$ an n-aperiodic tree automaton. Then any subautomaton or homomorphic image of $\mathbb{A}$ is also n-aperiodic.*

Before proving that taking generalized cascade product also preserves n-aperiodicity, we make the following observation:

Lemma 2. *Let $\mathbb{A}$ be a Σ-tree automaton, $\mathbb{B}$ a Δ-tree automaton and $\mathbb{C} = \mathbb{A} \times_\alpha \mathbb{B}$ a generalized cascade product of $\mathbb{A}$ and $\mathbb{B}$. Then for any integer $n \geq 0$, (nontrivial) tree $t \in T_\Sigma(X_n)$ and states $a_1, \ldots, a_n \in A$ there exists a (nontrivial) tree $s \in T_\Delta(X_n)$ such that*

$$t^{\mathbb{C}}\big((a_1, b_1), \ldots, (a_n, b_n)\big) = \big(t^{\mathbb{A}}(a_1, \ldots, a_n), s^{\mathbb{B}}(b_1, \ldots, b_n)\big)$$

for any $(a_1, b_1), \ldots, (a_n, b_n) \in C$.

Lemma 3. *Let $n > 0$ be an integer, $\mathbb{A}$ an n-aperiodic Σ-tree automaton, $\mathbb{B}$ an n-aperiodic Δ-tree automaton, and $\mathbb{C} = \mathbb{A} \times_\alpha \mathbb{B}$ a generalized cascade product of $\mathbb{A}$ and $\mathbb{B}$. Then $\mathbb{C}$ is also n-aperiodic.*

Proof. We know that there exist integers N_A and N_B such that $f^{N_A} = f^{N_A+1}$ and $g^{N_B} = g^{N_B+1}$ for all proper term functions $f : A^n \to A^n$ and $g : B^n \to B^n$ of $\mathbb{A}$ and $\mathbb{B}$, respectively. We show that this implies that $h^{N_A+N_B} = h^{N_A+N_B+1}$ for all proper term functions $h : (A \times B)^n \to (A \times B)^n$ of $\mathbb{C}$.

Let $\underline{t} = (t_1, \ldots, t_n)$ be an n-tuple of nontrivial trees in $T_\Sigma(X_n)$. Let us denote by $((a'_1, b'_1), \ldots, (a'_n, b'_n))$ the n-tuple $(\underline{t}^{\mathbb{C}})^{N_A}((a_1, b_1), \ldots, (a_n, b_n))$. Applying Lemma 2 to the trees $(t_1, \ldots, t_n)$ and the states $a'_1, \ldots, a'_n \in A$ we get that there exist nontrivial trees $s_1, \ldots, s_n \in T_\Delta(X_n)$ such that for any $(a'_1, x_1), \ldots, (a'_n, x_n) \in C$ the value of $\underline{t}^{\mathbb{C}}((a'_1, x_1), \ldots, (a'_n, x_n))$ can be written as

$$\big((t_1^{\mathbb{A}}(a'_1, \ldots, a'_n), s_1^{\mathbb{B}}(x_1, \ldots, x_n)), \ldots, (t_n^{\mathbb{A}}(a'_1, \ldots, a'_n), s_n^{\mathbb{B}}(x_1, \ldots, x_n))\big).$$

From the definition of $a'_1, \ldots, a'_n$ we get that $t^{\mathbb{A}}_i(a'_1, \ldots, a'_n) = a'_i$ for all $i \in [n]$, so

$$\underline{t}^{\mathbb{C}}((a'_1, x_1), \ldots, (a'_n, x_n)) = ((a'_1, s^{\mathbb{B}}_1(x_1, \ldots, x_n)), \ldots, (a'_n, s^{\mathbb{B}}_n(x_1, \ldots, x_n))).$$

Iterating this from $(a'_1, b'_1), \ldots, (a'_n, b'_n)$ it follows that for any $k \geq 0$ we can write $(\underline{t}^{\mathbb{C}})^k((a'_1, b'_1), \ldots, (a'_n, b'_n))$ as

$$((a'_1, (s^{\mathbb{B}}_1)^k(b'_1, \ldots, b'_n)), \ldots, (a'_n, (s^{\mathbb{B}}_n)^k(b'_1, \ldots, b'_n))).$$

But we know that $(s^{\mathbb{B}}_i)^{N_B} = (s^{\mathbb{B}}_i)^{N_B+1}$ holds for all $i \in [n]$. Summing up we get the following equality:

$$
\begin{aligned}
& (\underline{t}^{\mathbb{C}})^{N_A+N_B}((a_1, b_1), \ldots, (a_n, b_n)) \\
={}& (\underline{t}^{\mathbb{C}})^{N_B}((a'_1, b'_1), \ldots, (a'_n, b'_n)) \\
={}& ((a'_1, (s^{\mathbb{B}}_1)^{N_B}(b'_1, \ldots, b'_n)), \ldots, (a'_n, (s^{\mathbb{B}}_n)^{N_B}(b'_1, \ldots, b'_n))) \\
={}& ((a'_1, (s^{\mathbb{B}}_1)^{N_B+1}(b'_1, \ldots, b'_n)), \ldots, (a'_n, (s^{\mathbb{B}}_n)^{N_B+1}(b'_1, \ldots, b'_n))) \\
={}& (\underline{t}^{\mathbb{C}})^{N_B+1}((a'_1, b'_1), \ldots, (a'_n, b'_n)) \\
={}& (\underline{t}^{\mathbb{C}})^{N_A+N_B+1}((a_1, b_1), \ldots, (a_n, b_n)),
\end{aligned}
$$

so $\mathbb{C}$ is indeed n-aperiodic. $\blacksquare$

For each n, let $\mathbf{SAper}_n$ denote the class of all n-aperiodic Σ-tree automata, where Σ ranges over all ranked alphabets (of rank type R). Thus, $\mathbf{SAper} = \bigcap_{n \geq 1} \mathbf{SAper}_n$ is the class of all strongly aperiodic tree automata. Moreover, let $\mathbf{CAper}$ denote the class of all context aperiodic tree automata. From the above facts we get the following:

Theorem 1. *For any integer $n > 0$, $\mathbf{SAper}_n$ is a generalized cascade variety. Hence, $\mathbf{SAper}$ is also a generalized cascade variety.*

In a similar way, we have:

Theorem 2. $\mathbf{CAper}$ *is a generalized cascade variety.*

A Σ-tree automaton is called *definite* [10] if there exists an integer $k \geq 0$ such that $t^{\mathbb{A}} = s^{\mathbb{A}}$ whenever $t, s \in T_\Sigma(X_n)$ agree up to "depth k". We let $\mathbf{D}$ denote the class of definite tree automata.

Corollary 2. $\mathbf{D} \subseteq \mathbf{SAper}$.

Proof. It was shown in [5] that $\mathbf{D}$ is the least cascade variety containing the two-state Σ-tree automaton $\mathbb{D}_0$ such that Σ_n contains two symbols for each $n \in R$, inducing respectively the two constant valued operations in n variables. It is easy to check that $\mathbb{D}_0$ is strongly aperiodic. $\blacksquare$

4   Strict Containments

By Propositions 1, 2 and Corollary 2,

$$\mathbf{CAper} \supseteq \mathbf{SAper}_1 \supseteq \mathbf{SAper}_2 \supseteq \ldots \supseteq \mathbf{SAper} \supseteq \mathbf{D} \qquad (1)$$

is a decreasing chain. In this section we prove that when R contains an integer > 1, then each of the containments in (1) is strict. But first we treat the case when $R = \{1\}$ or $R = \{0,1\}$. (The case $R = \{0\}$ is trivial.)

Proposition 7. *When $R = \{1\}$ or $R = \{0,1\}$, it holds that*

$$\mathbf{CAper} = \mathbf{SAper}_1 \supset \mathbf{SAper}_2 = \mathbf{SAper} = \mathbf{D}.$$

Proof. So let $R = \{1\}$ or $R = \{0,1\}$. The first equality is clear. We also know that $\mathbf{SAper} \supseteq \mathbf{D}$. If we can show that $\mathbf{SAper}_2 \subseteq \mathbf{D}$, then it follows that $\mathbf{SAper}_2 = \mathbf{SAper} = \mathbf{D}$. Moreover, $\mathbf{SAper}_1 \supset \mathbf{SAper}_2$ since there exists a counter-free automaton which is not definite.

To prove that $\mathbf{SAper}_2 \subseteq \mathbf{D}$, suppose that $\mathbb{A}$ is *not* definite. In the classical case, i.e., when $R = \{1\}$, it is known that $\mathbb{A}$ has a proper term function having two or more fixed points, cf. [4]. Thus, by Proposition 5, $\mathbb{A}$ is not 2-aperiodic. If $0 \in R$ the same reasoning applies, since if $\mathbb{A}$ is not definite, then the algebra $\mathbb{A}'$ obtained from $\mathbb{A}$ by removing the constants (so that the rank type of $\mathbb{A}'$ is $\{1\}$) is also not definite. $\blacksquare$

Thus, when R does not contain any integer > 1, then the hierarchy (1) collapses. In the remaining part of this section we will show that when R contains an integer > 1, then the hierarchy is proper.

Proposition 8. *If R contains an integer $n > 1$ then $\mathbf{SAper}_1 \subset \mathbf{CAper}$.*

Proof. Let Σ contain the symbol σ of rank > 1. Define the Σ-tree automaton $\mathbb{A}$ on the set of states $A = \{0,1,2\}$ as follows:

$$\sigma^{\mathbb{A}}(x_1, \ldots, x_n) = \begin{cases} 0 \text{ if } x_1 = \ldots = x_n = 1; \\ 1 \text{ if } x_1 = \ldots = x_n = 0; \\ 2 \text{ otherwise.} \end{cases}$$

Let the interpretation of any other symbol be a constant function. (We may add all constant functions if Σ is large enough.) This automaton is not 1-aperiodic, since the term function $A \to A$ induced by the term $\sigma(x_1, \ldots, x_1)$ maps 0 to 1 and 1 to 0.

However, $\mathbb{A}$ is context aperiodic: For any proper term function $f : A \to A$ induced by a nontrivial context it holds that f^2 is a constant function, so that $f^2 = f^3$. $\blacksquare$

Proposition 9. *Suppose that R contains an integer > 1. Then for each $n > 1$ there exists an $(n-1)$-aperiodic tree automaton which is not n-aperiodic.*

Proof. Given n, we will first prove the claim assuming that the rank type contains an integer $m \geq n$.

We are going to construct a tree automaton $\mathbb{A}$ on the set

$$\{0, 1, \ldots, n, 1', \ldots, n'\}$$

having at least the operations $\sigma_i^{\mathbb{A}} : A^m \to A$, $i \in [n]$. For each $i \in [n]$, we define

$$\sigma_i^{\mathbb{A}}(1, \ldots, n, \ldots, n) = i'$$
$$\sigma_i^{\mathbb{A}}(1', \ldots, n', \ldots, n') = i.$$

In all remaining cases, the operations $\sigma_i^{\mathbb{A}}$ return 0. For each integer $k \in R$, $k \neq m$, we take a single operation symbol in Σ_k and interpret it as the constant function $A^k \to A$ with value 0. This defines the tree automaton $\mathbb{A}$.

For each $i \in [n]$, let $s_i = \sigma_i(x_1, \ldots, x_n, \ldots, x_n) \in T_\Sigma(X_n)$. If $f = \langle s_1^{\mathbb{A}}, \ldots, s_n^{\mathbb{A}} \rangle$, then $f(1, \ldots, n) = (1', \ldots, n')$ and $f(1', \ldots, n') = (1, \ldots, n)$ showing that $\mathbb{A}$ is not n-aperiodic.

To prove that $\mathbb{A}$ is $(n-1)$-aperiodic, consider any proper tree $t \in T_\Sigma(X_{n-1})$. We show that $t^{\mathbb{A}} : A^{n-1} \to A$ is constant with value 0. This is clearly true when an operation symbol different from one of the σ_i appears in t. Suppose now all operations symbols appearing in t are in the set $\{\sigma_1, \ldots, \sigma_n\}$. Then t has a subtree of the form $s = \sigma_i(x_{j_1}, \ldots, x_{j_m})$. Since the first n variables $x_{j_1}, \ldots, x_{j_n}$ cannot be all distinct, $s^{\mathbb{A}}$ is the constant function $A^{n-1} \to A$ with value 0. It follows now that $t^{\mathbb{A}}$ is also this function.

We show how to modify the above construction when each integer in R is less than n. Let m denote the maximal integer in R, so that $m > 1$. Let k denote the least integer with $1 + k(m-1) \geq n$, so that $k \geq 2$. We take symbols $\sigma_{i,1}, \ldots, \sigma_{i,k}$ of rank m, for all $i \in [n]$. Consider the trees in $T_\Sigma(X_n)$,

$$s_{i,1} = \sigma_{i,1}(x_1, \ldots, x_m)$$
$$s_{i,2} = \sigma_{i,2}(s_{i,1}, x_{m+1}, \ldots, x_{m+m-1})$$

$$\vdots$$

$$s_{i,k-1} = \sigma_{i,k-1}(s_{i,k-2}, x_{m+(k-2)(m-1)+1}, \ldots, x_{m+(k-1)(m-1)})$$
$$s_{i,k} = \sigma_{i,k}(s_{i,k-1}, x_{m+(k-1)m+1}, \ldots, x_n, \ldots, x_n)$$

Let $s_i = s_{i,k}$, $i \in [n]$. Now define the interpretation of the $\sigma_{i,j}$ so that

$$s_i^{\mathbb{A}}(1, \ldots, n) = i'$$
$$s_i^{\mathbb{A}}(1', \ldots, n') = i,$$

for all i, moreover, for each i and $j < k$,

$$s_{i,j}^{\mathbb{A}}(1, \ldots, n) \quad \text{and} \quad s_{i,j}^{\mathbb{A}}(1', \ldots, n')$$

are *new* elements. The set A is the union of the set $\{0, 1, \ldots, n, 1', \ldots, n'\}$ with these new elements. In all other cases, the operations $\sigma_{i,j}$ return 0. As before, for all $p \neq m$, we take a single operation symbol in Σ_p whose interpretation is a constant function with value 0. We omit the formal verification of the correctness of this construction.   ∎

Proposition 10. *If R contains an integer $k > 1$, then $\mathbf{D} \subset \mathbf{SAper}$.*

Proof. Let Σ be a ranked alphabet of rank type R and $\sigma \in \Sigma_k$ a function symbol with $k > 1$. Define the Σ-tree automaton $\mathbb{A}$ as follows. Let $A = \{0, 1, 2\}$ and

$$\sigma^{\mathbb{A}}(a_1, \ldots, a_k) = \begin{cases} 1 \text{ if } 0 \in \{a_1, \ldots, a_k\} \subseteq \{0, 1\}; \\ 2 \text{ otherwise,} \end{cases}$$

for all $a_1, \ldots, a_k \in A$. All other symbols in Σ are interpreted as a constant function with value 2.

It is clear that for any nontrivial tree $t \in T_\Sigma(X_n)$, $t^{\mathbb{A}}(a_1, \ldots, a_n) \in \{1, 2\}$ holds; moreover, if $a_1, \ldots, a_n \in \{1, 2\}$, then $t^{\mathbb{A}}(a_1, \ldots, a_n) = 2$. It follows that for any proper term function $f : A^n \to A^n$, each component of f^2 is a constant function with value 2.

We show that $\mathbb{A}$ is not definite. Consider the following sequences s_i, t_i of trees in $T_\Sigma(X_2)$: $s_0 = \sigma(x_1, \ldots, x_1)$, $s_{n+1} = \sigma(x_1, \ldots, x_1, s_n)$ and $t_0 = \sigma(x_2, \ldots, x_2)$, $t_{n+1} = \sigma(x_1, \ldots, x_1, t_n)$. It is easy to check that $s_n^{\mathbb{A}}(0, 1) = 1$ and $t_n^{\mathbb{A}}(0, 1) = 2$ holds for any integer $n \geq 0$, hence—since the trees s_n and t_n agree up to depth n—$\mathbb{A}$ is indeed not definite.   ∎

5   Decidability and Complexity

Since for any Σ-tree automaton $\mathbb{A}$ and for any $n \geq 1$, $S_n(\mathbb{A})$ is finite, it is clear that there exists an algorithm to decide, given a Σ-tree automaton $\mathbb{A}$ and an integer n, whether $\mathbb{A}$ is n-aperiodic. Also, context aperiodicity is decidable. However, strong aperiodicity of a tree automaton is not immediately decidable, since the definition of strong aperiodicity involves a condition for each n. (However, see Remark 1.) In this section, we show that strong aperiodicity is decidable in polynomial time. It is known that deciding aperiodicity of classical automata (the case when $R = \{1\}$) is PSPACE-complete, cf. [2,3]. We use this fact to show that for any fixed n, deciding whether a tree automaton belongs to $\mathbf{SAper}_n$ is PSPACE-hard.

Let $\mathbb{A}$ be a Σ-tree automaton and consider the direct product $\mathbb{A} \times \mathbb{A}$ of $\mathbb{A}$ with itself. Let $B \subseteq A \times A$ a set of state pairs. We denote by $[B]$ the following set containing state pairs:

$$\{t^{\mathbb{A} \times \mathbb{A}}((a_1, b_1), \ldots, (a_n, b_n)) : n \geq 0,\ t \in T_\Sigma(X_n) \text{ nontrivial}, (a_i, b_i) \in B\}.$$

Note that for any $\mathbb{A}$ and $B \subseteq A \times A$, the set $[B]$ is computable in time polynomial in $|\mathbb{A}|$, the *size* of $\mathbb{A}$ defined as $n + \sum_{i \in R} s_i n^i$, where $n = |A|$ and $s_i = |\Sigma_i|$, for all $i \in R$. Before proving our PSPACE-hardness result for the complexity of strong aperiodicity, we make the following observation:

Lemma 4. *The following are equivalent for any Σ-tree automaton $\mathbb{A}$:*

i) *There exist an integer n and a proper term function $f : A^n \to A^n$ of $\mathbb{A}$ with at least two different fixed points.*

ii) *There exist an integer n and a proper term function $f : (A \times A)^n \to (A \times A)^n$ of the direct product $\mathbb{A} \times \mathbb{A}$ such that f has a fixed point of the form $((a_1, b_1), \ldots, (a_n, b_n))$ with $a_i \neq b_i$ for some $i \in [n]$.*

Proposition 11. *The following are equivalent for any Σ-tree automaton $\mathbb{A}$:*

i) *$\mathbb{A}$ is not strongly aperiodic;*

ii) *there exist an integer $n > 0$ and nontrivial trees $t_1, \ldots, t_n \in T_\Sigma(X_n)$ such that $(t_1, \ldots, t_n)^{\mathbb{A}}$ has at least two different fixed points;*

iii) *there exist a set $S \subseteq A \times A$ and states $a \neq b \in A$ such that $(a, b) \in S \subseteq [S]$;*

iv) *the mapping $P(A \times A) \to P(A \times A)$ defined by $B \mapsto [B] \cap B$ has a fixed point S which contains a pair (a, b) with $a \neq b$;*

v) *the greatest fixed point of the mapping $B \mapsto [B] \cap B$ contains some pair (a, b) with $a \neq b$.*

Proof. i) $\to$ ii). We have already proved this (Corollary 1).

ii) $\to$ iii). Let $n > 0$ be an integer, $t_1, \ldots, t_n \in T_\Sigma(X_n)$ nontrivial trees and $(a_1, \ldots, a_n) \neq (b_1, \ldots, b_n)$ different state tuples such that both $(a_1, \ldots, a_n)$ and $(b_1, \ldots, b_n)$ are fixed points of $(t_1, \ldots, t_n)^{\mathbb{A}}$. Then $((a_1, b_1), \ldots, (a_n, b_n))$ is a fixed point of $(t_1, \ldots, t_n)^{\mathbb{A} \times \mathbb{A}}$. Hence, choosing $S = \{(a_i, b_i) : i \in [n]\}$ we get that $S \subseteq [S]$, and since $(a_1, \ldots, a_n) \neq (b_1, \ldots, b_n)$, there exists a state pair $(a_i, b_i) \in S$ with $a_i \neq b_i$.

iii) $\to$ iv). This is a simple reformulation.

iv) $\to$ v). The mapping $B \mapsto [B] \cap B$ is monotone, so it has a greatest fixed point.

v) $\to$ i). Suppose $S = \{(a_1, b_1), \ldots, (a_n, b_n)\}$ is the greatest fixed point. Since S is a fixed point, $S = [S] \cap S$. Moreover, by assumption, $a_i \neq b_i$ holds for some $i \in [n]$. It follows (from the definition of $[S]$) that there exist nontrivial trees $t_1, \ldots, t_n \in T_\Sigma(X_n)$ such that $((a_1, b_1), \ldots, (a_n, b_n))$ is a fixed point of $(t_1, \ldots, t_n)^{\mathbb{A} \times \mathbb{A}}$. Now we can apply Lemma 4 and Corollary 1 and get the result. ∎

Theorem 3. *It is decidable in polynomial time whether a given tree automaton $\mathbb{A}$ is strongly aperiodic.*

Proof. The condition v) of Proposition 11 is clearly decidable in polynomial time: we have to compute $F^{n^2}(A \times A)$, where $F : P(A \times A) \to P(A \times A)$ is defined by $F(B) = [B] \cap B$ and check whether the resulting set contains an ordered pair whose components are different. Since F is computable by a polynomial time procedure, its iteration for n^2 steps runs in polynomial time. ∎

It is clear that for any fixed n, testing whether a tree automaton is in **SAper**$_n$ can be done in exponential time. The following proposition establishes a PSPACE lower bound.

Proposition 12. *For each fixed n, it is* PSPACE-*hard to decide, given a tree automaton* $\mathbb{A}$, *whether* $\mathbb{A}$ *is n-aperiodic.*

Proof. Consider an ordinary automaton, i.e., a Σ-tree automaton $\mathbb{A}$, where Σ is of rank type $\{1\}$. We construct a Δ-tree automaton $\mathbb{B}$, for some Δ, which is n-aperiodic if and only if $\mathbb{A}$ is aperiodic. We define $B = \{0\} \cup \{(a,j) : a \in A, j \in [n]\}$ so that B has $|A| \times n + 1$ states. Then let $\Delta = \Delta_n = \{(\sigma, i) : \sigma \in \Sigma, i \in [n]\}$. The operations $(\sigma, i)^{\mathbb{A}}$ are defined as follows. Suppose that $\sigma^{\mathbb{A}}(a) = b$, for some $a, b \in A$ and $\sigma \in \Sigma$. Then we let $(\sigma, i)^{\mathbb{A}}((a,1), \ldots, (a,n)) = (b,i)$. In all remaining cases, the operation returns 0. For each Σ-term $t \in T_\Sigma(X_n)$, define $\underline{t} = (t_1, \ldots, t_n) \in T_\Delta(X_n)^n$ as follows. If $t = x_1$ then $t_i = x_i$, for each i. If $t = \sigma(s)$, then $t_i = (\sigma, i)(\underline{s})$, for each i. Now if $t^{\mathbb{A}}(a) = b$, then $\underline{t}^{\mathbb{B}}(\underline{a}) = \underline{b}$, where $\underline{a} = (a_1, \ldots, a_n)$ and $\underline{b}$ is defined in the same way. It follows that if $\mathbb{A}$ is not aperiodic, then $\mathbb{B}$ is not n-aperiodic.

We still have to show that if $\mathbb{B}$ is not n-aperiodic then $\mathbb{A}$ is not strongly aperiodic either. For this reason, we define the paths in a term $t \in T_\Delta(X_n)$ as follows. If $t = x_i$ then $\mathrm{paths}(t) = \{x_1\}$. If $t = (\sigma, i)(t_1, \ldots, t_n)$ then $\mathrm{paths}(t) = \{\sigma(s) : s \in \bigcup_{j=1}^n \mathrm{paths}(t_j)\}$. Note that $\mathrm{paths}(t) \subseteq T_\Sigma(X_1)$.

Now assume that $\mathbb{B}$ is not n-aperiodic, so that there exist some $q_1, \ldots, q_k \in A^n$, $k > 1$, a vector of nontrivial terms $\underline{t} = (t_1, \ldots, t_n) \in T_\Delta(X_n)^n$ such that $\underline{t}^{\mathbb{B}}(q_i) = q_{i+1}$, for all $i = 1, \ldots, k - 1$ and $\underline{t}^{\mathbb{B}}(q_k) = q_1$. Now each t_i contains a subtree of the form $(\sigma, j)(x_{k_1}, \ldots, x_{k_n})$. However, the variables $x_{k_1}, \ldots, x_{k_n}$ must be all distinct, since otherwise $t_i^{\mathbb{A}}$ would be constant with value 0. But then the components of each q_i must also be all distinct, moreover, the components of each q_j must give a permutation of the states $(a,1), \ldots, (a,n)$ for some $a \in A$, since otherwise we would have $t_i^{\mathbb{B}}(q_j) = 0$. Also, if $(\sigma, j)(s_1, \ldots, s_n)$ is any subtree of some component of $\underline{t}$, then for any q_i there must exist a state $a \in A$ such that $s_1^{\mathbb{A}}(q_i), \ldots, s_n^{\mathbb{A}}(q_i)$ is the sequence $(a,1), \ldots, (a,n)$. It then follows that there is a permutation π such that for each q_i there is some state $a_i \in A$ with $q_i = ((a, \pi(1)), \ldots, (a, \pi(n)))$. Of course, the sates a_i are all distinct. Thus, if we take any tree $s \in \mathrm{paths}(t_j)$, for any j, then we have $s^{\mathbb{A}}(a_i) = a_{i+1}$ for all $i = 1, \ldots, k - 1$ and $s^{\mathbb{A}}(a_k) = a_1$. This shows that $\mathbb{A}$ is not aperiodic. ∎

6   Aperiodicity and Logic

In this section we relate the above aperiodicity notions to formal logic. We establish the following facts for tree languages over ranked alphabets of rank type $R = \{0, 2\}$.

1. If a tree language is in **CTL**, then its minimal tree automaton is in **SAper₁**.
2. There exists a tree language in **CTL** whose minimal automaton is not in **SAper₂**.
3. There exists a tree language definable in first order logic equipped with the relation $<$ which is not 1-aperiodic.
4. There exists a 1-aperiodic tree language which is not definable in first order logic equipped with both $<$ and the successor relations.

The above results can all be extended to rank types containing 0 and at least one integer > 1. We recall that if $\mathbb{A}$ is a Σ-tree automaton, then the *tree language accepted by* $\mathbb{A}$ *with final states* $F \subseteq A$ is the set $\{t : t^{\mathbb{A}} \in F\}$. Languages acceptable by tree automata are called *regular*. It is known that each regular tree language can be accepted by a minimal tree automaton, unique up to isomorphism.

We start with the definitions of the above logics. Suppose that Σ is a ranked set of rank type $R = \{0, 2\}$. An *atomic formula* of the logic FO($<, S1, S2$) over Σ is of the form $P_\sigma(z)$, where $\sigma \in \Sigma$ and z is a *first order variable* ranging over the vertices of a tree in T_Σ. The meaning of this formula is that the vertex denoted by z is labeled σ. Further atomic formulas are $Si(z_1, z_2)$, $i = 1, 2$ and $z_1 < z_2$ expressing that the vertices z_1 and z_2 are related by the corresponding relation, i.e., z_2 is the ith *successor* of z_1 and z_2 is a descendant of z_1, respectively. The formulas of the logic FO($<, S1, S2$) are constructed from these atomic formulas by the Boolean connectives and existential and/or universal quantification. When φ is a closed formula of this logic, we let L_φ denote the set of all trees $t \in T_\Sigma$ satisfying φ, and call L_φ a *tree language definable in* FO($<, S1, S2$). The collection of all these languages for all ranked alphabets is denoted **FO**($<, S1, S2$). We let **FO**($<$) stand for the class of languages definable in the sublogic obtained by dropping the successor relations.

The other logic, CTL, has atomic formulas p_σ, $\sigma \in \Sigma$ expressing that the root of a tree is labeled σ. The set of formulas is the least set generated from the atomic formulas by the Boolean connectives and the X_i, $i = 1, 2$ and EU modalities. A tree t satisfies a formula $X_i\varphi$ if the ith successor of the root exists and satisfies φ. Moreover, t satisfies φ EU ψ if it has a path starting from the root which contains a vertex v such that the subtree rooted at this vertex satisfies ψ and each subtree rooted at a vertex preceding v satisfies φ. We let **CTL** denote the class of languages definable in CTL. In this logic, the EF modality is expressible, where a tree satisfies EFφ if it has a subtree satisfying φ.

It is known that **CTL** is a proper subclass of **FO**($<, S1, S2$) and that each language in **FO**($<, S1, S2$) is regular. Moreover, as shown in effect in [16], the minimal automaton [8] of each language in **FO**($<, S1, S2$) is context aperiodic.

In the rest of this section we fix $R = \{0, 2\}$.

Proposition 13. *If L is a tree language in* **CTL***, then the minimal automaton of L is in* **SAper**$_1$.

Proof. It was shown in [6,7] that a tree language is in **CTL** if and only if its minimal tree automaton is in the least generalized cascade variety containing the automaton over the two-element set $\{0, 1\}$ equipped with the binary **or** function and the two constants $0, 1$. Our claim thus follows from the facts that this tree automaton is 1-aperiodic and that 1-aperiodic tree automata form a generalized cascade variety, cf. Theorem 1.   ∎

Proposition 14. *There exists a tree language in* **CTL** *whose minimal tree automaton is not in* **SAper**$_2$.

Proof. Consider the Σ-tree automaton $\mathbb{A}$ on $\{0, 1\}$, where Σ contains the binary symbols σ_1 and σ_2 respectively interpreted as the binary **or** function and the binary constant function with value 1, and the constant symbol σ_0 interpreted as 0. Then $\mathbb{A}$ is the minimal tree automaton of the language $L \subseteq T_\Sigma$ consisting of all trees containing a vertex labeled σ_2. This language is defined by the formula $\mathbf{EF}p_{\sigma_2}$. On the other hand, the identity function is a proper term function having two fixed points. Thus $\mathbb{A} \notin \mathbf{SAper}_2$.   ∎

Proposition 15. *There exists a regular tree language in* $\mathbf{FO}(<)$ *whose minimal tree automaton is not 1-aperiodic.*

Proof. Let us define the ranked alphabet Σ with $\Sigma_2 = \{\sigma, \nu_1, \nu_2\}$ and $\Sigma_0 = \{c\}$. We construct a regular tree language $L \subseteq T_\Sigma$ in $\mathbf{FO}(<)$ whose minimal automaton is not 1-aperiodic. Let L be the following language:

$$L = \{t \in T_\Sigma : \mathrm{Paths}(t) \cap (\sigma\nu_1\sigma\nu_2)^* c \neq \emptyset\}.$$

Here, $\mathrm{Paths}(t)$ is the set of all words which are label sequences of maximal paths in t.

Now L is definable in $\mathrm{FO}(<)$: a tree $t \in T_\Sigma$ belongs to L if and only if there exists a vertex v such that the following hold:

1. the label of v is c (hence v is a leaf);
2. if y is a successor of x and $y < v$ then x is labeled σ if and only if y is labeled in $\{\nu_1, \nu_2\}$, and x is labeled in $\{\nu_1, \nu_2\}$ if and only if y is labeled σ;
3. if y is a second successor of x and $y < v$ then x is labeled ν_1 if and only if y is labeled ν_2, and x is labeled ν_2 if and only if y is labeled ν_1;
4. the root is labeled in $\{\sigma, c\}$ and if v is a successor of x then x is labeled ν_2.

It is clear that all these conditions are expressible in $\mathrm{FO}(<)$. Now consider the following infinite sequence of trees t_i: $t_0 = c$, and for all $i \geq 0$, $t_{i+1} = \sigma(\nu_1(t_i, t_i), \nu_2(t_i, t_i))$. It is clear that $t_i \in L$ if and only if i is even: hence the minimal automaton of L is not 1-aperiodic (since the sequence t_i corresponds to a sequence $f^i(a_0)$ in the minimal recognizer $\mathbb{A}$ of L with $a_0 = c^{\mathbb{A}}$, where f is the proper term function induced by the tree $\sigma(\nu_1(x_1, x_1), \nu_2(x_1, x_1))$).   ∎

Proposition 16. *There exists a tree language whose minimal automaton is 1-aperiodic which is not in* $\mathbf{FO}(<, S_1, S_2)$.

Proof. Let $\Sigma_0 = \{\mathbf{0}, \mathbf{1}\}$, $\Sigma_2 = \{\wedge, \vee\}$, and let us define the following tree automaton $\mathbb{A}$ on $A = \{0, 1, \bot\}$ that was introduced in [13]: $\mathbf{0}^{\mathbb{A}} = 0$, $\mathbf{1}^{\mathbb{A}} = 1$ and

$$\wedge^{\mathbb{A}}(x,y) = \begin{cases} 1 & \text{if } x = y = 1; \\ 0 & \text{if } \{x, y\} = \{0, 1\}; \\ \bot & \text{otherwise,} \end{cases} \qquad \vee^{\mathbb{A}}(x,y) = \begin{cases} 0 & \text{if } x = y = 0; \\ 1 & \text{if } \{x, y\} = \{0, 1\}; \\ \bot & \text{otherwise.} \end{cases}$$

Let $L = \{t \in T_\Sigma : t^{\mathbb{A}} = 1\}$. It has been shown in [13] that L is not contained in $\mathbf{FO}(<, S_1, S_2)$. It is easy to check that $\mathbb{A}$ is 1-aperiodic. We note that $\mathbb{A}$ is not 2-aperiodic, since both 1 and $\bot$ are fixed points of the term function $A \to A$ induced by $\wedge(x_1, x_1)$.   ∎

7   A Generalization

In this section, we provide a generalization of the aperiodicity notions studied above. The following definition is motivated by the Krohn-Rhodes decomposition theorem [4].

Definition 4. *Suppose that $\mathcal{G}$ is a class of finite simple groups closed under division. We say that a tree automaton $\mathbb{A}$ belongs to the class $\mathbf{Aut}(\mathcal{G}, n)$ for some positive integer n exactly when each simple group dividing $S_n(\mathbb{A})$ is in $\mathcal{G}$. The intersection of all classes $\mathbf{Aut}(\mathcal{G}, n)$ is $\mathbf{Aut}(\mathcal{G})$. Moreover, we let $\mathbf{Aut}_c(\mathcal{G})$ denote the class of all tree automata $\mathbb{A}$ such that every simple group divisor of $C(\mathbb{A})$ is in $\mathcal{G}$.*

When $\mathcal{G}$ contains only the trivial groups, then the above classes are just $\mathbf{SAper}_n$, $\mathbf{SAper}$ and $\mathbf{CAper}$, respectively. The following generalization of Theorems 1 and 2 holds.

Theorem 4. *For each n, $\mathbf{Aut}(\mathcal{G}, n)$ is a generalized cascade variety. $\mathbf{Aut}(\mathcal{G})$ and $\mathbf{Aut}_c(\mathcal{G})$ are generalized cascade varieties.*

We omit the proof which is based on the following facts. For the definition of the wreath product of transformation semigroups we refer to [4].

Lemma 5. *Suppose that $\mathbb{A}$ and $\mathbb{B}$ are Σ-tree automata such that $\mathbb{A}$ is a subautomaton or a homomorphic image of $\mathbb{B}$. Then for each n, $(A, S_n(\mathbb{A}))$ divides $(B, S_n(\mathbb{B}))$. Similarly, $(A, C(\mathbb{A}))$ divides $(B, C(\mathbb{B}))$.*

Lemma 6. *Suppose that $\mathbb{A}$ is a Σ-tree automaton, $\mathbb{B}$ is a Δ-tree automaton and consider a generalized cascade product $\mathbb{A} \times_\alpha \mathbb{B}$. Then for each n, $(A \times B, S_n(\mathbb{A} \times_\alpha \mathbb{B}))$ divides a wreath product of $(A, S(\mathbb{A}))$ and $(B, S(\mathbb{B}))$. Moreover, $(A \times B, C(\mathbb{A} \times_\alpha \mathbb{B}))$ divides a wreath product of $(A, C(\mathbb{A}))$ and $(B, C(\mathbb{B}))$.*

8   A Variant of Aperiodicity

In this section we study a variant of the aperiodicity notions resulting by considering polynomial functions instead of term functions. We will show that in this case the hierarchy collapses at $n = 2$.

Let $\mathbb{A}$ be a Σ-tree automaton of rank type R over the set A. By adding to Σ each element of A as a new constant symbol we obtain the ranked alphabet $\Sigma(A)$ of rank type $R \cup \{0\}$. (We may assume that Σ and A are disjoint.) We turn $\mathbb{A}$ into a $\Sigma(A)$-tree automaton by interpreting each element of A by itself. We let $\mathbb{A}^{(p)}$ denote the resulting $\Sigma(A)$-tree automaton. Below by a *proper translation* of $\mathbb{A}$ we shall mean a function induced by a nontrivial context in $T_{\Sigma(A)}(X_1)$. The following was proved in [5], Corollary 3.12.

Lemma 7. *The following are equivalent for any Σ-tree automaton $\mathbb{A}$:*

i) $\mathbb{A}$ is definite;
ii) Any proper translation of $\mathbb{A}$ has at most one fixed point.

We also note the obvious fact that $\mathbb{A}$ is definite if and only if $\mathbb{A}^{(p)}$ is.

For each n, we let $\mathbf{SAper}_n^{(p)}$ denote the class of all tree automata $\mathbb{A}$ such that $\mathbb{A}^{(p)}$ is in $\mathbf{SAper}_n$. We define the classes $\mathbf{SAper}^{(p)}$ and $\mathbf{CAper}^{(p)}$ in the same way.

Theorem 5. *The following are equivalent for any Σ-tree automaton $\mathbb{A}$:*

   i) $\mathbb{A}$ is definite;
   ii) $\mathbb{A} \in \mathbf{SAper}^{(p)}$;
   iii) $\mathbb{A} \in \mathbf{SAper}_2^{(p)}$;
   iv) Each proper translation of $\mathbb{A}$ has at most one fixed point.

Proof. The first and last conditions are equivalent by Lemma 7, and the first condition implies the second by Corollary 2. It is clear that the second condition implies the third. Finally, the third implies the fourth by Proposition 5. ■

Corollary 3. *It is decidable in polynomial time whether a tree automaton is in $\mathbf{SAper}^{(p)}$.*

Proof. A polynomial time algorithm is implicit in [5]. ■

Proposition 17. $\mathbf{SAper}_2^{(p)} \subset \mathbf{SAper}_1^{(p)} \subset \mathbf{CAper}^{(p)}$.

Proof. The inclusions are clear. The second inclusion is strict by the proof of Proposition 8. The first is strict since the tree automaton on $\{0,1\}$ with the or function as the single operation is in $\mathbf{SAper}_1^{(p)} \setminus \mathbf{D}$. ■

Proposition 18. *Suppose the rank type R contains an integer > 1. Then for each n there exist tree automata $\mathbb{A}$ and $\mathbb{B}$ such that both $\mathbb{A}$ and $\mathbb{B}$ are $(n-1)$-aperiodic, neither of them is n-aperiodic, moreover, $\mathbb{A}$ is contained in $\mathbf{SAper}_1^{(p)}$ and $\mathbb{B}$ is not.*

Proof. First we deal with the case when R contains an integer $k \geq n$.

Consider the automaton $\mathbb{A}$ constructed in the proof of Proposition 9. We have proved that $\mathbb{A}$ is $(n-1)$-aperiodic and not n-aperiodic. It is easy to see that for every non-constant proper term function $f : A \to A$ of $\mathbb{A}^{(p)}$ there exists *at most* one state a with $f(a) \neq 0$, moreover, this state a is different from 0. From this we can conclude that $f^2 = f^3$ holds for any such function. Thus, $\mathbb{A}$ is contained in $\mathbf{SAper}_1^{(p)}$.

Now we construct the automaton $\mathbb{B}$ by modifying the construction of $\mathbb{A}$ as follows. The state set of $\mathbb{B}$ is $B = \{0, \dots, n\} \cup \{1'\}$. For each $i \in [n]$, we define $\sigma_i^{\mathbb{B}}$ as follows:

$$\sigma_i^{\mathbb{B}}(1, \dots, n, \dots, n) = \sigma_i^{\mathbb{B}}(1', 2, \dots, n, \dots, n) = i \text{ if } i > 1;$$

$$\sigma_1^{\mathbb{B}}(1, \dots, n, \dots, n) = 1' \text{ and } \sigma_1^{\mathbb{B}}(1', 2, \dots, n, \dots, n) = 1.$$

In all remaining cases the functions $\sigma_i^{\mathbb{B}}$ return 0, and all other function symbols are interpreted as a constant function with value 0.

Now by the same argument as in the proof of Proposition 9 we get that $\mathbb{B}$ is $(n-1)$-aperiodic but not n-aperiodic. Also, since for $t = \sigma_1(x_1, 2, \ldots, n)$ in $T_{\Sigma(A)}(X_1)$ we have $t^{\mathbb{B}}(1) = 1'$ and $t^{\mathbb{B}}(1') = 1$, $\mathbb{B}$ is not contained in $\mathbf{SAper}_1^{(p)}$.

If R contains only integers less than n, then we can modify $\mathbb{B}$ as in the proof of Proposition 9. ∎

Proposition 19. *Suppose that R contains some integer $k > 1$. Then there exists an automaton $\mathbb{A}$ contained in $\mathbf{SAper} \backslash \mathbf{SAper}_1^{(p)}$.*

Proof. We define the following Σ-tree automaton $\mathbb{A}$, where $\Sigma_i = \{\sigma_i\}$ for each $i \in R$ and $A = \{0, 1, 2\}$. For each $i \in R$ with $i \neq k$, let $\sigma_i^{\mathbb{A}}$ be the constant function with value 2 and let us define $\sigma_k^{\mathbb{A}}$ as follows:

$$\sigma_k^{\mathbb{A}}(a_1, \ldots, a_k) = \begin{cases} 1 & \text{if } \{a_1, \ldots, a_k\} = \{0, 2\}; \\ 2 & \text{otherwise.} \end{cases}$$

Now it is easy to see that $t^{\mathbb{A}}(a_1, \ldots, a_n) \in \{1, 2\}$ for all proper trees $t \in T_{\Sigma}(X_n)$, $n \geq 1$, and for all $a_1, \ldots, a_n \in A$. Moreover, if each a_i is 1 or 2, then $t^{\mathbb{A}}(a_1, \ldots, a_n) = 2$. It follows that for any tuple of proper trees $\underline{t} = (t_1, \ldots, t_n) \in \left(T_{\Sigma}(X_n)\right)^n$, each component of $(\underline{t}^{\mathbb{A}})^2 : A^n \to A^n$ is constant with value 2, hence $\mathbb{A}$ is strongly aperiodic.

However, since the function induced by $\sigma_k(0, \ldots, 0, x_1)$ maps 1 to 2 and 2 to 1, $\mathbb{A}$ is not contained in $\mathbf{SAper}_1^{(p)}$. ∎

Corollary 4. *As shown in the the Figure, when R contains an integer > 1, the class $\mathbf{SAper}_1^{(p)}$ nontrivially intersects $\mathbf{SAper}$ and each member of the hierarchy $\mathbf{SAper}_n$ for $n \geq 2$.*

9 Conclusions

Given a tree automaton $\mathbb{A}$ (i.e., a finite algebra), two natural categories associated with $\mathbb{A}$ are the category $\mathcal{T}^+(\mathbb{A})$ of proper term functions[1] and the category $\mathcal{P}ol^+(\mathbb{A})$ of proper polynomial functions of $\mathbb{A}$. In both categories, the objects may be taken as the positive integers. Now $\mathbb{A}$ is n-aperiodic if and only if the hom-set $\mathcal{T}^+(\mathbb{A})(n, n)$ is group-free (which yields that each $\mathcal{T}^+(\mathbb{A})(m, m)$ with $m < n$ is also group-free). And $\mathbb{A}$ is strongly aperiodic if each hom-set $\mathcal{T}^+(\mathbb{A})(n, n)$ is group-free. Thus these notions are very natural from the mathematical point of view. Similarly, $\mathbb{A}$ is n-aperiodic with respect to polynomials if and only if the hom-set $\mathcal{P}ol^+(\mathbb{A})(n, n)$ is group-free, and strongly aperiodic with respect to polynomials if and only if each $\mathcal{P}ol^+(\mathbb{A})(n, n)$ is group-free.

We have fully described the relationship between the above notions and established decidability and complexity results for them. Aperiodicity captures some nice combinatorial properties of regular tree languages. For example, a regular

---

[1] In categorical algebra, one usually takes the category $\mathcal{T}(\mathbb{A})$ of all term functions of $\mathbb{A}$, called the *Lawvere theory* of term functions.

tree language $L \subseteq T_\Sigma$ is n-aperiodic (i.e., its minimal automaton is n-aperiodic) if and only if there is some k such that for all $r \in T_\Sigma(X_n)$, $t \in T_\Sigma(X_n)^n$ and $s \in T_\Sigma^n$,

$$r \cdot t^k \cdot s \in L \quad \Leftrightarrow \quad r \cdot t^{k+1} \cdot s \in L. \tag{2}$$

The languages satisfying this condition form a *tree language variety* as defined in [6], a notion closely related to Almeida's and Steinby's tree language varieties, cf. [1,14,15]. The operation of product in (2) is of course defined by tree substitution.

Aperiodicity is also related to logical definability, though probably not very closely. The decidability of the membership of a regular tree language in **FO**$(<, S1, S2)$ or **CTL** is open. According to Proposition 13, **CTL** is contained in the class of 1-aperiodic tree languages. The containment is strict by Proposition 16, so that in order to obtain a decidable characterization one needs to look for additional conditions.

Open problem. Give a decidable characterization of **CTL**.

References

1. Almeida, J.: On pseudovarieties, varieties of languages, filters of congruences, pseudoidentities and related topics. Algebra Universalis 27, 333–350 (1990)
2. Bernátsky, L.: Regular expression star-freeness is PSPACE-complete. Acta Cybernetica 13, 1–21 (1997)
3. Cho, S., Huynh, D.T.: Finite-automaton aperiodicity is PSPACE-complete. Theoretical Computer Science 88, 99–116 (1991)
4. Eilenberg, S.: Automata, Languages, and Machines, vol. A&B. Academic Press, London (1974/1976)
5. Ésik, Z.: Definite tree automata and their cascade compositions. Publ. Math. 48, 243–262 (1996)
6. Ésik, Z.: An algebraic characterization of temporal logics on finite trees. Parts I, II, III. In: 1st International Conference on Algebraic Informatics, 2005, pp. 53–77, 79–99, 101–110, Aristotle Univ. Thessaloniki, Thessaloniki (2005)
7. Ésik, Z.: Characterizing CTL-like logics on finite trees. Theoretical Computer Science 356, 136–152 (2006)
8. Gécseg, F., Steinby, M.: Tree Automata. Akadémiai Kiadó, Budapest (1984)
9. Grätzer, G.: Universal Algebra, 2nd edn. Springer, Heidelberg (1979)
10. Heuter, U.: Definite tree languages. Bulletin of the EATCS 35, 137–142 (1988)
11. Heuter, U.: First-order properties of trees, star-free expressions, and aperiodicity. RAIRO Inform. Théor. Appl. 25, 125–145 (1991)
12. McNaughton, R., Papert, S.: Counter-free Automata. MIT Press, Cambridge, MA (1971)
13. Potthoff, A.: Modulo-counting quantifiers over finite trees. Theoretical Computer Science 126, 97–112 (1994)
14. Steinby, M.: A theory of tree language varieties. In: Tree Automata and Languages, North-Holland, Amsterdam, pp. 57–81 (1992)
15. Steinby, M.: General varieties of tree languages. Theoretical Computer Science 205, 1–43 (1998)
16. Thomas, W.: Logical aspects in the study of tree languages. In: Ninth colloquium on trees in algebra and programming (Bordeaux, 1984), pp. 31–49. Cambridge Univ. Press, Cambridge (1984)

Appendix

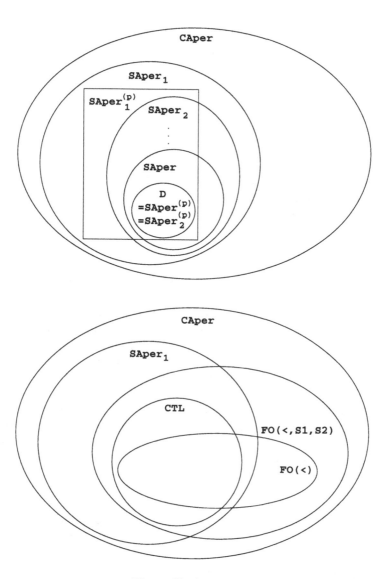

Fig. 1. The hierarchies

The Syntactic Complexity of Eulerian Graphs[*]

Antonios Kalampakas

Technical Institute of Kavala,
Department of Exact Sciences,
65404, Kavala, Greece
akalamp@math.auth.gr

Abstract. In this paper we prove that the set of directed Eulerian graphs is not recognizable. On the other hand, the set of directed graphs with an Eulerian underlying graph is shown to be recognizable. Furthermore, we compute the syntactic complexity of this language and we compare it with that of connected graphs.

1 Introduction

The notion of graph language recognizability, by virtue of the syntactic magmoid, was investigated in [5]. In this setup, the syntactic complexity of a given recognizable graph language can be determined, giving rise to a syntactic classification inside the class of recognizable graph languages. In this paper we investigate the set of Eulerian graphs and that of graphs with an Eulerian underlying graph. The syntactic complexity of the latter is determined, and we see that it is placed higher than that of connected graphs in the syntactic classification of recognizable graph languages.

A hypergraph consists of a set of nodes and a set of hyperedges, just as an ordinary (directed) graph except that a hyperedge may have an arbitrary sequence of sources and an arbitrary sequence of targets. Each hyperedge is labelled with a symbol from a doubly ranked alphabet Σ in such a way that the first (second) rank of its label equals the number of its sources (targets respectively). Also, every hypergraph is multi-pointed in the sense that it has a sequence of m "begin" and n "end" nodes, $m, n \geq 0$.

From now on a hypergraph will also be called a graph, and its hyperedges edges; furthermore, to specify the number of begin and end nodes, it will be called an (m, n)-graph. We denote by $GR_{m,n}(\Sigma)$ the set of all (m, n)-graphs labelled over Σ.

If G is an (m, n)-graph and H is an (n, k)-graph then their *product* $G \circ H$ is the (m, k)-graph obtained by taking the disjoint union of G and H and then identifying the ith end node of G with the ith begin node of H, for every $i \in \{1, ..., n\}$; also, the sequence of begin nodes of $G \circ H$ is the one of G, and its sequence of end nodes the one of H.

The *sum* $G \square H$ of arbitrary graphs G and H is their disjoint union with their sequences of begin nodes concatenated and similarly for their end nodes.

---

[*] This research was partially supported by I.K.Y.

S. Bozapalidis and G. Rahonis (Eds.): CAI 2007, LNCS 4728, pp. 208–217, 2007.
© Springer-Verlag Berlin Heidelberg 2007

The family $GR(\Sigma) = (GR_{m,n}(\Sigma))_{m,n\in\mathbb{N}}$ with the operations $\circ$ and $\square$ forms a magmoid in the sense of [1,2], that is, a strict monoidal category (or x-category) whose objects are the natural numbers (see e.g. [8,7]). In [4], $GR(\Sigma)$ is characterized as the quotient of the free magmoid generated by Σ together with a set of five elements, and divided by a finite set of equations.

Magmoids simulate the ordinary monoid structure and a natural regularity notion derives from this simulation (cf. [5]). Precisely, we say that $L \subseteq M$ is recognizable whenever there exist a locally finite magmoid N (i.e., $N_{m,n}$ is finite for all $m, n \geq 0$) and a morphism of magmoids $h : M \to N$, so that $L = h^{-1}(P)$ for some $P \subseteq N$. Note that, $L \subseteq M$ means that L is a doubly ranked set $L = (L_{m,n})$, and for all $m, n \geq 0$, $L_{m,n} \subseteq M_{m,n}$.

An advantage of investigating graph languages within magmoids is that in this setup it is possible to classify them according to their syntactic complexity. We say that two elements of a magmoid M are equivalent modulo the *syntactic congruence* $\sim_L$ of a subset $L \subseteq M$, whenever they have the same set of contexts with respect to L. The syntactic magmoid of L, denoted M_L, is then the quotient of M by $\sim_L$, which is actually the smallest quotient of M recognizing L. In [5] it is shown that a subset $L \subseteq M$ is recognizable if and only if its syntactic magmoid is locally finite.

The *syntactic complexity* of a recognizable set $L \subseteq M$ is measured by a function, called the syntax complexity function of L, mapping any pair (m, n) of natural numbers to the number of syntactic classes of M_L at the rank (m, n). It is proved in [5] that the syntactic complexity of the set $Con(\Sigma)$ of connected graphs is:

$$SC_{Con(\Sigma)}(m, n) = B_{m+n} + 1,$$

where B_k denotes the k-th bellian number i.e., the number of all partitions of the set $\{1, 2, ..., k\}$, cf. [3]. In this paper we see that the set of directed Eulerian graphs (multiple edges allowed) is not recognizable. Furthermore, we prove that the set $EU(\Sigma)$ of directed graphs with an Eulerian underlying graph (which contains the set of connected graphs) is recognizable. The syntax complexity function of $EU(\Sigma)$ is calculated and it turns out that, for $m + n > 2$, it holds

$$SC_{Con(\Sigma)}(m, n) < SC_{EU(\Sigma)}(m, n).$$

The paper is divided into 4 sections. The notion of a magmoid, together with some preliminary matter, are presented in Section 2. We particularly insist in the construction of the *magmoid of hypergraphs* by recalling the definition of hypergraphs introduced in [6] together with the operations product and sum. The notions of magmoid recognizability and syntactic complexity of a graph language are presented in Section 3. In the last section we study the set of directed Eulerian graphs and that of directed graphs with an Eulerian underlying graph.

2 Magmoids and Graphs

Recall that a doubly ranked set (or a doubly ranked alphabet) $(A_{m,n})_{m,n\in\mathbb{N}}$ is a set A together with a function $rank : A \to \mathbb{N} \times \mathbb{N}$, where $\mathbb{N}$ is the set of natural numbers. For $m, n \in \mathbb{N}$, $A_{m,n}$ is the set $\{a \in A \mid rank(a) = (m,n)\}$. In what follows we will drop the subscript $m, n \in \mathbb{N}$ and denote a doubly ranked set simply by $(A_{m,n})$.

Definition 1. *A* magmoid *is a doubly ranked set* $M = (M_{m,n})$ *equipped with two operations*

$$\circ : M_{m,n} \times M_{n,k} \to M_{m,k}, \qquad m, n, k \geqslant 0$$

$$\Box : M_{m,n} \times M_{m',n'} \to M_{m+m',n+n'}, \qquad m, n, m', n' \geqslant 0$$

which are associative in the obvious way and satisfy the distributivity law

$$(f \circ g) \Box (f' \circ g') = (f \Box f') \circ (g \Box g')$$

whenever all the above operations are defined. Moreover, both the operations $\circ$ *and* $\Box$ *are unitary, i.e.,* M *is equipped with a sequence of constants* $e_n \in M_{n,n}$ $(n \geqslant 0)$, *called* units, *such that*

$$e_m \circ f = f = f \circ e_n, \quad e_0 \Box f = f = f \Box e_0$$

for all $f \in M_{m,n}$ *and all* $m, n \geqslant 0$, *and the additional condition*

$$e_m \Box e_n = e_{m+n}, \qquad \text{for all } m, n \geqslant 0$$

holds.

Submagmoids, morphisms, congruences and quotients of magmoids are defined in the obvious way.

Next we construct the magmoid of hypergraphs. Given a finite alphabet X, we denote by X^* the set of all words over X and for every word $w \in X^*$, $|w|$ denotes its length. Formally a *concrete* (m,n)-*graph* over a doubly ranked alphabet $\Sigma = (\Sigma_{m,n})$ is a tuple

$$G = (V, E, s, t, l, begin, end)$$

where V is the finite set of nodes, E is the finite set of hyperedges, $s : E \to V^*$ is the source function, $t : E \to V^*$ is the target function, $l : E \to \Sigma$ is the labelling function such that $rank(l(e)) = (|s(e)|, |t(e)|)$ for every $e \in E$, $begin \in V^*$ with $|begin| = m$ is the sequence of begin nodes and $end \in V^*$ with $|end| = n$ is the sequence of end nodes. For an edge e of a hypergraph G we simply write $rank(e)$ to denote $rank(l(e))$.

The specific sets V and E chosen to define a concrete graph G are actually irrelevant. We shall not distinguish between two isomorphic graphs. Hence the following definition of an abstract graph. Two concrete (m,n)-graphs $G = (V, E, s, t, l, begin, end)$ and $G' = (V', E', s', t', l', begin', end')$ over Σ are isomorphic iff there exist two bijections $h_V : V \rightarrow V'$ and $h_E : E \rightarrow E'$ commuting with source, target, labelling, *begin* and *end* in the usual way.

An *abstract* (m, n)-*graph* is defined to be the equivalence class of a concrete (m, n)-graph with respect to isomorphism. We denote by $GR_{m,n}(\Sigma)$ the set of all abstract (m, n)-graphs over Σ. Since we shall mainly be interested in abstract graphs we shall simply call them graphs except when it is necessary to emphasize that they are defined up to an isomorphism.

Any graph $G \in GR_{m,n}(\Sigma)$ having no edges, is called a *discrete* (m, n)-*graph*. Given an edge label $\sigma \in \Sigma_{m,n}$, we denote by $G(\sigma)$ the (m, n)-graph such that $V = \{x_1, \ldots, x_m, y_1, \ldots, y_n\}$, $E = \{e\}$ with $l(e) = \sigma$, $begin = s(e) = x_1 \cdots x_m$ and $end = t(e) = y_1 \cdots y_n$.

If G is an (m, n)-graph represented by $(V, E, s, t, l, begin, end)$ and H is an (n, k)-graph represented by $(V', E', s', t', l', begin', end')$ then their *product* $G \circ H$ is the (m, k)-graph represented by the concrete graph obtained by taking the disjoint union of G and H and then identifying the *i*th end node of G with the *i*th begin node of H, for every $i \in \{1, \ldots, n\}$; also, $begin(G \circ H) = begin(G)$ and $end(G \circ H) = end(H)$.

The *sum* $G \square H$ of arbitrary graphs G and H is their disjoint union with their sequences of begin nodes concatenated and similarly for their end nodes.

For instance let $\Sigma = \{a, b, c, d\}$, with $rank(a) = (2, 1)$, $rank(b) = (1, 1)$, $rank(c) = (2, 2)$ and $rank(d) = (1, 2)$. In the following pictures, edges are represented by boxes, nodes by dots, and the sources and targets of an edge by directed lines that enter and leave the corresponding box, respectively. The order of the sources and targets of an edge is the vertical order of the directed lines as drawn in the pictures. We display two graphs $G \in GR_{3,2}(\Sigma)$ and $H \in GR_{2,2}(\Sigma)$, where the *i*th begin node is indicated by b_i, and the *i*th end node by e_i.

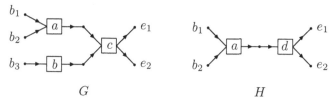

$$G \qquad\qquad\qquad H$$

Then their product $G \circ H$ is the $(3, 2)$-graph

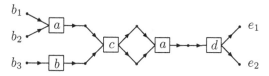

and, their sum $G \square H$ is the $(5, 4)$-graph

For every $n \in \mathbb{N}$ the unit E_n of rank (n, n) is the discrete graph with nodes $x_1, \ldots, x_n$ and $begin(E_n) = end(E_n) = x_1 \cdots x_n$. Note that E_0 is the empty graph.

It is straightforward to verify that $GR(\Sigma) = (GR_{m,n}(\Sigma))$ with the operations defined above is a magmoid, see Lemma 6 of [6].

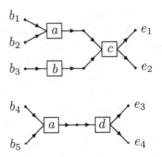

3   Recognizability and Syntactic Complexity

An elegant characterization of a congruence can be achieved by means of the notion of the context. In a magmoid M an *(m,n)-context* is a 4-tuple $\omega = (g_1, f_1, f_2, g_2)$, with $f_i \in M_{m_i, n_i}$ $(i = 1, 2)$, $g_1 \in M_{a, m_1+m+m_2}$, $g_2 \in M_{n_1+n+n_2, b}$, where $a, b \in N$.

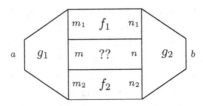

The set of all (m, n)-contexts is denoted $Cont_{m,n}(M)$. For any $f \in M_{m,n}$ and $\omega = (g_1, f_1, f_2, g_2)$ as above, we write $\omega[f] = g_1 \circ (f_1 \,\Box\, f \,\Box\, f_2) \circ g_2$; note that $\omega[f] \in M_{a,b}$. Let L be a subset of the magmoid M and $f \in M_{m,n}$, we set

$$C_L(f) = \{\omega \mid \omega \in Cont_{m,n}(M), \; \omega[f] \in L\}.$$

Proposition 1 (cf. [5]). *The equivalence $\sim_L$ on M defined by*

$$f \sim_{L,m,n} g, \quad \text{whenever } C_L(f) = C_L(g)$$

is a congruence.

Given a magmoid M and a set $L \subseteq M$, $\sim_L$ is called the *syntactic congruence* of L and the quotient magmoid $M_L = M/\sim_L$ is the *syntactic magmoid* of L. Thus, for all $m, n \geqslant 0$, the set $(M_L)_{m,n}$ can be identified with the set consisting of all distinct contexts of the elements of $M_{m,n}$, i.e., we may write

$$(M_L)_{m,n} = \{C_L(f) \mid f \in M_{m,n}\}$$

whereas, the operations of M_L are given by the next formulas:

$$C_L(f) \circ C_L(g) = C_L(f \circ g), \quad C_L(f) \,\Box\, C_L(g) = C_L(f \,\Box\, g).$$

Let $M = (M_{m,n})$ be a magmoid. A subset L of M is called *recognizable* if there exists a locally finite magmoid $N = (N_{m,n})$ (i.e., $N_{m,n}$ finite for all $m, n \in \mathbb{N}$) and a morphism $H : M \to N$, so that $L = H^{-1}(P)$, for some $P \subseteq N$.

Theorem 1 (cf. [5]). *Let M be a magmoid and L its subset. The following conditions are equivalent:*

1. *L is recognizable;*
2. *L is saturated by a congruence of a locally finite index;*
3. *$\sim_L$ has locally finite index;*
4. *the set $card\{C_L(f) \mid f \in M_{m,n}\}$ is finite for all $m, n \in \mathbb{N}$;*
5. *the syntactic magmoid M_L is locally finite.*

The *syntactic complexity* of a recognizable subset L of a magmoid M can be measured by the function $SC_L : \mathbb{N} \times \mathbb{N} \to \mathbb{N}$, called the *syntax complexity function* of L, mapping any pair (m, n) of natural numbers to the number of syntactic classes of M_L at the rank (m, n), that is

$$SC_L(m, n) = card(M_L)_{m,n}, \qquad m, n \in \mathbb{N}.$$

Theorem 2 (cf. [5]). *The syntactic complexity of the graph language $Con(\Sigma)$ of all connected graphs is*

$$SC_{Con(\Sigma)}(m, n) = B_{m+n} + 1, \quad \text{for all } m, n \in \mathbb{N}.$$

Recall that B_k is the k-th bellian number i.e., the number of all partitions of the set $\{1, 2, ..., k\}$, cf. [3].

4   The Syntactic Complexity of Eulerian Graphs

In this section we see that the set of directed Eulerian graphs (with multiple edges allowed) is not recognizable, while the set of directed graphs with Eulerian underlying graph is proved to be recognizable. Furthermore, we calculate the syntax complexity function of this graph language and we compare it with that of connected graphs. In what follows, we deal only with ordinary graphs i.e., all edges will be of rank $(1, 1)$.

We start with the necessary definitions. A directed trail inside a graph is an alternating sequence of nodes and edges $v_0, e_1, \ldots, e_n v_n$ where the edges align source to target, so that each node (except v_0) is the target of the preceding edge and (except v_n) the source of the subsequent one and also no edge is repeated. A directed Euler tour is a directed trail that contains all the edges of the graph and its origin and terminus are the same. A directed graph is Eulerian if it has a directed Euler tour.

A well known characterization for directed Eulerian graphs states that a connected directed graph is Eulerian if and only if the in-degree (i.e., the number of incoming edges) of every node equals its out-degree (i.e., the number of outgoing edges). With the help of this result we prove the following.

Proposition 2. *The set $E(\Sigma)$ of Eulerian graphs over a doubly ranked alphabet $\Sigma = \Sigma_{1,1}$ is not recognizable.*

Proof. First a notation, for a node v of a graph F we denote by $[v]$ the difference between the number of the incoming and the outgoing edges of v. Thus, a graph F is Eulerian if and only if it is connected and it holds $[v] = 0$ for every node v of F. Given a connected graph $F \in GR_{1,1}(\Sigma)$ let v_1 be the begin node and v_2 the end node of F. Then there exists a $(1,1)$-context ω such that $\omega[F] \in E(\Sigma)$, if and only if

$$[v] = 0, \quad \text{for all } v \neq v_1, v_2 \quad \text{and} \quad [v_1] = -[v_2].$$

Indeed, in order to turn F into an Eulerian graph we only have to add $[v_1]$ edges from v_1 to v_2 (or $[v_2]$ edges from v_2 to v_1). In any other case, F cannot be transformed into an Eulerian graph since we cannot add any edges to the nodes v ($v \neq v_1, v_2$).

Now let $F, F' \in GR_{1,1}(\Sigma)$ be two different connected graphs with the above property and also $[v_1] \neq [v_1']$. Hence, for the two graphs it holds $C_{E(\Sigma)}(F) \neq C_{E(\Sigma)}(F')$. This way we can construct infinitely many different syntactic classes of $E(\Sigma)$ at the rank $(1,1)$, and by virtue of Theorem 1 we get that this language is not recognizable. □

Notice that in the above proposition the graphs of $E(\Sigma)$ are allowed to have multiple edges and that is a critical point of the proof.

We proceed by studying the set of directed graphs with Eulerian underlying graph. The underlying graph of a directed graph is the graph obtained by replacing every directed edge by an undirected one. Thus, an undirected trail is an alternating sequence of nodes and edges $v_0, e_1, \ldots, e_n v_n$ where the edges align source to target or target to source and also no edge is repeated. An undirected Euler tour is an undirected trail that contains all the edges of the graph and its origin and terminus are the same. A directed graph has an Eulerian underlying graph if it has an undirected Euler tour.

It is known that a connected graph is Eulerian if and only if every node has an even degree. Thus, a connected directed graph has an Eulerian underlying graph if and only if every node has an even degree (the degree in this case is the sum of the numbers of the incoming and the outgoing edges of the node). By virtue of this characterization we shall prove the next theorem.

Theorem 3. *The set $EU(\Sigma)$ of directed graphs with an Eulerian underlying graph over an alphabet $\Sigma = \Sigma_{1,1}$ is recognizable. The syntax complexity function of this set is*

$$SC_{EU(\Sigma)}(m,n) = 1 + \sum \binom{m+n}{k} S(k)$$

where the sum runs over all evens $0 \leq k \leq m+n$ and $S(k) =$

$$\sum \frac{k!}{(k_1!)^{s_1} \cdot s_1! \cdot (k_2!)^{s_2} \cdot s_2! \cdots (k_t!)^{s_t} \cdot s_t!} \binom{m+n-k}{r} (s_1 + \cdots + s_t)^r B_{m+n-k-r}$$

where the sum runs over all decompositions $s_1 \cdot k_1 + \cdots + s_t \cdot k_t = k$, with $0 \leq k_1, \ldots, k_t \leq k$ even numbers, and over all $0 \leq r \leq m+n-k$.

Proof. For $m, n \in \mathbb{N}$ we should calculate the distinct syntactic classes $C_{EU(\Sigma)}(F)$, with $F \in GR_{m,n}(\Sigma)$. Let $v_1, v_2, \ldots, v_m$ and $u_1, u_2, \ldots, u_n$ be respectively the m begin and n end nodes of F. Each one of these nodes has either an even or odd degree. Notice that if the degree of any one of the rest nodes is different from zero then $C_{EU(\Sigma)} = \emptyset$, since the degree of these nodes cannot be changed. Hence, we may assume that the degree of all $v \neq v_1, v_2, \ldots, v_m, u_1, u_2, \ldots, u_n$ is even. Moreover, the graph F consists of one or more connected components, thus we divide the begin and end nodes of F into distinct classes according to which connected component they belong. We also observe that each connected component can have an even number of nodes with an odd degree since the sum of all the degrees of its nodes should be an even number (it is twice the number of its edges). For the same reason, the total number of odd degrees in the begin and end nodes of the graph F should also be an even number.

Now we see that for different distributions of even and odd degrees to the nodes $v_1, \ldots, v_m, u_1, u_2, \ldots, u_n$ of F and different divisions of them into classes, with an even number of nodes with an odd degree in each class, we get different syntactic classes. Hence, in order to calculate the number of different syntactic classes of an (m, n)-graph we should count the different ways we can assign even labels and (an even number of) odd labels into $m + n$ objects and then multiply this with the number of different divisions of these $m + n$ objects into classes that have an even number of objects with odd labels.

The first task consists in calculating the sum of the number of ways we can choose k objects, with $k = 0, 2, \ldots, m + n$ if k is an even number, or $k = 0, 2, \ldots, m + n - 1$ if k is odd, from a set of $m + n$ objects. Hence, there are

$$\binom{m+n}{0} + \binom{m+n}{2} + \cdots + \binom{m+n}{m+n}$$

or

$$\binom{m+n}{0} + \binom{m+n}{2} + \cdots + \binom{m+n}{m+n-1}$$

ways we can do that.

Next, for the second task, suppose that we have assigned k odd labels (k even) and of course the rest of the $m + n$ objects are labelled even. First we count the number of ways we can divide the set of k distinct objects into classes with an even number of elements. This can be done in

$$\sum \frac{k!}{(k_1!)^{s_1} \cdot s_1! \cdot (k_2!)^{s_2} \cdot s_2! \cdots (k_t!)^{s_t} \cdot s_t!}$$

different ways, where the sum runs over all decompositions $s_1 \cdot k_1 + \cdots + s_t \cdot k_t = k$, with $0 \leq k_1, \ldots, k_t \leq k$ even numbers. For example, for $k = 6$, we have

$$\frac{6!}{(2!)^3 \cdot 3!} + \frac{6!}{4! \cdot 2!} + \frac{6!}{6!} = 31.$$

Now, we need to distribute the rest $m + n - k$ objects, with the even labels, either in these $s_1 + \cdots + s_t$ sets or into an arbitrary number of new ones. Suppose that

r $(0 \leq r \leq m+n-k)$ of these $m+n-k$ objects are going into the old sets and hence $m+n-k-r$ are partitioned into new sets. Then for each one of the

$$\frac{k!}{(k_1!)^{s_1} \cdot s_1! \cdot (k_2!)^{s_2} \cdot s_2! \cdots (k_t!)^{s_t} \cdot s_t!}$$

divisions of k into $s_1 + \cdots + s_t$ sets there are

$$(s_1 + \cdots + s_t)^r$$

ways that we can distribute the r objects. Furthermore, there are $B_{m+n-k-r}$ ways that we can partition the set of $m+n-k-r$ objects. Altogether now, the second task can be done in $S(k) =$

$$\sum \frac{k!}{(k_1!)^{s_1} \cdot s_1! \cdot (k_2!)^{s_2} \cdot s_2! \cdots (k_t!)^{s_t} \cdot s_t!} \binom{m+n-k}{r} (s_1 + \cdots + s_t)^r B_{m+n-k-r}$$

ways, where the sum runs over all decompositions $s_1 \cdot k_1 + \cdots + s_t \cdot k_t = k$, with $0 \leq k_1, \ldots, k_t \leq k$ even numbers and over all $0 \leq r \leq m+n-k$. Notice that for $k = 0$, i.e., the case where all labels are even, it holds $S(0) = B_{m+n}$, which represents the number of partitions of the $m+n$ objects with the even labels.

Taking into account the above calculations, the different syntactic classes at the rank (m, n) are

$$1 + \sum \binom{m+n}{k} S(k)$$

where the sum runs over all evens $0 \leq k \leq m+n$, and we also added the empty class $(C_{EU(\Sigma)} = \emptyset)$.   □

Remark 1. The intuitive expectation that the set of graphs with an Eulerian underlying graph should be syntactically more complicated than that of connected graphs is validated from the above result. Indeed, the formula of $SC_{EU(\Sigma)}(m, n)$ for $k = 0$ is equal with $B_{m+n} + 1$, which is the syntax complexity of the set of connected graphs at the rank (m, n). Hence, it holds,

$$SC_{Con(\Sigma)}(m, n) \leq SC_{EU(\Sigma)}(m, n), \qquad \text{for all } m, n \in \mathbb{N},$$

where the equality holds only for $m + n < 2$.

References

1. Arnold, A., Dauchet, M.: Théorie des magmoides. I. RAIRO Inform. Théor. 12(3), 235–257 (1978)
2. Arnold, A., Dauchet, M.: Théorie des magmoides. II. RAIRO Inform. Théor. 13(2), 135–154 (1979)
3. Bell, E.T.: Exponential numbers. Amer. Math. Monthly 41, 411–419 (1934)
4. Bozapalidis, S., Kalampakas, A.: An Axiomatization of Graphs. Acta Informatica 41, 19–61 (2004)

5. Bozapalidis, S., Kalampakas, A.: Recognizability of graph and pattern languages. Acta Informatica 42, 553–581 (2006)
6. Engelfriet, J., Vereijken, J.J.: Context-free graph grammars and concatenation of graphs. Acta Informatica 34, 773–803 (1997)
7. Hotz, G.: Eine Algebraisierung des Syntheseproblems von Schaltkreisen. EIK 1, 185-205, 209-231 (1965)
8. MacLane, S.: Categories for the working mathematician. Springer, Heidelberg (1971)

Learning Deterministically Recognizable Tree Series — Revisited

Andreas Maletti

Institute of Theoretical Computer Science, Faculty of Computer Science
Technische Universität Dresden
maletti@tcs.inf.tu-dresden.de

Abstract. We generalize a learning algorithm originally devised for deterministic all-accepting weighted tree automata (wta) to the setting of arbitrary deterministic wta. The learning is exact, supervised, and uses an adapted minimal adequate teacher; a learning model introduced by Angluin. Our algorithm learns a minimal deterministic wta that recognizes the taught tree series and runs in polynomial time in the size of that wta and the size of the provided counterexamples. Compared to the original algorithm, we show how to handle non-final states in the learning process; this problem was posed as an open problem in [Drewes, Vogler: *Learning Deterministically Recognizable Tree Series*, J. Autom. Lang. Combin. 2007].

1 Introduction

We devise a supervised learning algorithm for deterministically recognizable tree series. Learning algorithms for formal languages have a long and studied history (see the seminal and survey papers [16,1,3,4]). The seminal paper [16] reports first results on identification in the limit; in particular it shows that every recursively enumerable language can be learned from a teacher. Here we study series; quantitative versions of languages. In particular, a tree series associates to each tree of T_Σ (the set of all well-formed expressions over the ranked alphabet Σ) a coefficient. Thus, it is nothing else than a mapping $\psi\colon T_\Sigma \to A$ for some suitable set A. It depends on A whether the coefficient represents, e.g., a probability, a count, a string, etc. For the moment, we assume that $(A, +, \cdot, 0, 1)$ is a field. A tree language $L \subseteq T_\Sigma$ can then be identified with the tree series that maps the elements of L to 1 and remaining elements of T_Σ to 0.

Angluin [2] proposed *query learning*, a model of interactive learning. In this learning model, the learner can question a teacher (or oracle). The teacher will answer predetermined types of questions. For example, the *minimally adequate teacher* [2,12] for a tree series $\psi\colon T_\Sigma \to A$ answers only two types of questions about ψ: *coefficient* and *equivalence* queries. A coefficient query asks for the coefficient of a certain tree t in the tree series ψ. The teacher truthfully supplies $\psi(t)$. Second, the learner can query the teacher whether his learned tree series φ coincides with ψ. The teacher either returns the special token $\perp$ to signal equality

S. Bozapalidis and G. Rahonis (Eds.): CAI 2007, LNCS 4728, pp. 218–235, 2007.
© Springer-Verlag Berlin Heidelberg 2007

(i.e., $\varphi = \psi$) or he supplies a counterexample. Such a counterexample is a tree t on which φ and ψ disagree (i.e., $\varphi(t) \neq \psi(t)$).

Certainly, we need to be able to finitely represent the learned tree series. To this end, we use an automaton model called *(bottom-up) weighted tree automaton* (for short: wta; see [8] and the references therein). These devices are classical bottom-up tree automata [14,15] with transition weights. The weights are elements of A and are combined using the operations $+$ and $\cdot$ of the field (see Definition 3). In [17], a learning algorithm based on the introduced minimally adequate teacher is presented for wta over fields. Here we will restrict ourselves to deterministic wta [8] and their recognized series, which are called deterministically recognizable. Since no general determinization procedure for wta over fields is known, this task is not encompassed by the result of [17].

For deterministic wta over fields (actually, semifields), the learning algorithm [12] was proposed. It is based on a restricted MYHILL-NERODE theorem [12] for the series recognized by deterministic *all-accepting* (i.e., all states are final) wta (for short: aa-wta). Consequently, this algorithm learns the minimal deterministic aa-wta that recognizes ψ (which is unique up to renaming of states) provided that any deterministic aa-wta recognizing ψ exists. We extend this algorithm to arbitrary deterministic wta and solve the open problem of [12].

Let us discuss the main differences. First, an aa-wta M makes no distinction between final and non-final states because all of its states are final. In essence, the internal working of M is completely exposed to the outside. It yields that the recognized series ψ is *subtree-closed* [12]. This property demands that with every tree t such that $\psi(t) \neq 0$, also all of its subtrees are mapped (under ψ) to some nonzero weight. With this property, the weight of the last transition that is used to accept $t = \sigma(t_1, \ldots, t_k)$ can simply be computed as $\psi(t) \cdot \prod_{i=1}^{k} \psi(t_i)^{-1}$ (see Definition 11). Consequently, the minimal deterministic aa-wta recognizing ψ is unique (up to renaming of states).

On the contrary, there exists no unique minimal deterministic wta recognizing ψ because the weights on transitions that lead to non-final states can be varied (occasionally called *pushing* [13]). In summary, we need to (i) distinguish final and non-final states and (ii) use a more complicated mechanism to compute the transition weights (because some $\psi(t_i)$ might be 0 in the above expression). The basis for our generalized learning will be the general MYHILL-NERODE theorem [5], which provides a characterization of the deterministically recognizable tree series by means of finite-index congruences of the initial term algebra (T_Σ, Σ). We then follow the approach of [5] and introduce a helping tree series (see Definition 10). The exact changes to the learning algorithm are discussed in the main body. Our new algorithm runs in time $\mathcal{O}(sm^2 nr)$ where s is the size of the largest counterexample supplied by the teacher, m and n are the number of transitions and the number of states of the returned automaton, respectively, and r is the maximal rank of the input symbols.

Including this Introduction, the paper comprises 6 sections. The second section recalls basic notions and notations. In the next section, we recall wta and the MYHILL-NERODE theorem [5]. In Sect. 4, we present the main contribution of this paper, which is the generalized learning algorithm. Moreover, we prove its correctness and continue in Sect. 5 with an elaborated example run of the algorithm. In the last section, we discuss the runtime complexity of our new algorithm and compare it to the learning algorithm of [12].

2   Preliminaries

We write $\mathbb{N}$ to represent the nonnegative integers. Further, we write $[l, u]$ for $\{n \in \mathbb{N} \mid l \leqslant n \leqslant u\}$. Any nonempty and finite set Σ is an *alphabet*. A *ranked alphabet* is a partition $(\Sigma_k)_{k \in \mathbb{N}}$ of an alphabet Σ. For every ranked alphabet $\Sigma = (\Sigma_k)_{k \in \mathbb{N}}$, the set of Σ-*trees*, denoted by T_Σ, is inductively defined to be the smallest set T such that for every $\sigma \in \Sigma_k$ and $t_1, \ldots, t_k \in T$ also $\sigma(t_1, \ldots, t_k) \in T$. We write α instead of $\alpha()$ if $\alpha \in \Sigma_0$. Given a set $T \subseteq T_\Sigma$, the set $\{\sigma(t_1, \ldots, t_k) \mid \sigma \in \Sigma_k, t_1, \ldots, t_k \in T\}$ is denoted by $\Sigma(T)$. The size of a tree $t \in T_\Sigma$, denoted by $\mathrm{size}(t)$, is the number of occurrences of symbols of Σ in t. Let $\Box \notin \Sigma$ be a distinguished nullary symbol. Let $\Sigma'_k = \Sigma_k$ for every $k > 0$ and $\Sigma'_0 = \Sigma_0 \cup \{\Box\}$. A Σ-*context* c is a tree of $T_{\Sigma'}$ such that $\Box$ occurs exactly once in c. The set of all Σ-contexts is denoted by C_Σ, and we write $c[t]$ for the tree that is obtained by replacing in $c \in C_\Sigma$ the occurrence of $\Box$ with $t \in T_\Sigma$.

Let $\equiv$ be an equivalence on a set S. We write $[s]_\equiv$ for the equivalence class of $s \in S$ and $(S/\equiv)$ for $\{[s]_\equiv \mid s \in S\}$. We drop the subscript from $[s]_\equiv$ if $\equiv$ is clear. Finally, if $S = T_\Sigma$, then $\equiv$ is a *congruence* if for every $\sigma \in \Sigma_k$ and $t_1, \ldots, t_k, u_1, \ldots, u_k \in T_\Sigma$ such that $t_i \equiv u_i$ for every $i \in [1, k]$ also $\sigma(t_1, \ldots, t_k) \equiv \sigma(u_1, \ldots, u_k)$.

A *(commutative) semiring* is an algebraic structure $(A, +, \cdot, 0, 1)$ comprising two commutative monoids $(A, +, 0)$ and $(A, \cdot, 1)$ such that $\cdot$ distributes over $+$ and 0 is absorbing for $\cdot$. A semiring $(A, +, \cdot, 0, 1)$ is a *semifield* if for every $a \in A \setminus \{0\}$ there exists an $a^{-1} \in A$ such that $a \cdot a^{-1} = 1$. A *tree series* ψ is a mapping $\psi \colon T_\Sigma \to A$; the set of all such mappings is denoted by $\mathcal{A}\langle\!\langle T_\Sigma \rangle\!\rangle$. Given $t \in T_\Sigma$, we denote $\psi(t)$ also by (ψ, t). The HADAMARD product of two tree series $\psi, \varphi \in \mathcal{A}\langle\!\langle T_\Sigma \rangle\!\rangle$ is denoted by $\psi \cdot \varphi$ and given by $(\psi \cdot \varphi, t) = (\psi, t) \cdot (\varphi, t)$ for every $t \in T_\Sigma$. Finally, a series $\psi \in \mathcal{A}\langle\!\langle T_\Sigma \rangle\!\rangle$ is *subtree-closed* if for every $t \in T_\Sigma$ with $(\psi, t) \neq 0$ also $(\psi, u) \neq 0$ for every subtree u of t.

3   Weighted Tree Automaton

In this section, we recall from [8,5] the central notions of this contribution: deterministic weighted tree automata (wta) and deterministically recognizable tree series. For the rest of the paper, let $\mathcal{A} = (A, +, \cdot, 0, 1)$ be a commutative semifield; in examples we will use the field $\mathbb{R} = (\mathbb{R}, +, \cdot, 0, 1)$ of real numbers. In Sect. 4 we show how to learn a deterministic wta from a teacher using the characterization given by the MYHILL-NERODE theorem [5].

Definition 1 (see [8, Definitions 3.1 and 3.3]). *A weighted tree automaton* M *is a tuple* $(Q, \Sigma, \mathcal{A}, F, \mu)$ *with*

- *a finite set* Q *of* states;
- *a ranked alphabet* Σ *of* input symbols;
- *a set* $F \subseteq Q$ *of* final states; *and*
- *a tree representation* $(\mu_k)_{k \in \mathbb{N}}$ *such that* $\mu_k \colon \Sigma_k \to \mathcal{A}^{Q^k \times Q}$.

We call M *(bottom-up)* deterministic *if for every symbol* $\sigma \in \Sigma_k$ *and* $w \in Q^k$ *there exists at most one* $q \in Q$ *such that* $\mu_k(\sigma)_{w,q} \neq 0$.

Note that a wta model with final weights (i.e., with $F \colon Q \to \mathcal{A}$ instead of $F \subseteq Q$) is considered in [6]. However, for every deterministic "final weight" wta an equivalent deterministic wta can be constructed [7, Lemma 6.1.4]. Instead of $\mu_0(\alpha)_{\varepsilon,q}$ with $\alpha \in \Sigma_0$ and ε the empty word, we commonly write $\mu_0(\alpha)_q$.

Let us present our running-example wta. It is supposed to assign a probability to (simplified) syntax trees of simple English sentences. If the tree is ill-formed, then the assigned probability shall be 0. This shall signal that it is rejected. Moreover, the probability shall diminish with the length of the input sentence.

Example 2. Let $\Sigma = (\Sigma_k)_{k \in \mathbb{N}}$ with $\bigcup_{k \in \mathbb{N} \setminus \{0,2\}} \Sigma_k = \emptyset$ and $\Sigma_2 = \{\sigma\}$ and $\Sigma_0 = \{\text{Alice, Bob, loves, hates, ugly, nice, mean, tall}\}$. Moreover, let $(Q, \Sigma, \mathbb{R}, F, \mu)$ be the deterministic wta with $\{\text{NN, VB, ADJ, NP, VP, S}\}$ as set Q of states and $F = \{\text{S}\}$ and the nonzero tree representation entries

$$0.5 = \mu_0(\text{Alice})_{\text{NN}} = \mu_0(\text{Bob})_{\text{NN}} = \mu_0(\text{loves})_{\text{VB}} = \mu_0(\text{hates})_{\text{VB}}$$
$$0.25 = \mu_0(\text{ugly})_{\text{ADJ}} = \mu_0(\text{nice})_{\text{ADJ}} = \mu_0(\text{mean})_{\text{ADJ}} = \mu_0(\text{tall})_{\text{ADJ}}$$
$$0.5 = \mu_2(\sigma)_{\text{NN VP,S}} = \mu_2(\sigma)_{\text{NP VP,S}} = \mu_2(\sigma)_{\text{VB NN,VP}} = \mu_2(\sigma)_{\text{VB NP,VP}}$$
$$0.5 = \mu_2(\sigma)_{\text{ADJ NN,NP}} = \mu_2(\sigma)_{\text{ADJ NP,NP}} \; . \qquad\blacksquare$$

In the sequel, we will sometimes abbreviate the nullary symbols used in Example 2 to just their initial letter. Let us continue with the semantics of wta.

Definition 3 (see [8, Definition 3.3]). *Let* $M = (Q, \Sigma, \mathcal{A}, F, \mu)$ *be a wta. The mapping* $h_\mu \colon T_\Sigma \to \mathcal{A}^Q$ *is given by*

$$h_\mu(\sigma(t_1, \ldots, t_k))_q = \sum_{q_1 \cdots q_k \in Q^k} \mu_k(\sigma)_{q_1 \cdots q_k, q} \cdot h_\mu(t_1)_{q_1} \cdot \ldots \cdot h_\mu(t_k)_{q_k}$$

for every $\sigma \in \Sigma_k$, $q \in Q$, *and* $t_1, \ldots, t_k \in T_\Sigma$. *The tree series that is* recognized *by* M, *denoted by* $S(M)$, *is defined for every* $t \in T_\Sigma$ *by* $(S(M), t) = \sum_{q \in F} h_\mu(t)_q$.

We note that deterministic wta do not essentially use the additive operation. A tree series $\psi \in \mathcal{A}\langle\!\langle T_\Sigma \rangle\!\rangle$ is *deterministically recognizable* if there exists a deterministic wta M such that $S(M) = \psi$. Let us illustrate the definition of the semantics on a small example.

Example 4. Recall the deterministic wta M of Example 2. Then

$$(S(M), \sigma(\text{Alice}, \sigma(\text{loves}, \text{Bob}))) = 3.125 \cdot 10^{-2}$$
$$(S(M), \sigma(\sigma(\text{mean}, \text{Bob}), \sigma(\text{hates}, \sigma(\text{ugly}, \text{Alice})))) = 4.8828125 \cdot 10^{-4}$$
$$(S(M), \sigma(\sigma(\text{Alice}, \text{loves}), \text{Bob})) = 0 \ .$$

Let us illustrate the computation of the last coefficient. To this end, let $t = \sigma(\sigma(A, l), B)$. Since S is the only final state of M, we obtain that $(S(M), t) = h_\mu(t)_S$. We continue with

$$h_\mu(\sigma(\sigma(A, l), B))_S$$
$$= \sum_{q_1 q_2 \in Q^2} \mu_2(\sigma)_{q_1 q_2, S} \cdot h_\mu(\sigma(A, l))_{q_1} \cdot h_\mu(B)_{q_2}$$
$$= \sum_{q_1 \in Q} 0.5 \cdot h_\mu(\sigma(A, l))_{q_1} \cdot \mu_0(B)_{VP} = 0 \ .$$

We showed two parse trees for the sentence "Alice loves Bob". One of them is ill-formed and the other is assigned a positive probability. Thus, the sentence would not be considered ill-formed because a parse tree with nonzero weight exists. ∎

Let us conclude this section with the MYHILL-NERODE theorem [5] for deterministically recognizable tree series. Let $\psi \in A\langle\langle T_\Sigma \rangle\rangle$. The MYHILL-NERODE congruence relation $\equiv_\psi \subseteq T_\Sigma \times T_\Sigma$ is given by $t \equiv_\psi u$ if and only if there exists a coefficient $a \in A \setminus \{0\}$ such that $(\psi, c[t]) = a \cdot (\psi, c[u])$ for every $c \in C_\Sigma$. Finally, by L_ψ we denote $\{t \in T_\Sigma \mid \forall c \in C_\Sigma \colon (\psi, c[t]) = 0\}$.

Theorem 5 (see [5, Theorem 2]). *A tree series $\psi \in A\langle\langle T_\Sigma \rangle\rangle$ is deterministically recognizable if and only if $\equiv_\psi$ has finite index. Moreover, every minimal deterministic wta recognizing ψ has* $\text{card}((T_\Sigma \setminus L_\psi)/\equiv_\psi)$ *states.*

4 Learning Algorithm

Next, we show how to learn a minimal deterministic wta for a given deterministically recognizable tree series ψ with the help of a teacher. To this end, we now fix a tree series $\psi \in A\langle\langle T_\Sigma \rangle\rangle$. Let us clarify the role of the teacher. He is able to answer two types of questions:

1. *Coefficient queries:* Given $t \in T_\Sigma$, the teacher supplies (ψ, t).
2. *Equivalence queries:* Given a wta M, he answers whether $S(M) = \psi$. If so, he returns the special token $\bot$. Otherwise he returns a counterexample; i.e., some tree $t \in T_\Sigma$ such that $(S(M), t) \neq (\psi, t)$.

This straightforward adaptation of the *minimally adequate teacher* [2] was proposed in [12] and is based on the adaptation for tree languages [9,11]. Equivalence queries might be considered unrealistic in a fully automatic setting and might there be replaced by tests that check a predetermined number of trees in applications. We will, however, not investigate the ramifications of this approximation.

At this point, we will only note that equivalence of deterministic wta is decidable [6]. So in the particular case, that the teacher uses a deterministic wta to represent ψ, both types of queries can automatically be answered.

The following development is heavily inspired by the learning algorithm devised in [12]; in its turn an extension of the learning algorithm of [11] to deterministic *all-accepting* [12] wta. It was argued in [12] that the all-accepting property is no major restriction because any deterministically recognizable tree series ψ can be presented as the HADAMARD product of a series ψ' recognized by a deterministic all-accepting wta and a series ψ'' recognized by a deterministic BOOLEAN (i.e., only weights 0 and 1) wta (for the latter class, learning algorithms are known [9,11]).

Let us discuss the problems of this approach. First, the decomposition is not unique; in general, we need to guess coefficients in ψ' (namely the ones where ψ is 0). The guessed coefficients affect the size of the minimal deterministic wta recognizing ψ'. Second, we learn minimial deterministic wta M' and M'' recognizing ψ' and ψ'', respectively, however, the HADAMARD product of M' and M'' is not necessarily a minimal deterministic wta recognizing $\psi = \psi' \cdot \psi''$. Third, we run two very similar algorithms and then perform a HADAMARD product construction; this is most likely not the most efficient solution.

The first problem (the completion of ψ to a subtree-closed ψ') can indeed be easily solved, provided that a representation of ψ by a deterministic wta is available (on the other hand, provided that a representation as deterministic wta is available, we could also just minimize the available representation). If no such representation is available, then the problem is far more complicated, and we will now show that very simple completions can even lead to deterministically non-recognizable tree series.

Example 6. Recall the wta M from Example 2. Clearly, $S(M)$ is not yet subtree-closed, so we complete it to $\psi' \in \mathbb{R}\langle\langle T_\Sigma \rangle\rangle$ (cf. Definition 10) by

$$(\psi', t) = \begin{cases} (S(M), t) & \text{if } (S(M), t) \neq 0 \\ 1 & \text{otherwise} \end{cases}$$

for every $t \in T_\Sigma$. We consider trees of the form $\sigma(\text{m}, \sigma(\text{m}, \ldots \sigma(\text{m}, \text{B}) \ldots))$. For every $n \in \mathbb{N}$, let t_n be the such obtained tree with n occurrences of m. Clearly, $(\psi', t_n) = 1$ for every $n \in \mathbb{N}$. Thus, for every $i, j \in \mathbb{N}$, we have $t_i \equiv_{\psi'} t_j$ if and only if $(\psi', c[t_i]) = (\psi', c[t_j])$ for every $c \in C_\Sigma$ because $(\psi', t_i) = (\psi', t_j)$. Now, we consider the context $c = \sigma(\text{A}, \sigma(1, \square))$. An easy computation shows that $(\psi', c[t_n]) = 0.5^5 \cdot (0.5 \cdot 0.25)^n$ for every $n \in \mathbb{N}$. Consequently, $t_i \not\equiv_{\psi'} t_j$ whenever $i \neq j$. Consequently, $\equiv_{\psi'}$ has infinite index and thus ψ' is not deterministically recognizable by Theorem 5. ∎

Our main contribution is a slightly modified learning algorithm that is not restricted to deterministic *all-accepting* wta. To this end, we first define a restriction of the MYHILL-NERODE congruence [5]. Henceforth, we will drop the index ψ from $\equiv_\psi$ and L_ψ.

Definition 7 (cf. [5, Sect. 5]). *Let $C \subseteq C_\Sigma$. The relation $\equiv_C$ contains all $(t, u) \in T_\Sigma \times T_\Sigma$ for which there exists an $a \in A \setminus \{0\}$ such that for every context $c \in C$ the equality $(\psi, c[t]) = a \cdot (\psi, c[u])$ holds.*

Clearly, $\equiv_C$ is an equivalence for every $C \subseteq C_\Sigma$. Moreover, the relation $\equiv_{C_\Sigma}$ coincides with the MYHILL-NERODE congruence [5] and for every $t, u \in T_\Sigma$ and $c \in C_\Sigma$ it holds that $t \equiv_{\{c\}} u$ if and only if $(\psi, c[t]) \neq 0$ precisely when $(\psi, c[u]) \neq 0$ (cf. Condition (MN2) in [5]). In particular, for $c = \square$, we have that $t \equiv_{\{\square\}} u$ if and only if both (ψ, t) and (ψ, u) are nonzero or both zero. Consequently, the context $\square$ will allow us to distinguish final and non-final states. Finally, let

$$L_C = \{t \in T_\Sigma \mid \forall c \in C \colon (\psi, c[t]) = 0\}$$

for every $C \subseteq C_\Sigma$. Note that $L = L_{C_\Sigma}$.

An important observation is that there exists a finite set C of contexts such that $\equiv_C$ and $\equiv$ coincide, if $\equiv$ has finite index. Moreover, we note that for every $C \subseteq C_\Sigma$ the index of $\equiv_C$ is at most as large as the index of $\equiv$. Consequently, if $\equiv$ has finite index, then also $\equiv_C$ has finite index. Our learning strategy is to learn a set C of contexts such that $\equiv_C$ and $\equiv$ coincide. Next, we present our main data structure.

Definition 8 (cf. [12, Definition 4.3]). *We call a triple (E, T, C) an observation table if*

1. *E and T are finite subsets of T_Σ such that $E \subseteq T \subseteq \Sigma(E)$;*
2. *C is a subset of C_Σ with $\square \in C$ and $\mathrm{card}(C) \leq \mathrm{card}(E) + \mathrm{card}(T) + 1$;*
3. *$T \cap L_C = \emptyset$; and*
4. *$\mathrm{card}(E) = \mathrm{card}(E/\equiv_C)$.*

If, additionally, $\mathrm{card}(E) = \mathrm{card}(T/\equiv_C)$, then we call (E, T, C) complete.

The only major difference to [12] is found in Condition 2. First, the presence of the context $\square$ in C basically enables us to distinguish final and non-final states. There is no need for $\square$ in [12] because all states will be final. Second, we changed the size restriction on C from $\mathrm{card}(C) \leq \mathrm{card}(E)$ (as in [12]) to $\mathrm{card}(C) \leq \mathrm{card}(E) + \mathrm{card}(T) + 1$. In [12], for every $e, e' \in E$, the coefficient a of Definition 7 (required to show that $e \equiv_C e'$) can always be determined with the help of the context $\square$. Clearly, $\mathrm{card}(E)$ contexts are then sufficient to separate the elements of E. In our more general setting, we cannot always determine the coefficient a of Definition 7 with the help of the context $\square$. Rather, the contexts of C shall not only separate the elements of E, but shall also serve as explicit evidence that no tree in T (and thus also in E) is in L_C. This evidence is needed to determine the right coefficient in Definition 7 and is, consequently, used in Definition 10 to fix the right weight.

The third condition encodes the avoidance of trees t such that no supertree of t can be accepted (dead states; see [12]). This condition is only checked for those contexts that we accumulated in C.

Proposition 9. *Let $L \neq \emptyset$, and let $C \subseteq C_\Sigma$ be such that $\equiv_C$ coincides with $\equiv$. Then $L_C = L$.*

Proof. The direction $L \subseteq L_C$ is trivial. We prove the remaining direction by contradiction. Let $t \in L_C \setminus L$. Thus, $(\psi, c[t]) = 0$ for every $c \in C$, and clearly, $t \equiv_C u$ for every $u \in L$. However, there exists a context $c \in C_\Sigma \setminus C$ such that $(\psi, c[t]) \neq 0$ because $t \notin L$. This yields that $t \not\equiv u$ for every $u \in L$. Thus, we can derive the contradiction that $\equiv_C$ and $\equiv$ do not coincide because $L \neq \emptyset$. □

The condition $L \neq \emptyset$ is necessary in the above statement because the partition induced by $\equiv$ (and thus also $\equiv_C$) does not distinguish between an equivalence class containing only one tree, which happens to be in L, and an equivalence class containing only one tree, which is not in L.

The fourth and completeness condition in Definition 8 are equivalent to: $e \not\equiv_C e'$ for every two distinct $e, e' \in E$, and for every $t \in T$ there exists an $e \in E$ such that $t \equiv_C e$, respectively. Clearly, such an element e is uniquely determined by the former condition. In the sequel, given a complete observation table $\mathcal{T} = (E, T, C)$ and $t \in T$ we write $\mathcal{T}(t)$ for the unique $e \in E$ such that $e \equiv_C t$. Clearly, $\mathcal{T}(e) = e$ for every $e \in E$ (see [12, Lemma 4.4]). Next we show how to construct a deterministic wta given a complete observation table. To achieve this, we modify the construction [5] of a deterministic wta from the MYHILL-NERODE congruence $\equiv$.

Definition 10 (cf. [5, Lemma 8 and Page 9]). *Let $\mathcal{T} = (E, T, C)$ be a complete observation table. Let $\psi(\mathcal{T}) : T_\Sigma \to A \setminus \{0\}$ be such that for every $t \in T_\Sigma$*

$$(\psi(\mathcal{T}), t) = \begin{cases} (\psi, t) & \text{if } (\psi, t) \neq 0 \\ (\psi, c[t]) \cdot (\psi, c[\mathcal{T}(t)])^{-1} & \text{if } (\psi, t) = 0 \text{ and } t \in T \text{ and} \\ & \quad (\psi, c[t]) \neq 0 \text{ for some } c \in C \\ 1 & \text{otherwise.} \end{cases}$$

Here, we only consider a baseline implementation; an efficient implementation could avoid many queries to the teacher [10] and store the required information in an extended observation table. Note that, for example, a suitable context for the second case in Definition 10 is observed by our algorithm (see Algorithms 1 and 3) when t is added to the observation table; it could thus be stored for efficient retrieval.

Some notes on the well-definedness of $\psi(\mathcal{T})$ are necessary. First, the condition $(\psi, t) \neq 0$ can be checked easily by a coefficient query. Second, $t \in T$ implies $t \notin L_C$ by the third condition of Definition 8. Thus, there trivially exists a context $c \in C$ such that $(\psi, c[t]) \neq 0$. It follows that $(\psi, c[\mathcal{T}(t)]) \neq 0$ because $t \equiv_C \mathcal{T}(t)$ and hence $t \equiv_{\{c\}} \mathcal{T}(t)$. Consequently, the inverse is well-defined. It remains to show that the result is independent of the selection of the context c. To this end, let $c' \in C$ be another context such that $(\psi, c'[t]) \neq 0$. Following the above argumentation, $(\psi, c'[\mathcal{T}(t)]) \neq 0$. Since $t \equiv_C \mathcal{T}(t)$, there exists a coefficient $a \in A \setminus \{0\}$ such that $(\psi, c''[t]) = a \cdot (\psi, c''[T(t)])$ for every $c'' \in C$. It follows that

$$(\psi, c[t]) \cdot (\psi, c[T(t)])^{-1} = a = (\psi, c'[t]) \cdot (\psi, c'[T(t)])^{-1} .$$

Definition 11 (cf. [5, Definition 4]). *Let $T = (E, T, C)$ be a complete observation table. We construct the wta $\mathcal{M}(T) = (E, \Sigma, A, F, \mu)$ such that*

- $F = \{e \in E \mid (\psi, e) \neq 0\}$;
- *for every $\sigma \in \Sigma_k$ and $e_1, \ldots, e_k \in E$ such that $\sigma(e_1, \ldots, e_k) \in T$*

$$\mu_k(\sigma)_{e_1 \cdots e_k, T(\sigma(e_1, \ldots, e_k))} = (\psi(T), \sigma(e_1, \ldots, e_k)) \cdot \prod_{i=1}^{k} (\psi(T), e_i)^{-1}$$

- *and all remaining entries in μ are 0.*

Let us immediately observe some properties of the constructed wta. Clearly, $\mathcal{M}(T)$ is deterministic. Moreover, $S(\mathcal{M}(T))$ coincides with ψ on all trees of T.

Lemma 12 (cf. [12, Lemma 4.5]). *Let $T = (E, T, C)$ be a complete observation table. Then $(S(\mathcal{M}(T)), t) = (\psi, t)$ for every $t \in T$.*

Proof. Suppose that $\mathcal{M}(T) = (E, \Sigma, A, F, \mu)$. We first prove that

$$h_\mu(t)_{T(t)} = (\psi(T), t) \tag{1}$$

for every $t \in T$. Let $t = \sigma(t_1, \ldots, t_k)$ for some $\sigma \in \Sigma_k$ and $t_1, \ldots, t_k \in E$. By the induction hypothesis, we have $h_\mu(t_i)_{T(t_i)} = (\psi(T), t_i)$ for every $i \in [1, k]$. Clearly, $h_\mu(t_i)_e = 0$ for all states $e \in E$ with $e \neq T(t_i)$ because $\mathcal{M}(T)$ is deterministic (see [8, Lemma 3.6]). Then

$$h_\mu(\sigma(t_1, \ldots, t_k))_{T(\sigma(t_1, \ldots, t_k))}$$

$$= \sum_{e_1, \ldots, e_k \in E} \mu_k(\sigma)_{e_1 \cdots e_k, T(\sigma(t_1, \ldots, t_k))} \cdot \prod_{i=1}^{k} h_\mu(t_i)_{e_i}$$

$$= \mu_k(\sigma)_{T(t_1) \cdots T(t_k), T(\sigma(t_1, \ldots, t_k))} \cdot \prod_{i=1}^{k} (\psi(T), t_i)$$

$$= \mu_k(\sigma)_{t_1 \cdots t_k, T(\sigma(t_1, \ldots, t_k))} \cdot \prod_{i=1}^{k} (\psi(T), t_i)$$

$$= (\psi(T), \sigma(t_1, \ldots, t_k)) \cdot \prod_{i=1}^{k} (\psi(T), t_i)^{-1} \cdot \prod_{i=1}^{k} (\psi(T), t_i)$$

$$= (\psi(T), \sigma(t_1, \ldots, t_k))$$

where the second equality is by the induction hypothesis; the third is due to the fact that $t_1, \ldots, t_k \in E$; and the fourth is by the definition of μ (see Definition 11).

Algorithm 1. Learn a minimal deterministic wta recognizing ψ

$\mathcal{T} \leftarrow (\emptyset, \emptyset, \{\Box\})$ {initial observation table}
2: **loop**
      $M \leftarrow \mathcal{M}(\mathcal{T})$ {construct new wta}
4:      $t \leftarrow$ EQUAL?(M) {ask equivalence query}
      if $t = \bot$ **then**
6:         **return** M {return the approved wta}
      else
8:         $\mathcal{T} \leftarrow$ EXTEND$(\mathcal{T}, t)$ {extend the observation table}

Thus, $h_\mu(t)_{\mathcal{T}(t)} \neq 0$. We now return to the main statement and complete the proof by distinguishing two cases: $(\psi, t) = 0$ and $(\psi, t) \neq 0$. In the former case, $(\psi, \mathcal{T}(t)) = 0$ because $\mathcal{T}(t) \equiv_C t$ and thus $\mathcal{T}(t) \equiv_{\{\Box\}} t$ (since $\Box \in C$). Consequently, $\mathcal{T}(t) \notin F$ and $(S(\mathcal{M}(\mathcal{T})), t) = 0$. In the latter case, an analoguous reasoning leads to $(\psi, \mathcal{T}(t)) \neq 0$ and $\mathcal{T}(t) \in F$. Consequently, $(S(\mathcal{M}(\mathcal{T})), t) = h_\mu(t)_{\mathcal{T}(t)} = (\psi(\mathcal{T}), t) = (\psi, t)$. $\qquad\square$

In Algorithm 1 we show the principal structure of the learner. The bulk of work is done in EXTEND, which is shown in Algorithm 3. We start with the initial empty observation table $(\emptyset, \emptyset, \{\Box\})$ and iteratively query the teacher for counterexamples and update our complete observation table with the returned counterexample. Once the teacher approves our wta, we simply return it. Clearly, the returned wta recognizes ψ because the teacher certifies this. In Sect. 5 we show an example application of the learning algorithm to learn the series recognized by the wta of Example 2. We say that an algorithm works *correctly* if whenever the pre-conditions (Require) are met at the beginning of the algorithm, then the algorithm terminates and the post-conditions (Ensure) hold at the point of return.

Theorem 13 (see [12, Theorem 5.4]). *Let us suppose that* EXTEND *works correctly and ψ is deterministically recognizable. Then Algorithm 1 terminates and returns a minimal deterministic wta recognizing ψ.*

Proof. Let ψ be deterministically recognizable. Then $\equiv$ has finite index by Theorem 5. Let $l = \mathrm{card}(T_\Sigma/\equiv)$. We already remarked that, for every $C \subseteq C_\Sigma$, the index of $\equiv_C$ is at most l. This yields that for every observation table (E, T, C) we have $\mathrm{card}(E) \leqslant l$ because

$$\mathrm{card}(E) = \mathrm{card}(E/\equiv_C) \leqslant \mathrm{card}(T_\Sigma/\equiv_C) \leqslant \mathrm{card}(T_\Sigma/\equiv) = l \ .$$

It is easily checked that EXTEND is always called with a complete observation table and a counterexample as parameters. Since $\mathrm{card}(E)$ and $\mathrm{card}(T)$ are bounded, there can only be finitely many calls to EXTEND. Thus, Algorithm 1 terminates. Moreover, the returned wta, say $\mathcal{M}(\mathcal{T})$, is approved by the teacher, so we have $S(\mathcal{M}(\mathcal{T})) = \psi$. By the construction of $\mathcal{M}(\mathcal{T})$, we know that $\mathcal{M}(\mathcal{T})$ has at most l states. Consequently, $\mathcal{M}(\mathcal{T})$ is a minimal deterministic wta recognizing ψ by Theorem 5. $\qquad\square$

Algorithm 2. The COMPLETE function

Require: an observation table (E, T, C)

Ensure: return a complete observation table (E', T, C) such that $E \subseteq E'$

    for all $t \in T$ **do**
2:      **if** $t \not\equiv_C e$ for every $e \in E$ **then**
          $E \leftarrow E \cup \{t\}$
4: **return** (E, T, C)

Next, we describe the functionality of COMPLETE, which is shown in Algorithm 2. This function takes an observation table (E, T, C) and returns a complete observation table (E', T, C) with $E \subseteq E'$. We simply check for every $t \in T$ whether there exists an $e \in E$ such that $t \equiv_C e$. If this is not the case, then we add t to E. It is clear that COMPLETE works correctly.

Finally, let us discuss the EXTEND function, which is shown in Algorithm 3. We search for a minimal subtree that is still a counterexample using a technique called *contradiction backtracking* [18]. Let $T = (E, T, C)$ be a complete observation table, $\mathcal{M}(T) = (E, \Sigma, \mathcal{A}, F, \mu)$ be the constructed wta, and $t \in T_\Sigma$ be a counterexample; i.e., a tree t such that $(S(\mathcal{M}(T)), t) \neq (\psi, t)$. We first decompose t into a context $c \in C_\Sigma$ and a tree u that is not in E but whose direct subtrees are all in E. In some sense, this is a minimal offending subtree because the wta works correctly on all trees of T by Lemma 12. Moreover, such a subtree must exist because t is a counterexample.

Now we distinguish two cases. If u was already seen (i.e., $u \in T$), then $u \equiv_C T(u)$. By Lemma 12, the wta $\mathcal{M}(T)$ works correctly on u. Thus the error is made when processing the context c. We test whether c separates u and $T(u)$. Provided that $u \equiv_{C \cup \{c\}} T(u)$, then u and $T(u)$ behave equally in all contexts of $C \cup \{c\}$ and we continue our search for the counterexample with $c[T(u)]$.

In all other cases, either u and $T(u)$ should be separated or u was not seen before (i.e., is not already present in T). In the latter case, $h_\mu(u)_e = 0$, and consequently, also $h_\mu(c[u])_e = 0$ for every $e \in E$ (see [8, Lemma 3.7]). Hence $(S(\mathcal{M}(T)), c[u]) = 0$ and $(\psi, c[u]) \neq 0$. Thus we claim that in $T' = (E, T \cup \{u\}, C \cup \{c\})$ is an observation table and return the completion of T'. If $u \notin T$, then $(\psi, c[u]) \neq 0$ and thus $u \notin L_{C \cup \{c\}}$. Moreover, $\equiv_{C \cup \{c\}} \subseteq \equiv_C$ and either we add u to T (if $u \notin T$) or we add u to E (if $u \in T$ but $u \not\equiv_{C \cup \{c\}} T(u)$). Thus the post-condition of the algorithm and the size restriction on the set of contexts are met.

The next lemma will rely on two straightforward lemmata; their proofs offer little insight and can thus be skipped on first reading.

Lemma 14 (see [5, Theorem 1]). *Let $M = (Q, \Sigma, \mathcal{A}, F, \mu)$ be a deterministic wta, and let $t, u \in T_\Sigma$ be such that $h_\mu(t)_p \neq 0$ and $h_\mu(u)_p \neq 0$ for some state $p \in Q$. Then for every context $c \in C_\Sigma$ and state $q \in Q$*

$$h_\mu(c[t])_q \cdot h_\mu(t)_p^{-1} = h_\mu(c[u])_q \cdot h_\mu(u)_p^{-1} \ .$$

Algorithm 3. The EXTEND function

Require: a complete observation table $\mathcal{T} = (E, T, C)$ and a counterexample $t \in T_\Sigma$

Ensure: return a complete observation table $\mathcal{T}' = (E', T', C')$
  such that $E \subseteq E'$ and $T \subseteq T'$ and one inclusion is strict

    Decompose t into $t = c[u]$ where $c \in C_\Sigma$ and $u \in \Sigma(E) \setminus E$
2: **if** $u \in T$ and $u \equiv_{C \cup \{c\}} \mathcal{T}(u)$ **then**
        return EXTEND$(\mathcal{T}, c[\mathcal{T}(u)])$ {normalize and continue}
4: **else**
        return COMPLETE$(E, T \cup \{u\}, C \cup \{c\})$ {add u and c}

Proof. We prove the statement by induction on the context c. Let $c = \square$. Then $h_\mu(c[t])_q = h_\mu(t)_q$ and $h_\mu(c[u])_q = h_\mu(u)_q$. We now distinguish two cases: (i) $q = p$ and (ii) $q \neq p$. In the former case, we immediately obtain

$$h_\mu(t)_p \cdot h_\mu(t)_p^{-1} = 1 = h_\mu(u)_p \cdot h_\mu(u)_p^{-1} \; .$$

In the latter case, $h_\mu(t)_q = 0 = h_\mu(u)_q$ because M is deterministic (see [8, Lemma 3.6]). Consequently,

$$h_\mu(t)_q \cdot h_\mu(t)_p^{-1} = 0 = h_\mu(u)_q \cdot h_\mu(u)_p^{-1} \; .$$

In the induction step we assume that $c = \sigma(t_1, \ldots, t_{i-1}, c', t_{i+1}, \ldots, t_k)$ for some $\sigma \in \Sigma_k$, context $c' \in C_\Sigma$, position $i \in [1, k]$, and $t_1, \ldots, t_k \in T_\Sigma$. Then

$$h_\mu(\sigma(t_1, \ldots, t_{i-1}, c', t_{i+1}, \ldots, t_k)[t])_q \cdot h_\mu(t)_p^{-1}$$
$$= h_\mu(\sigma(t_1, \ldots, t_{i-1}, c'[t], t_{i+1}, \ldots, t_k))_q \cdot h_\mu(t)_p^{-1}$$
$$= \Big(\sum_{q_1, \ldots, q_k \in Q} \mu_k(\sigma)_{q_1 \cdots q_k, q} \cdot h_\mu(c'[t])_{q_i} \cdot \prod_{i \in [1,k] \setminus \{i\}} h_\mu(t_i)_{q_i} \Big) \cdot h_\mu(t)_p^{-1}$$
$$= \Big(\sum_{q_1, \ldots, q_k \in Q} \mu_k(\sigma)_{q_1 \cdots q_k, q} \cdot h_\mu(c'[u])_{q_i} \cdot \prod_{i \in [1,k] \setminus \{i\}} h_\mu(t_i)_{q_i} \Big) \cdot h_\mu(u)_p^{-1}$$
$$= h_\mu(\sigma(t_1, \ldots, t_{i-1}, c'[u], t_{i+1}, \ldots, t_k))_q \cdot h_\mu(u)_p^{-1}$$
$$= h_\mu(\sigma(t_1, \ldots, t_{i-1}, c', t_{i+1}, \ldots, t_k)[u])_q \cdot h_\mu(u)_p^{-1}$$

where the third equality holds by the induction hypothesis and distributivity.
$\square$

Lemma 15. *Let $\mathcal{T} = (E, T, C)$ be an observation table, and let $t, u \in T$ be such that $t \equiv_C u$. For every $c \in C$*

$$(\psi, c[t]) \cdot (\psi(\mathcal{T}), t)^{-1} = (\psi, c[u]) \cdot (\psi(\mathcal{T}), u)^{-1} \; .$$

Proof. By $t \equiv_C u$ there exists an $a \in A \setminus \{0\}$ such that for every context $c' \in C$ we have $(\psi, c'[t]) = a \cdot (\psi, c'[u])$. Consequently,

$$(\psi, t) = a \cdot (\psi, u) \quad \text{and} \quad (\psi, c[t]) = a \cdot (\psi, c[u]) \; . \tag{2}$$

1. First, let $(\psi, c[t]) = 0$. By (2) also $(\psi, c[u]) = 0$, which proves the statement.
2. Second, let $(\psi, c[t]) \neq 0$ and $(\psi, t) \neq 0$. Then we can again conclude with the help of (2) that $(\psi, c[u]) \neq 0$ and $(\psi, u) \neq 0$. Further,

$$(\psi, c[t]) \cdot (\psi(\mathcal{T}), t)^{-1} = (\psi, c[t]) \cdot (\psi, t)^{-1} = (\psi, c[u]) \cdot (\psi, u)^{-1}$$
$$= (\psi, c[u]) \cdot (\psi(\mathcal{T}), u)^{-1}$$

where the second equality holds by (2).

3. Finally, let $(\psi, c[t]) \neq 0$ and $(\psi, t) = 0$. We again immediately note that $(\psi, c[u]) \neq 0$ and $(\psi, u) = 0$ by (2). Since $t, u \notin L_C$,

$$(\psi, c[t]) \cdot (\psi(\mathcal{T}), t)^{-1} = (\psi, c[t]) \cdot \left((\psi, c[t]) \cdot (\psi, c[\mathcal{T}(t)])^{-1} \right)^{-1}$$
$$= (\psi, c[\mathcal{T}(t)]) = (\psi, c[\mathcal{T}(u)])$$
$$= (\psi, c[u]) \cdot \left((\psi, c[u]) \cdot (\psi, c[\mathcal{T}(u)])^{-1} \right)^{-1} = (\psi, c[u]) \cdot (\psi(\mathcal{T}), u)^{-1}$$

where the third equality holds because $t \equiv_C u$.   □

Now we are ready with the two auxiliary lemmata. It remains to prove that the recursive call of EXTEND meets the pre-conditions of EXTEND. It is clear, that $\mathcal{T}$ is a complete observation table, but we need to prove that $c[\mathcal{T}(u)]$ is also a counterexample. This is achieved in the next lemma.

Lemma 16. Let $\mathcal{T} = (E, T, C)$ be a complete observation table, $u \in T$, and $c \in C_\Sigma$ such that $u \equiv_{C \cup \{c\}} \mathcal{T}(u)$. If $(S(\mathcal{M}(\mathcal{T})), c[u]) \neq (\psi, c[u])$, then also $(S(\mathcal{M}(\mathcal{T})), c[\mathcal{T}(u)]) \neq (\psi, c[\mathcal{T}(u)])$.

Proof. Let $\mathcal{M}(\mathcal{T}) = (E, \Sigma, \mathcal{A}, F, \mu)$. We distinguish two cases: First, let $h_\mu(c[u])_q = 0$ for every $q \in Q$. Then also $h_\mu(c[\mathcal{T}(u)])_q = 0$ for every $q \in Q$ because $\mathcal{M}(\mathcal{T})$ is deterministic and $h_\mu(u)_{\mathcal{T}(u)} \neq 0$ and $h_\mu(\mathcal{T}(u))_{\mathcal{T}(u)} \neq 0$ by Lemma 12 (see also Lemma 14). Clearly, $(S(\mathcal{M}(\mathcal{T})), c[u]) = 0$ and $(S(\mathcal{M}(\mathcal{T})), c[\mathcal{T}(u)]) = 0$. Consequently, $(\psi, c[u]) \neq 0$ and $(\psi, c[\mathcal{T}(u)]) \neq 0$ because $u \equiv_{C \cup \{c\}} \mathcal{T}(u)$. This proves the statement in the first case.

Second, let $q \in Q$ be such that $h_\mu(c[u])_q \neq 0$. Note that $h_\mu(u)_{\mathcal{T}(u)} \neq 0$ and $h_\mu(\mathcal{T}(u))_{\mathcal{T}(u)} \neq 0$ by Lemma 12. Then

$$(S(\mathcal{M}(\mathcal{T})), c[u]) \cdot h_\mu(u)_{\mathcal{T}(u)}^{-1} = \sum_{p \in \{q\} \cap F} h_\mu(c[u])_p \cdot h_\mu(u)_{\mathcal{T}(u)}^{-1}$$

$$= \sum_{p \in \{q\} \cap F} h_\mu(c[\mathcal{T}(u)])_p \cdot h_\mu(\mathcal{T}(u))_{\mathcal{T}(u)}^{-1}$$

$$= (S(\mathcal{M}(\mathcal{T})), c[\mathcal{T}(u)]) \cdot h_\mu(\mathcal{T}(u))_{\mathcal{T}(u)}^{-1} \qquad (3)$$

where the second equality is by Lemmata 12 and 14. We now reason as follows.

$$(S(\mathcal{M}(\mathcal{T})), c[\mathcal{T}(u)])$$
$$= (S(\mathcal{M}(\mathcal{T})), c[u]) \cdot h_\mu(\mathcal{T}(u))_{\mathcal{T}(u)} \cdot h_\mu(u)_{\mathcal{T}(u)}^{-1} \qquad \text{by (3)}$$

$$= (S(\mathcal{M}(\mathcal{T})), c[u]) \cdot (\psi(\mathcal{T}), \mathcal{T}(u)) \cdot (\psi(\mathcal{T}), u)^{-1} \qquad \text{by (1)}$$
$$\neq (\psi, c[u]) \cdot (\psi(\mathcal{T}), \mathcal{T}(u)) \cdot (\psi(\mathcal{T}), u)^{-1}$$
$$= (\psi, c[\mathcal{T}(u)]) \qquad\qquad\qquad\qquad\qquad \text{by Lemma 15} \qquad \square$$

The previous lemma justifies the recursive call of EXTEND. It remains to check whether the recursion terminates (see [12, Lemma 5.3]). For this we consider a call EXTEND($\mathcal{T}, t$) triggered in Line 8 of Algorithm 1. Since the recursive call of EXTEND also has $\mathcal{T}$ as parameter, we now fix a complete observation table $\mathcal{T} = (E, T, C)$ for all invocations of EXTEND that are triggered by the considered call EXTEND($\mathcal{T}, t$). Moreover, let $v : T_\Sigma \to \mathbb{N}$ be the mapping that assigns to every $u \in T_\Sigma$ the number of occurrences of subtrees of u that are not in E. Next we show that every call in our chain of invocations strictly decreases $v(t)$ where t is the second parameter of the call to EXTEND. Suppose we now consider the call EXTEND($\mathcal{T}, t$), and let $t = c[u]$ be the decomposition as given in Line 1 of Algorithm 3. Without regard of the occurrence of the recursive call to EXTEND, it is of the form EXTEND($\mathcal{T}, c[\mathcal{T}(u)]$). By Line 1 in Algorithm 3 we have $u \in \Sigma(E) \setminus E$. So $v(t) = \text{size}(c)$ and $v(c[\mathcal{T}(u)]) = \text{size}(c) - 1$ because $\mathcal{T}(u) \in E$ and E is trivially subtree-closed (i.e., if $e \in E$ then also all subtrees of e are in E). Thus the recursion must terminate and hence each call of EXTEND terminates.

Corollary 17 (of Theorem 13). *Provided that ψ is deterministically recognizable, Algorithm 1 terminates and returns a minimal deterministic wta recognizing ψ.*

5 An Example

Let us show, how the algorithm learns the tree series ψ recognized by the wta of Example 2. We start (Line 1) with the initial empty observation table $\mathcal{T}_0 = (\emptyset, \emptyset, \{\square\})$. The constructed (Line 3) wta $\mathcal{M}_0 = (\emptyset, \Sigma, \mathcal{A}, \emptyset, \mu)$ recognizes the tree series that maps every tree to 0. We have seen in Example 4 that $(\psi, \sigma(A, \sigma(1, B))) = 3.125 \cdot 10^{-2}$, so suppose the equivalence query (Line 4) is answered with $t_1 = \sigma(A, \sigma(1, B))$. Consequently, we will call EXTEND($\mathcal{T}_0, t_1$). Inside the call, we first decompose t_1 into $c_1 = \sigma(\square, \sigma(1, B))$ and $u_1 = A$. Consequently, we return

$$\text{COMPLETE}(\emptyset, \{u_1\}, \{\square, c_1\}) = (\{u_1\}, \{u_1\}, \{\square, c_1\}) = \mathcal{T}_1$$

in Line 8 of Algorithm 3. We thus finished the first loop in Algorithm 1.

The wta $\mathcal{M}(\mathcal{T}_1)$ will only have the non-final state A and the nonzero tree representation entry $\mu_0(A)_A = 1$. Hence t_1 is still a counterexample, and we might assume that t_1 is returned by the teacher. Thus we call EXTEND($\mathcal{T}_1, t_1$). There we first decompose t_1 into the context $c_2 = \sigma(A, \sigma(\square, B))$ and $u_2 = 1$. Consequently, the call returns

$$\text{COMPLETE}(\{u_1\}, \{u_1, u_2\}, \{\square, c_1, c_2\}) = (\{u_1, u_2\}, \{u_1, u_2\}, \{\square, c_1, c_2\})$$

because $(\psi, \sigma(1, \sigma(1, B))) = 0$. Let T_2 be the complete observation table displayed above. This concludes the second iteration.

In the third iteration, we can still use t_1 as counterexample and the decomposition $c_3 = \sigma(A, \sigma(1, \Box))$ and $u_3 = B$. The call to EXTEND then returns $T_3 = (\{u_1, u_2\}, \{u_1, u_2, u_3\}, \{\Box, c_1, c_2, c_3\})$ because we have $A \equiv_{\{\Box, c_1, c_2, c_3\}} B$. Another iteration with the counterexample t_1 again yields the decomposition c_3 and u_3. Now u_3 was already seen before and $A \equiv_{\{\Box, c_1, c_2, c_3\}} B$, so we return $\text{EXTEND}(T_3, \sigma(A, \sigma(1, A)))$. In that call, we decompose the second argument into $c_4 = \sigma(A, \Box)$ and $u_4 = \sigma(1, A)$ and return

$$\text{COMPLETE}(\{u_1, u_2\}, \{u_1, \ldots, u_4\}, \{\Box, c_1, \ldots, c_4\})$$
$$= (\{u_1, u_2, u_4\}, \{u_1, \ldots, u_4\}, \{\Box, c_1, \ldots, c_4\}) = T_4 \ .$$

We will not demonstrate the construction of the wta $\mathcal{M}(T_4)$ but will give an elaborate example at the end. For the moment, rest assured that t_1 is still a counterexample (because $\mathcal{M}(T_4)$ has no final states). The decomposition of t_1 will be c_3 and u_3. As previously, this yields the recursive call $\text{EXTEND}(T_4, \sigma(A, \sigma(1, A)))$. Now the decomposition will be $c_5 = \Box$ and $u_5 = \sigma(A, \sigma(1, A))$ and EXTEND will return

$$\text{COMPLETE}(\{u_1, u_2, u_4\}, \{u_1, \ldots, u_5\}, \{\Box, c_1, \ldots, c_4\})$$
$$= (\{u_1, u_2, u_4, u_5\}, \{u_1, \ldots, u_4\}, \{\Box, c_1, \ldots, c_4\}) = T_5 \ .$$

Note that u_5 is a final state of $\mathcal{M}(T_5)$ and that t_1 is no longer a counterexample. If we continue with $t_2 = \sigma(A, \sigma(h, \sigma(u, B)))$ until it is no longer a counterexample, then we obtain

$$T_8 = (\{u_1, u_2, u_4, u_5, u\}, \{u_1, \ldots, u_5, h, u, \sigma(u, A)\},$$
$$\{\Box, c_1, \ldots, c_4, \sigma(u_1, \sigma(\Box, \sigma(u, u_3))), \sigma(u_1, \sigma(u_2, \sigma(\Box, u_3)))\}) \ .$$

Next we select $t_3 = \sigma(\sigma(t, \sigma(m, A)), \sigma(1, \sigma(n, B)))$ as counterexample and continue in the same manner. We obtain T_{11} as

$$(\{A, 1, \sigma(1, A), \sigma(A, \sigma(1, A)), u\}, \{u_1, \ldots, u_5, h, u, \sigma(u, A), m, t, n\}, C')$$

for some $C' \subseteq C_\Sigma$. At last, let us construct the wta $\mathcal{M}(T_{11})$. By Definition 11 we obtain the wta $(Q, \Sigma, \mathcal{A}, F, \mu)$ with

- $Q = \{A, 1, \sigma(1, A), \sigma(A, \sigma(1, A)), u\}$
- $F = \{\sigma(A, \sigma(1, A))\}$; and
- the nonzero tree representation entries

$$1 = \mu_0(A)_A = \mu_0(B)_A = \mu_0(1)_1 = \mu_0(h)_1$$
$$1 = \mu_0(n)_u = \mu_0(t)_u = \mu_0(u)_u = \mu_0(m)_u$$
$$1 = \mu_2(\sigma)_{1\,A, \sigma(1, A)}$$
$$0.125 = \mu_2(\sigma)_{u\,A, A}$$
$$0.03125 = \mu_2(\sigma)_{A\,\sigma(1, A), \sigma(A, \sigma(1, A))} \ .$$

Clearly, $\mathcal{M}(\mathcal{T}_{11})$ recognizes exactly ψ. In the next iteration, the teacher thus approves $\mathcal{M}(\mathcal{T}_{11})$. The returned wta has only 5 states (compared to the 6 states of the wta in Example 2). By Corollary 17 the returned wta is minimal. Thus, the learning algorithm might also be used to minimize deterministic wta but it is rather inefficient at that task.

6 Complexity Analysis

Our formal runtime complexity analysis follows the approach of [11]. In [12] a similar analysis is outlined but not actually shown. Our computation model will be the random access machine and we assume that the multiplicative semifield operations (including taking the inverse and equality tests) and the queries to the teacher can be performed in constant time. Finally, we assume that the algorithm terminates with the deterministic wta $(Q, \Sigma, \mathcal{A}, F, \mu)$. In the sequel, let

$$m = \mathrm{card}(\{(\sigma, q, q_1, \ldots, q_k) \mid \mu_k(\sigma)_{q_1 \cdots q_k, q} \neq 0\})$$

and $n = \mathrm{card}(Q)$. Let $r = \max\{k \mid \Sigma^{(k)} \neq \emptyset\}$, and let $\mathcal{T} = (E, T, C)$ be a complete observation table encountered during the run of the algorithm. Let us start with the complexity of COMPLETE.

Proposition 18 (cf. [11, Lemma 4.7]). *Within time $\mathcal{O}(mn)$ the call Complete $(E, T \cup \{u\}, C \cup \{c\})$ returns.*

Proof. First we check for each $t \in T \setminus E$ whether the new context c splits t and $T(t)$; i.e., whether $t \equiv_{C \cup \{c\}} T(t)$. Suppose that with each $t \in T \setminus E$ we store the coefficient a required in Definition 7 for $t \equiv_C T(t)$. Now we simply need to check whether this coefficient also qualifies for $t \equiv_{\{c\}} T(t)$. These simple checks require $\mathcal{O}(m)$ because $\mathrm{card}(T) \leq m$.

Should the check fail for some t_1 and t_2 that previously have been in the same equivalence class, then we need to compare them to each other. For each t these comparisons can amount up to $\mathcal{O}(n)$ because $\mathrm{card}(E) \leq n$.

Now it only remains to classify the new tree u provided that $u \notin T$. We simply compare u to each identified representative. Clearly, this requires us to check all contexts $C \cup \{c\}$. This takes $\mathcal{O}(n(m + n))$, which can also be given as $\mathcal{O}(2mn)$ because $n \leq m$. Thus the overall complexity is $\mathcal{O}(mn)$.   □

With the previous proposition we can state the complexity of a call to EXTEND.

Proposition 19 (cf. [11, Lemma 4.6]). *The call $\mathrm{EXTEND}(\mathcal{T}, t)$ returns in time $\mathcal{O}(\mathrm{size}(t)mnr)$.*

Proof. We already argued that at most $\mathrm{size}(t)$ recursive calls might be triggered by this call. In each invocation, we need to perform the decomposition into $c[u]$. In [11, Lemma 4.5], it is shown how this can be achieved in time $\mathcal{O}(nr)$. Thus it is also in $\mathcal{O}(mr)$. Using a similar technique, we can also test whether $u \in T$ in time $\mathcal{O}(mr)$. Finally, if $u \in T$, then the check $u \equiv_{C \cup \{c\}} T(u)$ can be performed in constant time because we can assume that a pointer to $\mathcal{T}(u)$ and the required coefficient for Definition 7 is stored with u. Thus, we only need to confirm that coefficient for the new context c. Altogether, this yields that the call to EXTEND returns in time $\mathcal{O}(\text{size}(t)mnr)$. □

Proposition 20 (cf. [11, Lemma 4.7]). *The wta $\mathcal{M}(\mathcal{T})$ can be constructed in time $\mathcal{O}(mr)$.*

Let s be the size of the largest counterexample returned by the teacher. Our simple and straightforward complexity analysis yields the following overall complexity (cf. $\mathcal{O}(mn^2(n+s)r)$ for the algorithm of [12]).

Theorem 21. *Our devised learning algorithm runs in time $\mathcal{O}(sm^2nr)$.*

Proof. We already saw that at most $m + n \leq 2m$ calls to EXTEND can happen before termination. Thus, we obtain the statement. □

Acknowledgements

The author would like to thank Heiko Vogler and Frank Drewes for lively discussions. Further, the author wants to express cordial thanks to the referees of the draft version of this paper. Their insight and criticism enabled the author to improve the paper.

References

1. Angluin, D.: Inductive inference of formal languages from positive data. Inform. and Control 45(2), 117–135 (1980)
2. Angluin, D.: Learning regular sets from queries and counterexamples. Inform. and Comput. 75(2), 87–106 (1987)
3. Angluin, D.: Queries and concept learning. Machine Learning 2(4), 319–342 (1987)
4. Angluin, D.: Queries revisited. In: Abe, N., Khardon, R., Zeugmann, T. (eds.) ALT 2001. LNCS (LNAI), vol. 2225, pp. 12–31. Springer, Heidelberg (2001)
5. Borchardt, B.: The Myhill-Nerode theorem for recognizable tree series. In: Ésik, Z., Fülöp, Z. (eds.) DLT 2003. LNCS, vol. 2710, pp. 146–158. Springer, Heidelberg (2003)
6. Borchardt, B.: A pumping lemma and decidability problems for recognizable tree series. Acta Cybernet. 16(4), 509–544 (2004)
7. B. Borchardt. The Theory of Recognizable Tree Series. PhD thesis, Technische Universität Dresden (2005)
8. Borchardt, B., Vogler, H.: Determinization of finite state weighted tree automata. J. Autom. Lang. Combin. 8(3), 417–463 (2003)
9. Drewes, F., Högberg, J.: Learning a regular tree language from a teacher. In: Ésik, Z., Fülöp, Z. (eds.) DLT 2003. LNCS, vol. 2710, pp. 279–291. Springer, Heidelberg (2003)
10. Drewes, F., Högberg, J.: Extensions of a MAT learner for regular tree languages. In: Proc. 23rd Annual Workshop of the Swedish Artificial Intelligence Society, Umeå University, pp. 35–44 (2006)

11. Drewes, F., Högberg, J.: Query learning of regular tree languages: How to avoid dead states. Theory of Comput. Syst. 40(2), 163–185 (2007)
12. Drewes, F., Vogler, H.: Learning deterministically recognizable tree series. J. Automata, Languages and Combinatorics, (to appear, 2007)
13. Eisner, J.: Simpler and more general minimization for weighted finite-state automata. In: Human Language Technology Conf. of the North American Chapter of the Association for Computational Linguistics, pp. 64–71 (2003)
14. Gécseg, F., Steinby, M.: Tree Automata. Akadémiai Kiadó, Budapest (1984)
15. Gécseg, F., Steinby, M.: Tree languages. In: Handbook of Formal Languages, ch. 1, vol. 3, pp. 1–68. Springer, Heidelberg (1997)
16. Gold, E.M.: Language identification in the limit. Inform. and Control 10(5), 447–474 (1967)
17. Habrard, A., Oncina, J.: Learning multiplicity tree automata. In: Sakakibara, Y., Kobayashi, S., Sato, K., Nishino, T., Tomita, E. (eds.) ICGI 2006. LNCS (LNAI), vol. 4201, pp. 268–280. Springer, Heidelberg (2006)
18. Shapiro, E.Y.: Algorithmic Program Debugging. ACM Distinguished Dissertation. MIT Press, Cambridge (1983)

The Second Eigenvalue of Random Walks On Symmetric Random Intersection Graphs[*]

Sotiris Nikoletseas[1,2], Christoforos Raptopoulos[1,2], and Paul G. Spirakis[1,2]

[1] Computer Technology Institute, P.O. Box 1122, 26110 Patras, Greece
[2] University of Patras, 26500 Patras, Greece
nikole@cti.gr, raptopox@ceid.upatras.gr, spirakis@cti.gr

Abstract. In this paper we examine spectral properties of random intersection graphs when the number of vertices is equal to the number of labels. We call this class symmetric random intersection graphs. We examine symmetric random intersection graphs when the probability that a vertex selects a label is close to the connectivity threshold τ_c. In particular, we examine the size of the second eigenvalue of the transition matrix corresponding to the Markov Chain that describes a random walk on an instance of the symmetric random intersection graph $G_{n,n,p}$. We show that with high probability the second eigenvalue is upper bounded by some constant $\zeta < 1$.

1 Introduction

Random graphs are interesting combinatorial objects that were introduced by P. Erdös and A. Rényi and still attract a huge amount of research in the communities of Theoretical Computer Science, Algorithms, Graph Theory and Discrete Mathematics. This continuing interest is due to the fact that, besides their mathematical beauty, such graphs are very important, since they can model interactions and faults in networks and also serve as typical inputs for an average case analysis of algorithms.

There exist various models of random graphs. The most famous is the $G_{n,p}$ random graph, a sample space whose points are graphs produced by randomly sampling the edges of a graph on n vertices independently, with the same probability p. Other models have also been quite a lot investigated: $G_{n,r}$ (the "random regular graphs", produced by randomly and equiprobably sampling a graph from all regular graphs of n vertices and vertex degree r) and $G_{n,M}$ (produced by randomly and equiprobably selecting an element of the class of graphs on n vertices having M edges). For an excellent survey of these models, see [1,2].

[*] This work was partially supported by the IST Programme of the European Union under contact number IST-2005-15964 (AEOLUS) and by the Programme PENED under contact number 03ED568, co-funded 75% by European Union – European Social Fund (ESF), 25% by Greek Government – Ministry of Development – General Secretariat of Research and Technology (GSRT), and by Private Sector, under Measure 8.3 of O.P. Competitiveness – 3rd Community Support Framework (CSF).

S. Bozapalidis and G. Rahonis (Eds.): CAI 2007, LNCS 4728, pp. 236–246, 2007.

In this work we give a probabilistic (i.e. that holds with high probability) upper bound on the second eigenvalue of the Markov Chain describing the random walk on an instance of a relatively recent model of random graphs, namely the random intersection graphs model introduced by Karoński, Sheinerman and Singer-Cohen [11,20]. Also, Godehardt and Jaworski [9] considered similar models. In the $G_{n,m,p}$, to each of the n vertices of the graph, a random subset of a universal set of m elements is assigned, by independently choosing elements with the same probability p. Two vertices u, v are then adjacent in the $G_{n,m,p}$ graph if and only if their assigned sets of elements have at least one element in common.

Importance and Motivation. First of all, we note that (as proved in [12]) any graph is an intersection graph. Thus, the $G_{n,m,p}$ model is very general. Furthermore, for some ranges of the parameters m, p ($m = n^\alpha, \alpha > 6$) the spaces $G_{n,m,p}$ and $G_{n,\hat{p}}$ are equivalent (as proved by Fill, Sheinerman and Singer-Cohen [8], showing that in this range, the total variation distance between the graph random variables has limit 0).

Second, random intersection graphs may model real-life applications more accurately (compared to the $G_{n,\hat{p}}$ Bernoulli random graphs case). In fact, there are practical situations where each communication agent (e.g. a wireless node) gets access only to some ports (statistically) out of a possible set of communication ports. When another agent also selects a communication port, then a communication link is implicitly established and this gives rise to communication graphs that look like random intersection graphs. Even epidemiological phenomena (like spread of disease) tend to be more accurately captured by this "proximity-sensitive" random intersection graphs model. Other applications may include oblivious resource sharing in a distributed setting, interactions of mobile agents traversing the web etc.

Finally, we have to mention that the second eigenvalue of a random walk on some graph is a very important quantity. More specifically, it is well known that random walks whose second largest eigenvalue is sufficiently less than 1 are "rapidly mixing", i.e. they get close (in terms of the variation distance) to the steady state distribution after only a polylogarithmic (in the number of vertices/states) number of steps (see e.g. [19]); this has important algorithmic applications e.g. in efficient random generation and counting of combinatorial objects. Also, such graphs are expander graphs and can be used as basic building blocks in optimal network design.

Related Work. Random intersection graphs have recently attracted a growing research interest. The question of how close $G_{n,m,p}$ and $G_{n,\hat{p}}$ are for various values of m, p has been studied by Fill, Sheinerman and Singer-Cohen in [8]. In [14], new models of random intersection graphs have been proposed, along with an investigation of both the existence and efficient finding of close to optimal independent sets. The authors of [7] find thresholds (that are optimal up to a constant factor) for the appearance of hamilton cycles in random intersection graphs. The efficient construction of hamilton cycles is studied in [17]. Also, by using a sieve method, Stark [21] gives exact formulae for the degree distribution of an arbitrary fixed vertex of $G_{n,m,p}$ for a quite wide range of the parameters of the model.

Geometric proximity between randomly placed objects is also nicely captured by the model of random geometric graphs (see e.g. [4,5,16]) and important variations (like random scaled sector graphs, [6]). Other extensions of random graph models (such as random regular graphs) and several important combinatorial properties (connectivity, expansion, existence of a giant connected component) are performed in [13,15].

Our Contribution. As proved in [8], the spaces $G_{n,m,p}$ and $G_{n,\hat{p}}$ are equivalent when $m = n^{\alpha}$, with $\alpha > 6$, in the sense that their total variation distance tends to 0 as n goes to ∞. Also, the authors in [17] show that, when $\alpha > 1$, for any monotone increasing property there is a direct relation (including a multiplicative constant) of the corresponding thresholds of the property in the two spaces. So, it is very important to investigate what is happening when $\alpha \leq 1$ where the two spaces are statistically different. In this paper, we consider the regime $\alpha = 1$. We call this case *symmetric* because the number of labels m is equal to the number of vertices n. Let λ_1 be the second largest eigenvalue of the transition matrix corresponding to the Markov Chain that describes the random walk on an instance of the symmetric random intersection graph $G_{n,n,p}$, with $p = 4\frac{\ln n}{n}$ (i.e. p is close to the connectivity threshold in this case, which is shown to be $\tau_c \stackrel{def}{=} \frac{\ln n}{n}$ in [20]). We show here that, with high probability, λ_1 is upper bounded away from 1 by some constant. In order to prove this we use two related Markov Chains whose bounds on their second eigenvalues can be used to give our desired bound.

2   Notation and Definitions

Let $Bin(n,p)$ denote the Binomial distribution with parameters n and p. We first formally define the *random intersection graphs model*.

Definition 1 (Random Intersection Graph). *Consider a universe $\mathcal{M} = \{1, 2, \ldots, m\}$ of elements and a set of vertices $V(G) = \{v_1, v_2, \ldots, v_n\}$. If we assign independently to each vertex v_j, $j = 1, 2, \ldots, n$, a subset S_{v_j} of $\mathcal{M}$ choosing each element $i \in \mathcal{M}$ independently with probability p and put an edge between two vertices v_{j_1}, v_{j_2} if and only if $S_{v_{j_1}} \cap S_{v_{j_2}} \neq \emptyset$, then the resulting graph is an instance of the random intersection graph $G_{n,m,p}$. In this model we also denote by L_l the set of vertices that have chosen label $l \in \mathcal{M}$. The degree of $v \in V(G)$ will be denoted by $d_G(v)$. Also, the set of edges of $G_{n,m,p}$ will be denoted by $e(G)$.*

Consider now the bipartite graph with vertex set $V(G) \cup \mathcal{M}$ and edge set $\{(v_j, i) : i \in S_{v_j}\} = \{(v_j, i) : v_j \in L_i\}$. We will refer to this graph as the bipartite random graph $B_{n,m,p}$ assiaciated to $G_{n,m,p}$.

In this work we consider the *symmetric random intesection graph $G_{n,n,p}$* in which the number of labels m is equal to the number of vertices n. The conectivity threshold in this case is $\tau_c \stackrel{def}{=} \frac{\ln n}{n}$ (see [20]). More specifically, when $p \geq \frac{\ln n + g(n)}{n}$

for some $g(n)$ that goes to ∞ arbitrarily slowly, the $G_{n,n,p}$ is almost surely connected. Here we will take $p = 4\frac{\ln n}{n}$, i.e. p four times the connectivity threshold.

A *random walk* on a graph $G = (V(G), e(G))$ is a Markov Chain with state space $V(G)$ and transition probability matrix given by

$$P(x, y) = \begin{cases} \frac{1}{d_G(x)} & \text{if } (x, y) \in e(G) \\ 0 & \text{otherwise} \end{cases}$$

for all $x, y \in V(G)$. Hence, we will say that a particle performs a random walk on G if given that it occupies vertex x at some time step, it can equiprobably occupy any of the neighbours of x at the next time step. The steady state distribution for this chain is given by $\pi(x) = \frac{d_G(x)}{2|e(G)|}$, for all $x \in V(G)$. We here give an upper bound on the second eigenvalue of P, when G is an instance of the symmetric random intersection graph $G_{n,n,p}$, with $p = 4\tau_c$, which is a metric that shows how quickly the corresponding Markov Chain approaches (in terms of variation distance) its steady state distribution.

3   Some Useful Properties of $G_{n,n,p}$

In this section we prove some properties of symmetric random intersection graphs that will be used in the rest of the paper. The first Lemma concerns the cardinalities of the sets S_v, L_l.

Lemma 1. *The following hold with high probability in $G_{n,n,p}$ when $p = 4\tau_c$*

(i) *For every vertex $v \in V(G)$ we have $|S_v| \in (1 \pm \sqrt{4/5})4\ln n$.*
(ii) *For every label $l \in M$ we have $|L_l| \in (1 \pm \sqrt{4/5})4\ln n$.*
(iii) *There are no vertices $x \neq y \in V(G)$ that have more than 2 labels in common.*

Proof. (i) By the definition of the model $|S_v|$ follows $Bin(n, p)$ and its mean value is $np = 4\ln n$. Hence, using Boole's inequality and Chernoff bounds we get

$$\Pr(\exists v \in V(G) : ||S_v| - np| \geq \sqrt{4/5}np) \leq n\exp\left\{-\frac{16}{15}\ln n\right\} = o(1).$$

(ii) follows in exactly the same way, since $|L_l|$ follows $Bin(n, p)$.
(iii) A crude inequality indeed gives

$$\Pr(\exists x \neq y \in V(G) : |S_x \cap S_y| \geq 3) \leq \binom{n}{2}\binom{m}{3}p^6 \leq n^5\left(\frac{4\ln n}{n}\right)^6 = o(1).$$

◇

The following Lemma concerns the degrees of the vertices and the number of edges in $G_{n,n,p}$.

Lemma 2. *The following hold with high probability in $G_{n,n,p}$ when $p = 4\tau_c$*

(1) *For every vertex $x \in V(G)$ we have $d_G(x) = |S_x|\Theta(\ln n)$.*
(2) *The number of edges of $G_{n,n,p}$ satisfies $|e(G)| = \Theta(\ln n)\sum_{y \in V(G)}|S_y|$*

Proof. (1) By (iii) of Lemma 1 we have that whp *any* edge $(x, y) \in e(G)$ is generated by (at least 1 and) at most 2 labels. Hence, whp for any vertex $x \in V(G)$, the formula $\sum_{l \in S_x}|L_l|$ is at most $2d_G(x)$ (and at least $d_G(x)$), since it counts every edge adjacent to x at most twice (once for every label that generates it). But (ii) of Lemma 1 implies that $\sum_{l \in S_y}|L_l| = |S_y|\Theta(\ln n)$, for all $y \in V(G)$. So we must have $d_G(x) = |S_x|\Theta(\ln n)$.
   (2) This is obvious since $|e(G)| = \frac{1}{2}\sum_{y \in V(G)}d_G(y)$.   ◇

4   Bounds for the Second Eigenvalue and the Mixing Time

In this section we give an upper bound on the second eigenvalue of the random walk on an instance of $G_{n,n,p}$, with $p = 4\tau_c$, that holds for almost every instance.
   Let $\tilde{W}$ be a Markov Chain on state space V (i.e. the vertices of $G_{n,n,p}$) and transition matrix given by

$$\tilde{P}(x, y) = \begin{cases} \sum_{l \in S_x \cap S_y} \frac{1}{|S_x| \cdot |L_l|} & \text{if } S_x \cap S_y \neq \emptyset \\ 0 & \text{otherwise.} \end{cases}$$

Note that this Markov Chain comes from observing the simple random walk on the $B_{n,n,p}$ graph associated with $G_{n,n,p}$ every two steps. This means that $\tilde{W}$ is reversible and we can easily verify that its stationary distribution is given by

$$\tilde{\pi}(x) = \frac{|S_x|}{\sum_{y \in V}|S_y|}, \qquad \text{for every } x \in V.$$

Now let W denote the random walk on $G_{n,n,p}$ and let P denote its transition probability matrix. It is known that W is reversible and its stationary distribution is given by $\pi(x) = \frac{d_G(x)}{2|e(G)|}$, for every $x \in V$.
   In order to compare the eigenvalues of W and $\tilde{W}$ we will use the following theorem of Diaconis and Saloff-Coste (see [3])

Theorem 1 ([3]). *Consider two reversible, irreducible markov chains on a finite set V with transition matrices $P, \tilde{P}$ respectively, stationary distributions $\pi, \tilde{\pi}$ respectively and eigenvalues $\beta_0 = 1 > \beta_1 \geq \beta_2 \geq \cdots \geq \beta_{|V|-1} \geq -1$ and $\tilde{\beta}_0 = 1 > \tilde{\beta}_1 \geq \tilde{\beta}_2 \geq \cdots \geq \tilde{\beta}_{|V|-1} \geq -1$ respectively.*

For each pair $x \neq y$ with $\tilde{P}(x, y) > 0$, fix a sequence of steps $x_0 = x, x_1, x_2, \ldots,$ $x_k = y$ with $P(x_i, x_{i+1}) > 0$. This sequence of steps is called a path γ_{xy} of length $|\gamma_{xy}| = k$. Let $\mathcal{E} = \{(x, y) : P(x, y) > 0)\}$, $\tilde{\mathcal{E}} = \{(x, y) : \tilde{P}(x, y) > 0\}$ and

$\tilde{\mathcal{E}}(z,w) = \{(x,y) \in \tilde{\mathcal{E}} : (z,w) \in \gamma_{xy}\}$, where $(z,w) \in \mathcal{E}$. Then, for $1 \le i \le |V|-1$,

$$\beta_i \le 1 - \frac{\delta}{A}(1 - \tilde{\beta}_i)$$

where δ is such that $\tilde{\pi}(x) \ge \delta\pi(x)$, for all $x \in V$, and

$$A = \max_{(z,w)\in\mathcal{E}} \left\{ \frac{1}{\pi(x)P(x,y)} \sum_{\tilde{\mathcal{E}}(z,w)} |\gamma_{xy}|\tilde{\pi}(x)\tilde{P}(x,y) \right\}.$$

$\diamond$

We now prove the following

Lemma 3. *Consider the two Markov Chains $W, \tilde{W}$ with transition matrices P and $\tilde{P}$ respectively. Let λ_1 and $\tilde{\lambda}_1$ be their second largest eigenvalues respectively. Then whp there is a constant $\gamma > 0$ such that*

$$\lambda_1 \le 1 - \gamma(1 - \tilde{\lambda}_1).$$

Proof. First note that, by Lemma 2, there is a constant $\delta > 0$ such that, for all $x \in V(G)$,

$$\tilde{\pi}(x) = \frac{|S_x|}{\sum_{y\in V}|S_y|} \ge \delta\frac{d_G(x)}{2|e(G)|} = \delta\pi(x).$$

Notice now that by the definition of the two Markov Chains W and $\tilde{W}$, and by the definition of $G_{n,n,p}$, we have that $P(x,y) > 0$ iff $\tilde{P}(x,y) > 0$. So, by considering the set of paths $\gamma_{x,y} = \{x_0 = x, x_1 = y\}$ for all (x,y) such that $\tilde{P}(x,y) > 0$, the quantity A of Theorem 1 becomes

$$A = \max_{(x,y):P(x,y)>0} \left\{ \frac{\tilde{\pi}(x)\tilde{P}(x,y)}{\pi(x)P(x,y)} \right\}.$$

It remains then to bound A by a constant. By the corresponding formulae for $\pi(x), P(x,y), \tilde{\pi}(x)$ and $\tilde{P}(x,y)$ we have that for all (x,y) such that $P(x,y) > 0$

$$\frac{\tilde{\pi}(x)\tilde{P}(x,y)}{\pi(x)P(x,y)} = \frac{\frac{|S_x|}{\sum_{y\in V}|S_y|} \cdot \sum_{l\in S_x\cap S_y} \frac{1}{|S_x|\cdot|L_l|}}{\frac{d_G(x)}{2|e(G)|} \cdot \frac{1}{d_G(x)}} = \frac{2|e(G)|\sum_{l\in S_x\cap S_y}\frac{1}{|L_l|}}{\sum_{y\in V}|S_y|}.$$

But by (ii) and (iii) of Lemma 1 and by (2) of Lemma 2 this fraction is whp $\Theta(1)$, which means that A is bounded by some constant.

Hence, by applying Theorem 1 and setting $\gamma = \frac{\delta}{A}$, we get the desired result.

$\diamond$

Because of Lemma 3, in order to give a bound on λ_1, we will first bound $\tilde{\lambda}_1$. To do this, we will define a third Markov Chain: Let $\hat{W}$ denote the random walk on the $B_{n,n,p}$ bipartite graph that is associated to $G_{n,n,p}$. Let also $\hat{P}$ denote its transition probability matrix and let $\hat{\lambda}_i, i = 1, \ldots, 2n$, its eigenvalues and $\hat{x}_i, i = 1, \ldots, n+n$, their corresponding eigenvectors. Also, let $\hat{\pi}$ be the stationary distribution of $\hat{W}$. Note that

$$\hat{P}^2 = \begin{bmatrix} \tilde{P} & \emptyset \\ \emptyset & Q \end{bmatrix}$$

where Q is some transition matrix. But for any $i = 1, \ldots, n+n$, we have $\hat{P}^2 \hat{x}_i = \hat{\lambda}_i^2 \hat{x}_i$. In particular, this means that $\tilde{\lambda}_1$ is at most $\hat{\lambda}_1^2$. So, in order to give an upper bound on $\tilde{\lambda}_1$, we need to give an upper bound on $\hat{\lambda}_1$ (that holds whp).In order to do so, we use the notion of *conductance* $\Phi_{\hat{W}}$ of the walk $\hat{W}$ that is defined as follows:

Definition 2. *Consider the bipartite random graph $B_{n,n,p}$ that is associated to $G_{n,n,p}$. The vertex set of $B_{n,n,p}$ is $V(B) = V(G) \cup \mathcal{M}$. For every $x \in V(B)$, let $d_B(x)$ be the degree of x in B. For any $S \subseteq V(B)$, let $e_B(S : \overline{S})$ be the set of neighbours of S with one end in S and the other in $\overline{S} = V(B) \backslash S$, let $d_B(S) = \sum_{v \in S} d_B(v)$ and $\hat{\pi}(S) = \sum_{v \in S} \hat{\pi}(v)$. Then*

$$\Phi_{\hat{W}} = \min_{\hat{\pi}(S) \leq 1/2} \frac{|e_B(S : \overline{S})|}{d_B(S)}.$$

Before dealing with the conductance of $B_{n,n,p}$ we need an auxiliary lemma.

Lemma 4. *With high probability there is no $S = V_1 \cup M_1$, with $V_1 \subseteq V(G)$, $M_1 \subseteq \mathcal{M}$, $|V_1| \geq (1 - 1/40)n$ and $|M_1| \geq (1 - 1/40)n$ such that $\hat{\pi}(S) \leq \frac{1}{2}$.*

Proof. Let $\overline{V}_1 = V(G) \backslash V_1$ and assume for contradiction that $|\overline{V}_1| \leq n/40$. By Lemma 1 we have that whp

$$\hat{\pi}(\overline{V}_1) = \sum_{x \in \overline{V}_1} \frac{|S_x|}{\sum_y |S_y|} \leq \frac{n}{40} \frac{1 + \sqrt{4/5}}{n(1 - \sqrt{4/5})} < \frac{1}{4}.$$

Since $\hat{\pi}(\overline{V}_1) + \hat{\pi}(V_1) = \frac{1}{2}$, this means that $\hat{\pi}(V_1) > \frac{1}{4}$.

Similarly, we can show that whp $\hat{\pi}(M_1) > \frac{1}{4}$. Hence, whp $\hat{\pi}(S) > \frac{1}{2}$.   ◇

We now prove the following

Lemma 5. *With high probability, the conductance of the random walk on $B_{n,n,p}$ satisfies $\Phi_{\hat{W}} \geq \phi$, where ϕ is some positive constant.*

Proof. Let $S = V_1 \cup M_1$, where $V_1 \subseteq V(G)$ and $M_1 \subseteq \mathcal{M}$. The cases $|V_1| = 0$ or $|M_1| = 0$ are trivial so we do not consider them.

Notice now that, because of Lemma 1, the number of edges coming out of V_1 is whp $|e_B(V_1 : \mathcal{M})| \in |V_1|(1 \pm \sqrt{4/5})4 \ln n$. Similarly, the number of edges coming out of M_1 is whp $|e_B(M_1 : V(G))| \in |M_1|(1 \pm \sqrt{4/5})4 \ln n$. It is then obvious that if $(1 + 1/2)|e_B(M_1 : V(G))| \leq |e_B(V_1 : \mathcal{M})|$ or $|e_B(V_1 : \mathcal{M})| \leq (1 - 1/2)|e_B(M_1 : V(G))|$, then we are done, since there will be at least a contant fraction of the number of edges spanned by S that "leave" S (which by the definition of conductance implies the lower bound on $\Phi_{\hat{W}}$). So we only have to deal with the case $(1 - 1/2)|e_B(M_1 : V(G))| \leq |e_B(V_1 : \mathcal{M})| \leq (1 + 1/2)|e_B(M_1 : V(G))|$. Given the relation of $|e_B(V_1 : \mathcal{M})|, |e_B(M_1 : V(G))|$ with $|V_1|, |M_1|$ respectively (see above), we will consider the broader (whp) case $\frac{1}{300}|M_1| \leq |V_1| \leq 300|M_1|$.

Note now that $|e_B(S : S)|$ follows $Bin(|V_1||M_1|, p)$. By using Chernoff bounds and Boole's inequality we can show that, for any constant $c_1 > e^{300}$ we have

$$Pr\left(\exists V_1, M_1 : \frac{1}{300}|M_1| \leq |V_1| \leq 300|M_1|, |e_B(S : S)| > \left(1 + c_1\frac{n}{|M_1|}\right)p|V_1||M_1|\right)$$

$$\leq \sum_{j=1}^{n} \sum_{i:\lfloor j/300\rfloor \leq i \leq 300j} \binom{n}{j}\binom{n}{i} \left(\frac{e^{c_1\frac{n}{j}}}{\left(1 + c_1\frac{n}{j}\right)^{\left(1 + c_1\frac{n}{j}\right)}}\right)^{pij}$$

$$\leq \sum_{j=1}^{n} \sum_{i:\lfloor j/300\rfloor \leq i \leq 300j} n^j n^i \left(\frac{e^{c_1\frac{n}{j}}}{\left(1 + c_1\frac{n}{j}\right)^{\left(1 + c_1\frac{n}{j}\right)}}\right)^{pij}$$

$$\leq \sum_{j=1}^{n} \sum_{i:\lfloor j/300\rfloor \leq i \leq 300j} \exp\left\{(j + i)\ln n + c_1 4i \ln n - c_1 4i \ln n \ln\left(1 + c_1\frac{n}{j}\right)\right\} = o(1)$$

Hence, whp there is no $S = V_1 \cup M_1$ such that $\frac{1}{300}|M_1| \leq |V_1| \leq 300|M_1|$ and the edges inside S surpass their mean value by more than $c_1\frac{n}{|M_1|}$ times.

We now consider two cases:

Case I: Assume that $|M_1| \geq \frac{n}{\ln n}$. Then it is easy to see that $e_B(S : \overline{S})$ has at least $\Omega(n^2/\ln n)$ possible edges. Since $|e_B(S : \overline{S})|$ follows $Bin(|V_1|(n - |M_1|) + |M_1|(n - |V_1|), p)$, the mean number of edges in $e_B(S : \overline{S})$ is clearly $\Omega(n)$. So, by using Chernoff bounds and Boole's inequality we conclude that the probability that there is some $S = M_1 \cup V_1$ satisfying $\frac{1}{300}|M_1| \leq |V_1| \leq 300|M_1|$, $|M_1| \geq \frac{n}{\ln n}$ and $|e_B(S : \overline{S})| \leq (1 - 1/2)E|e_B(S : \overline{S})|$ is at most

$$\sum_{j=n/\ln n} \sum_{i:j/300 \leq i \leq 300j} \binom{n}{j}\binom{n}{i} e^{-\Omega(n)}$$

$$\leq \sum_{j=n/\ln n}^{n} \sum_{i:j/300 \leq i \leq 300j} \binom{n}{n/\ln n}\binom{n}{\frac{n}{300 \ln n}} e^{-\Omega(n)}$$

$$\leq \sum_{j=n/\ln n}^{n} \sum_{i:j/300 \leq i \leq 300j} (e \ln n)^{\frac{n}{\ln n}} (300e \ln n)^{\frac{n}{300 \ln n}} e^{-\Omega(n)} = o(1) \quad (2)$$

where in the second inequality we used the fact that $\binom{n}{x}$ is decreasing for $x \geq n/2$ and in the third inequality we used the fact $\binom{n}{x} \leq \left(\frac{ne}{x}\right)^x$ for all x. Therefore, in this case, whp $|e_B(S : \overline{S})| \geq \frac{1}{2}E|e_B(S : \overline{S})| = \frac{\ln n}{2n}(|V_1|(n - |M_1|) + |M_1|(n - |V_1|))$. But Lemma 4 states that at least one of $(n - |M_1|)$ or $(n - |V_1|)$ is equal or greater to $n/40$. So, for any constant $c_2 < \frac{1}{4 \cdot 40 \cdot 300^2 \cdot c_1}$, we have that

$$c_2 \left(1 + c_1 \frac{n}{|M_1|}\right) p|V_1||M_1| \leq \frac{p}{2}(|V_1|(n - |M_1|) + |M_1|(n - |V_1|)).$$

By (1) and (2) above, this means that whp, in the case $|M_1| \geq \frac{n}{\ln n}$, there is at least a contant fraction $\gamma_1 \leq 1$ of the number of edges spanned by S that "leave" S (which by the definition of conductance implies the lower bound on $\Phi_{\hat{W}}$).

Case II: Assume that $|M_1| < \frac{n}{\ln n}$. As mentioned above, $|e(S : S)|$ follows $Bin(|V_1||M_1|, p)$. It is then easy to see that the probability that there is some $S = M_1 \cup V_1$ satisfying $\frac{1}{300}|M_1| \leq |V_1| \leq 300|M_1|$, $|M_1| < \frac{n}{\ln n}$ and $|e_B(S : S)| > 300^2|M_1| \ln \ln n$ is at most the probability that a set S' of k vertices, where $2 \leq k \leq 301n/\ln n$, of a Bernoulli random graph $G_{2n,p}$ has at least $k \ln \ln n$ edges that have both edges in S'. The last probability is at most

$$\sum_{k=2}^{301n/\ln n} \binom{2n}{k} \binom{\binom{k}{2}}{k \ln \ln n} p^{k \ln \ln n}$$

$$\leq \sum_{k=2}^{301n/\ln n} \left(\frac{2ne}{k}\right)^k \left(\frac{ke}{2 \ln \ln n}\right)^{k \ln \ln n} \left(\frac{4 \ln n}{n}\right)^{k \ln \ln n}$$

$$\leq \sum_{k=2}^{301n/\ln n} e^{k(\ln n - \ln k) + k \ln \ln n (\ln k - \ln \ln \ln n + \ln \ln k - \ln n) + 7k \ln \ln n} = o(1).$$

But Lemma 1 states that $d(S) = \Theta(|M_1| \ln n)$, which means that (whp), for every S in this case, there is at least a constant fraction $\gamma_2 \leq 1$ of the edges spanned by S that "leave" S (which by the definition of conductance implies the lower bound on $\Phi_{\hat{W}}$). This ends the proof of the lemma since we can take $\phi = \min\{\gamma_1, \gamma_2\}$.   ◇

By a result of [10,18], we know that $\hat{\lambda}_1 \leq 1 - \frac{\Phi_{\hat{W}}^2}{2}$ and so, by Lemma 5, $\hat{\lambda}_1$ is (upper) bounded away from 1. By the above discussion, we have proved the following

Theorem 2. *With high probability, the second largest eigenvalue of the random walk on $G_{n,n,p}$, with $p = 4\tau_c$, satisfies $\lambda_1 \leq \zeta$, where $\zeta \in (0, 1)$ is a constant that is bounded away from 1.*

5   Conclusions and Future Work

In this work, we proved that whp there is a constant gap between the second eigenvalue of the random walk on $G_{n,n,p}$, with $p = 4\tau_c$, and 1. Our analysis can be pushed further (although not without many technical difficulties) to provide tighter results (e.g. improve the multiplicative constant 4 on p). It is worth investigating other important properties of $G_{n,m,p}$, such as dominating sets, existence of vertex disjoint paths between pairs of vertices etc.

References

1. Alon, N., Spencer, J.: The Probabilistic Method, 2nd edn. John Wiley & Sons, Inc., Chichester (2000)
2. Bollobás, B.: Random Graphs, 2nd edn. Cambridge University Press, Cambridge (2001)
3. Diaconis, P., Saloff-Coste, L.: Comparison Theorems for Reversible Markov Chains. The Annals of Applied Probability 3(3), 696–730 (1993)
4. Díaz, J., Penrose, M.D., Petit, J., Serna, M.: Approximating Layout Problems on Random Geometric Graphs. Journal of Algorithms 39, 78–116 (2001)
5. Díaz, J., Petit, J., Serna, M.: Random Geometric Problems on $[0, 1]^2$. In: Rolim, J.D.P., Serna, M.J., Luby, M. (eds.) RANDOM 1998. LNCS, vol. 1518, pp. 294–306. Springer, Heidelberg (1998)
6. Díaz, J., Petit, J., Serna, M.: A Random Graph Model for Optical Networks of Sensors. In: The 1st International Workshop on Efficient and Experimental Algorithms (WEA) (2003), also in the IEEE Transactions on Mobile Computing Journal 2(3) 186–196 (2003)
7. Efthymiou, C., Spirakis, P.: On the Existence of Hamilton Cycles in Random Intersection Graphs. In: Caires, L., Italiano, G.F., Monteiro, L., Palamidessi, C., Yung, M. (eds.) ICALP 2005. LNCS, vol. 3580, pp. 690–701. Springer, Heidelberg (2005)
8. Fill, J.A., Sheinerman, E.R., Singer-Cohen, K.B.: Random Intersection Graphs when $m = \omega(n)$: An Equivalence Theorem Relating the Evolution of the $G(n, m, p)$ and $G(n, p)$ models, http://citeseer.nj.nec.com/fill98random.html
9. Godehardt, E., Jaworski, J.: Two models of Random Intersection Graphs for Classification. In: Opitz, O., Schwaiger, M. (eds.) Studies in Classification, Data Analysis and Knowledge Organisation, pp. 67–82. Springer, Heidelberg (2002)
10. Jerrum, M., Sinclair, A.: Approximate Counting, Uniform Generation and Rapidly Mixing Markov Chains. Information and Computation 82, 93–133 (1989)
11. Karoński, M., Scheinerman, E.R., Singer-Cohen, K.B.: On Random Intersection Graphs: The Subgraph Problem. Combinatorics, Probability and Computing journal 8, 131–159 (1999)
12. Marczewski, E.: Sur deux propriétés des classes d' ensembles. Fund. Math. 33, 303–307 (1945)
13. Nikoletseas, S., Palem, K., Spirakis, P., Yung, M.: Short Vertex Disjoint Paths and Multiconnectivity in Random Graphs: Reliable Network Computing. In: Shamir, E., Abiteboul, S. (eds.) ICALP 1994. LNCS, vol. 820, pp. 508–519. Springer, Heidelberg (1994), also in the Special Issue on Randomized Computing of the International Journal of Foundations of Computer Science (IJFCS) 11(2), 247–262 (2000)
14. Nikoletseas, S., Raptopoulos, C., Spirakis, P.: The Existence and Efficient Construction of Large Independent Sets in General Random Intersection Graphs. In: Díaz, J., Karhumäki, J., Lepistö, A., Sannella, D. (eds.) ICALP 2004. LNCS, vol. 3142, pp. 1029–1040. Springer, Heidelberg (2004), also in the Theoretical Computer Science (TCS) Journal (to appear, 2007)

15. Nikoletseas, S., Spirakis, P.: Expander Properties in Random Regular Graphs with Edge Faults. In: Mayr, E.W., Puech, C. (eds.) STACS 95. LNCS, vol. 900, pp. 421–432. Springer, Heidelberg (1995)
16. Penrose, M.: Random Geometric Graphs. Oxford Studies in Probability (2003)
17. Raptopoulos, C., Spirakis, P.: Simple and Efficient Greedy Algorithms for Hamilton Cycles in Random Intersection Graphs. In: Deng, X., Du, D.-Z. (eds.) ISAAC 2005. LNCS, vol. 3827, pp. 493–504. Springer, Heidelberg (2005)
18. Sinclair, A.: Algorithms for Random Generation and Counting: a Markov Chain Approach. PhD Thesis, University of Edimburg (1988)
19. Sinclair, A. (ed.): Algorithms for Random Generation and Counting. Birkhauser (1992)
20. Singer-Cohen, K.B.: Random Intersection Graphs. PhD thesis, John Hopkins University (1995)
21. Stark, D.: The Vertex Degree Distribution of Random Intersection Graphs. Random Structures & Algorithms 24(3), 249–258 (2004)

Verifying Security Protocols for Sensor Networks Using Algebraic Specification Techniques

Iakovos Ouranos[1] and Petros Stefaneas[2]

[1] School of Electrical and Computer Engineering
iouranos@central.ntua.gr
[2] School of Applied Mathematical and Physical Sciences
National Technical University of Athens
petros@math.ntua.gr

Abstract. Algebraic specification languages are formal methods that provide a rigorous basis for modeling of several systems. Security protocols are safety critical systems that need to be verified before their implementation. In this paper we have formally specified sensor network encryption protocol (SNEP) and a key agreement protocol for sensor networks, both from the SPINS protocol suite, with the OTS/CafeOBJ method, a well known formal specification technique applied not only in research, but also in industry. Based on this specification, we have proved that each protocol possesses an important safety(invariant) property.

Keywords: Algebraic Specification and Verification, CafeOBJ, Sensor Networks, Observational Transition Systems, SPINS Protocol suite.

1 Introduction

Formal Methods (FM) are techniques, languages and tools based on mathematics that provide a rigorous basis for specification and verification of software and hardware systems[1]. Using them during the design process of a system may prevent critical flaws that could lead to high cost. The OTS/CafeOBJ method [2][3] is one of the most promising approaches to the system verification. Protocols or systems are specified as observational transition systems, or OTSs, in CafeOBJ [4], an executable algebraic specification language. Next, the properties we want to verify, are expressed in CafeOBJ notation and the proof is done using induction and/or case analysis with the CafeOBJ system. This paper shows how the above method can be used for specifying security protocols for sensor networks and verifying properties of them. Since security in sensor networks is of major importance, it is very interesting to show how such safety critical protocols are verified using a rigorous algebraic technique. As a case study we have chosen protocols from the SPINS suite [5], that have not been analyzed before using such methods. The rest of the paper is organized as follows: section 2 gives some basic notation of CafeOBJ and OTSs. Section 3 describes the protocols to be modeled. Section 4 presents the algebraic specification of the protocols, while section 5 the verification of one invariant property for each protocol. Section 6 concludes the paper.

S. Bozapalidis and G. Rahonis (Eds.): CAI 2007, LNCS 4728, pp. 247–259, 2007.

2   CafeOBJ and Observational Transition Systems

2.1   CafeOBJ Basics

Initial and *hidden algebra*[6] are the logical basis of CafeOBJ [10]. The former is used to specify abstract data types such as integers, and the latter to specify objects. There are two kinds of sorts in CafeOBJ, *visible sorts* representing abstract data types and *hidden sorts* representing the set of states of an object. The operations to hidden sorts can be classified into *actions* and *observations*. An action can change a state of an object. It takes a state of an object and zero or more data, and returns another or the same state of the object. An observation is used to observe the value of a data component in an object. It takes a state of an object and zero or more data, and returns the value of a data component in the object.

Action and observation operators are declared by starting with bop, and others by starting with op. After bop or op, an operator name is written, followed by : and a list of sorts, and then, → and a sort are written. Operators are defined with equations. Equations are declared by starting with eq, and conditional ones by starting with ceq. After eq, two terms connected with = are written, ended with a full stop. After ceq, two terms connected with = are written, followed by if, and then a term denoting the condition and a full stop. The CafeOBJ system uses declared equations as left-to right rewrite rules and rewrites a given term.

Basic units of CafeOBJ specifications are modules. The CafeOBJ system provides built-in modules such as BOOL, where propositional logic is specified, or INT, where the set of integers together with their properties is specified. Below we present the specification of a counter of integers as a behavioural object in CafeOBJ.

```
mod* COUNTER {
pr(INT)
-- declaration of hidden sort
*[Counter]*
-- declaration of operators
op init : -> Counter              -- initial state
bop add : Int Counter -> Counter  -- action
bop read : Counter -> Int         -- observation
-- declaration of variables
vars I I' : Integer
var C : Counter
-- declaration of equations
eq read(init) = 0 .
eq read(add(I, C)) = I + read(C) .
}
```

The name of the module is COUNTER specified after the keyword mod* which is an abbreviation of module*. The module* declares a module with loose semantics, while for the modules with initial semantics the mod! is used. The visible sorts are declared within [] while for the hidden sorts the *[ ]* is used and the ordering of them by <. So, in the above example we have one hidden sort Counter. The lines beginning with the keyword - - are comments. Variables are declared by the keyword var (vars for more than one variables). Modules can be imported by using protecting, extending, or using. Protecting imports do not collapse elements or add new elements to the models of the imported module in contrast to the extending imports. Using imports provides no guaranty, so they might even collapse elements. The data of the above specification is INT which is a built-in module of the system. It is imported in the specification by protecting(INT). The init operator denotes any initial state, while add action adds an integer to the counter. Finally the read observation returns the value of the counter.

2.2   Observational Transition Systems

We suppose that there exists a universal state space called Y and that each data type used in OTSs is provided. A system is modeled by observing only quantities that are relevant to the system and how to change the quantities by state transition from the outside of each state of Y. An OTS can be used to model a system in this way. An OTS S is $\langle O, I, T \rangle$ such that:

- O: A finite set of observers. Each observer $o_{x_1:D_{o_1},...,x_m:D_{o_m}} : Y \rightarrow D_o$ is an indexed function that has m indexes $x_1, ..., x_m$ whose types are $D_{o1}, ..., D_{om}$. The equivalence relation $(u_1 =_s u_2)$ between two states $u_1, u_2 \in Y$ is defined as $\forall o_{x_1}, ..., x_m : O.(o_{x_1}, ..., x_m(u_1) = o_{x_1}, ..., x_m(u_2))$, where $\forall o_{x_1}, ..., x_m : O$ is the abbreviation of $\forall o_{x_1}, ..., x_m : O.\forall x_1 : D_{o1}...\forall x_m : D_{om}$.
- I: The set of initial states such that $I \subseteq \Upsilon$.
- T: A finite set of transitions. Each transition $t_{y_1:D_{t1},...,y_n:D_{tn}} : Y \rightarrow Y$ is an indexed function that has n indexes $y_1, ..., y_n$ whose types are $D_{t1}, ..., D_{tn}$ provided that $t_{y_1, ...,y_n}(u_1) =_s t_{y_1, ...,y_n}(u_2)$ for each $[u] \in Y/ =_s$, each $u_1, u_2 \in [u]$ and each $y_k : D_{tk}$ for $k = 1, ..., n$. $t_{y_1, ...,y_n}(u)$ is called the successor state of u wrt S. Each transition $t_{y_1, ...,y_n}$ has the condition $c - t_{y_1:D_{t1},...,y_n:D_{tn}} : Y \rightarrow$ Bool, which is called the effective condition of the transition. If $c - t_{y_1, ...,y_n}(u)$ does not hold, then $t_{y_1, ...,y_n}(u) =_s u$.

An OTS is described in CafeOBJ. Observers are denoted by CafeOBJ observation operations, and transitions by CafeOBJ action operations. Given an OTS S, reachable states wrt S are inductively defined as follows:

- Each $u_{init} \in I$ is reachable wrt S.
- For each $t_{y_1, ...,y_n} \in T$ and each $y_k : D_{tk}$ for k = 1,...,n, $t_{x_1, ...,x_n}(u)$ is reachable wrt S if $u \in Y$ is reachable wrt S.

Let R_s be the set of all reachable states wrt S.

Predicates whose types are $Y \rightarrow$ Bool are called state predicates. Any state predicate $p : Y \rightarrow$ Bool is called *invariant wrt S* if p holds in all reachable states wrt S, i.e. $\forall u : R_S.p(u)$. All properties considered in this paper are *invariants*.

We prove that an OTS S has an invariant property p mainly by induction on the number of transition rules applied (executed) as follows:

- Base case: For any state $u \in Y$ in which each observation $o \in O$ satisfies I, we show that $p(u)$ holds.
- Inductive step: Given any (reachable) state $u \in Y$ *wrt* S such that $p(u)$ holds, we show that, for any transition $t \in T$, $p(t(u))$ also holds.

3   The SPINS Protocol Suite

3.1   General

SPINS protocols were designed to support the security requirements of sensor networks. The limited computation resources of sensors, make it impossible to use asymmetric cryptography. As a result, the designers of SPINS protocols use purely symmetric cryptographic primitives. The security requirements of sensor networks include:

- Data Confidentiality
- Data Authentication
- Data Integrity
- Data Freshness

To achieve these security requirements, two protocols were designed and implemented: SNEP and μTESLA. SNEP provides data confidentiality, two-party data authentication, integrity, and freshness. μTESLA provides authentication for data broadcast. In addition, they built an authenticated routing application using the μTESLA and a two-party key agreement protocol, based on SNEP. In this paper we will analyze SNEP protocol and the two-party key agreement protocol based on it.

3.2   Notation

Following the notation of [5],

- A,B are principals, such as communicating nodes
- N_A is a nonce generated by A (to achieve freshness)
- $M_1 \mid M_2$ denotes the concatenation of messages M_1 and M_2
- K_{AB} denotes the secret (symmetric) key which is shared between A and B
- $\{M\}_{K_{AB}}$ is the encryption of message M with the symmetric key shared by A and B
- $\{M\}_{\langle K_{AB}, IV \rangle}$ denotes the encryption of message M, with key K_{AB}, and the initialization vector IV which is used in encryption modes such as cipher-block chaining (CBC), output feedback mode (OFB), or counter mode(CTR)
- $MAC(K_{AB}, M)$ denotes the message authentication code of message M, encrypted with key K_{AB}.

3.3 The SNEP Protocol

The entire Sensor Network Encryption Protocol (Fig.1) works as follows: If a node A wants to authenticate node B, it generates a nonce N_A and sends it to B. On receipt of the message, B obtains the nonce of A and sends its ID encrypted with shared key K_{encr}, and the counter C which is the initialization vector, along with the message authentication code encrypted with K_{mac} that contains the received nonce, the counter value, and the encrypted ID. On receipt of the message, A decrypts it, obtains the ID of B, and verifies the MAC, which contains the nonce. If the verification succeeds, then it is sure that B is really the sender of the message. We denote that K_{encr} and K_{mac} are derived from the master secret key K through a pseudo-random function.

Message 1. $A \rightarrow B : N_A$
Message 2. $B \rightarrow A : \{B\}_{\langle K_{encr}, C \rangle}, MAC(K_{mac}, N_A \mid C \mid \{B\}_{\langle K_{encr}, C \rangle})$

Fig. 1. The SNEP Protocol

SNEP provides a number of nice properties, such as:

1. *Semantic Security*: A strong security property which prevents eavesdroppers from inferring the message content from the encrypted message. The use of a shared counter between the sender and the receiver, as an initialization vector, provides a cryptographic mechanism with no additional sending overhead that achieves the property. The communicating parties share the counter and increment it after each block, without sending it with each message. Since the counter value is incremented after each message, the same message is encrypted differently each time.
2. *Replay Protection*: The counter value in the MAC prevents replaying old messages. If the counter were not present in the MAC, an adversary could easily replay messages.
3. *Strong Freshness*: If the MAC verifies correctly, node A knows that node B generated the response after it send the request.
4. *Data Authentication*: To achieve two-party authenticity and data integrity, the protocol uses a Message Authentication Code (MAC). If the MAC verifies correctly, the receiver can be assured that the message originated from the claimed sender.
5. *Low communication overhead*: The counter state is kept at each point and does not need to be sent in each message.

3.4 The Key Agreement Protocol

To bootstrap secure connections, a protocol for symmetric key setup is needed. The protocol has been constructed solely from symmetric key algorithms. It uses the base station as a trusted agent for key setup.

Assume that the node A wants to establish a shared secret session key SK_{AB} with node B. Since A and B do not share any secrets, they need to use a trusted third party S, which is the base station. In the trust setup, both A and B share a secret key with the base station, K_{AS} and K_{BS}, respectively. Figure 2 depicts the protocol that achieves secure key agreement as well as strong key freshness.

Message 1. $A \rightarrow B : N_A, A$
Message 2. $B \rightarrow S : N_A, N_B, A, B, \text{MAC}(K_{BS}, N_A \mid N_B \mid A \mid B)$
Message 3. $S \rightarrow A : \{SK_{AB}\}_{K_{AS}}, \text{MAC}(K_{AS}, N_A \mid B \mid \{SK_{AB}\}_{K_{AS}})$
Message 4. $S \rightarrow B : \{SK_{AB}\}_{K_{BS}}, \text{MAC}(K_{BS}, N_B \mid A \mid \{SK_{AB}\}_{K_{AS}})$

Fig. 2. The key agreement protocol

The protocol uses the SNEP protocol with strong freshness. The nonces N_A snd N_B ensure strong key freshness to both A and B. The SNEP protocol is responsible to ensure confidentiality of the established session key SK_{AB}, as well as message authenticity to make sure that the key was really generated by the base station. MAC in the second message helps defend the base station from denial-of-service attacks (DoS), so the base station only sends two messages to A and B if it received a legitimate request from one of the nodes.

4   Algebraic Specifications

4.1   SNEP Modeling

In the modeling of SNEP, we suppose that there exist untrustable nodes as well as trustable ones. Trustable nodes exactly follow the protocol, but untrustable ones may do something against the protocol as well, namely eavesdropping and/or faking of messages. The combination and cooperation of untrustable nodes is modeled as the most general intruder á la Dolev and Yao[7]. The intruder can do the following:

- Eavesdrop any message flowing in the network.
- Glean any nonce, cipher and mac from the message; however the intruder can decrypt a cipher or a mac only if he knows the corresponding key to decrypt.
- Fake and send messages based on the gleaned information; however the intruder cannot guess unknown nonces.

The basic data types i.e. the visible sorts and the corresponding data constructors, used to model the protocol, are as follows:

- **Node** denotes nodes. Constant **enemy** denotes the intruder.
- **Rand** denotes random numbers, which makes nonces unguessable and unique.
- **Nonce** denotes nonces. Given nodes a,b and a random number r, $n(a,b,r)$ denotes a nonce created by a for b. Projections **creator**, **forwhom** and **rand** return the first, second and third arguments.
- **Key** denotes symmetric keys. Given a node a, $k(a)$ denotes the key of a's group and operator p returns the argument of $k(a)$.

- MacKey denotes the keys used for creation of MACs.
- Counter denotes counters that are used for encrypting as IVs.
- Cipher denotes ciphertexts used in the protocol. Given a symmetric key k, a counter c and a node a, enc(k,c,a) denotes the ciphertext obtained by encrypting a with k and c. Operators k, c and p return the first, second and third arguments of enc(k,c,a).
- Mac denotes message authentication codes in protocol. Given a mac key k, a nonce n, a counter c and a cipher ci, mac(k,n,c,ci) denotes the message authentication code obtained by encrypting nonce, counter, and cipher with the key k. Operators n,k,c and ci return the first, second, third and fourth arguments of mac(k,n,c,ci).

In addition to those visible sorts, we use the visible sort Bool that denotes truth values, declared in the built-in module BOOL.

The two operators to denote the two kinds of messages are declared as follows:

```
op m1: Node Node Node Nonce -> Msg
op m2: Node Node Node Cipher Mac -> Msg
```

The visible sort Msg denotes messages. Projections crt, src and dst return the first (actual creator), second (sender) and third (receiver) arguments of each message. The first argument is meta-information that is only available to the outside observer and the principal that has sent the corresponding message, and that cannot be forged by the intruder, while the remaining arguments may be forged by the intruder. A predicate mi? checks if a given message is mi. The projection n returns the fourth argument of the first message, while projections c and m return the fourth and fifth argument of the second message correspondingly.

The network is modeled as a bag(multiset) of messages, which is used as the storage that the intruder can use. Any message that has been sent or put once into the network is supposed to be never deleted from the network. As a consequence the emptiness of the network means that no messages have been sent.

The enemy node tries to glean three kinds of quantities from the network. These are the nonces, the ciphers and the message authentication codes. The collections of those quantities gleaned by the enemy node are denoted my the operators:

```
op nonces  : Network -> ColNonces
op ciphers : Network -> ColCiphers
op macs    : Network -> ColMacs
```

The visible sort Network denotes networks and the visible sorts ColNonces, ColCiphers and ColMacs denotes collections of Nonces, Ciphers and Macs correspondingly.

For the case of ColMacs, the definition is:

```
eq MC \in macs(void) = false .
ceq MC \in macs(M,NW) = true if m2?(M) and MC = m(M) .
ceq MC \in macs(M,NW) = MC \in macs(NW)
                        if not(m2?(M) and MC = m(M)) .
```

The constant void denotes the empty bag and the operator _,_ denotes the data constructor of nonempty bags. Operator _\in_ is the membership predicate of bags. The equations say that no macs appearing in the second message are available if the network is empty, and the enemy node can glean such a mac MC from the network iff there exists a message m2 in the network which includes MC.

The OTS describing the protocol contains two observations and six kinds of transition rules:

```
bop nw : Snep -> Network   -- network
bop ur : Snep -> URands    -- used random numbers (Rands)
-- actions
   -- send message m1
   bop sdm1  : Snep Node Node Rand        -> Snep
   -- send message m2
   bop sdm2  : Snep Node Msg              -> Snep
   -- send fake m1
   bop fkm11 : Snep Node Node Rand        -> Snep
   -- send fake m1
   bop fkm12 : Snep Node Node Nonce       -> Snep
   -- send fake m2
   bop fkm21 : Snep Node Node Cipher Mac  -> Snep
   -- send fake m2
   bop fkm22 : Snep Node Node Rand        -> Snep
```

The hidden sort Snep denotes the state space. Observation ur is the set of used random numbers and the observation nw denotes the network. The sdm1 and sdm2 formalize sending messages according to the protocol, fkm11, fkm12, sending faking messages of first kind and fkm21, fkm22 sending faking messages of second kind. The equations to define fkm21 are:

```
op c-fkm21 : Snep Node Node Cipher Mac -> Bool
eq c-fkm21(S, P1, P2, CI, M) = (CI \in ciphers(nw(S)) and
                                M \in macs(nw(S))) .
ceq nw(fkm21(S, P1, P2, CI, M)) = (m2(enemy, P1, P2, CI, M),nw(S))
                        if c-fkm21(S, P1, P2, CI, M) .
eq ur(fkm21(S, P1, P2, CI, M)) = ur(S) .
ceq fkm21(S, P1, P2, CI, M) = S if not c-fkm21(S,P1,P2,CI,M) .
```

The operator c-fkm21 is the effective condition of any transition rule denoted by $fkm21.c\text{-}fkm21(s,p1,p2,ci,mc)$ means that in a state s, there exists a cipher ci and a mac mc in the network. In that state, an enemy node can use them and send a fake message m2.

4.2   Key Agreement Protocol Modeling

The key agreement protocol is modeled similarly. We distinguish two types of protocol agents, the nodes and the base station (visible sorts Node and Base). A malicious base station is denoted by the constant ibase. In addition:

- Sort Nkey denotes the key shared by two nodes. Given nodes n1 and n2 k(n1,n2) is the key shared by n1 and n2.
- Sort Bnkey denotes the key shared by a node and a base station. Given node n and base station bs, k(n,bs) denotes the key shared by n and bs.
- Sort Cipher denotes ciphers of the protocol. The encryption of the shared key of two nodes Nkey with a Bnkey form the ciphertext, i.e. enc*(bk,nk)*, with *bk* of the sort Bnkey and *nk* of the sort Nkey.
- Sort Mac1 denotes MACs of the second message while Mac2 denotes MACs of the third and fourth message.

Messages of the protocol are declared as visible sorts:

```
op m1 : Node Node Node Nonce            -> Msg
op m2 : Node Node Base Nonce Nonce Mac1 -> Msg
op m3 : Base Base Node Cipher Mac2      -> Msg
op m4 : Base Base Node Cipher Mac2      -> Msg
```

The enemy node can glean four kinds of quantities from the network. Nonces, Ciphers and two kinds of message authentication codes. The OTS that models the protocol contains two observations, four transition rules that formalize sending messages according to the protocol and twelve transition rules corresponding to the fake messages. Hidden sort Protocol models the state space.

```
-- reliable nodes
bop sdm1 : Protocol Node Node Rand      -> Protocol
bop sdm2 : Protocol Node Base Rand Msg -> Protocol
bop sdm3 : Protocol Base        Msg Msg -> Protocol
bop sdm4 : Protocol Base        Msg Msg -> Protocol
-- enemy nodes and base station
bop fkm11 : Protocol Node Node Rand               -> Protocol
bop fkm12 : Protocol Node Node Nonce              -> Protocol
bop fkm21 : Protocol Node Base Nonce Nonce Mac1 -> Protocol
bop fkm22 : Protocol Node Node Base Rand Rand    -> Protocol
bop fkm23 : Protocol Node Node Base Nonce Rand   -> Protocol
bop fkm24 : Protocol Node Node Base Nonce Rand   -> Protocol
bop fkm31 : Protocol Base Node Cipher Mac2        -> Protocol
bop fkm32 : Protocol Base Node Node Rand          -> Protocol
bop fkm33 : Protocol Base Node Node Nonce         -> Protocol
bop fkm41 : Protocol Base Node Cipher Mac2        -> Protocol
bop fkm42 : Protocol Base Node Node Rand          -> Protocol
bop fkm43 : Protocol Base Node Node Nonce         -> Protocol
```

The equations to define `sdm4` are as follows:

```
op c-sdm4 : Protocol Base Msg Msg -> Bool
eq c-sdm4(S,B,M1,M2) = (M1 \in nw(S) and m1?(M1) and
                        M2 \in nw(S) and m2?(M2) and B = dst-m2(M2)) .
ceq nw(sdm4(S, B, M1, M2)) = m4(B,B,src-m1m2(M2),enc(k(src-m1m2(M2),B),
                            k(src-m1m2(M1), dst-m1m3m4(M1))),
                            mac2(k(src-m1m2(M2), B),n(M2),src-m1m2(M1),
                            enc(k(src-m1m2(M2),B), k(src-m1m2(M1),
                            dst-m1m3m4(M1))))),nw(S) if c-sdm4(S,B,M1,M2) .
ceq ur(sdm4(S,B,M1,M2)) = ur(S) if c-sdm4(S,B,M1,M2) .
ceq sdm4(S,B,M1,M2) = S if not c-sdm4(S,B,M1,M2) .
```

c-sdm4(s,b,m,m') means that in a state s, there exists a message m of the kind m1 in the network, and a message m' of the kind m2 that is addressed to b. If this condition holds, the message m4(...) is put into the network.

5   Verification of Invariant Properties

Based on the specifications presented above, we have verified that each protocol possesses one safety(invariant) property. The property we verified for SNEP is called *authentication property* while the property we verified for key agreement protocol is called the *key agreement property*. Informally, they are:

Authentication Property. *Whenever node A receives a valid m2 message from node B, B is always the claimed node (i.e. not an enemy).*

Key Agreement Property. *Whenever node A(B) receives from base station a valid message m3(m4), then it is always true that the session key contained in the message is valid (i.e. is not the enemy's key).*

5.1   Proof Scores of Authentication Property

The property is expressed as an invariant: At any reachable state s, sensor nodes $n1,n2,n3$, key k, mac key k', nonce n and counter c,

invariant((not($k = $ k(enemy)) and not($k' = $ kmac(enemy)) and not($c = $ c(enemy)) and not(creator(n) = enemy) and (m2($p1,p2,p3$,enc($k,c,p2$),mac(k', n, c, enc(k, c, $p2$))) \in nw(s))) implies not($p2 = $ enemy)) .

To prove the property we follow the induction method as explained above. In the module INV the invariant is declared, while in module ISTEP is declared the induction step as follows:

```
op istep1 : Node Node Node Key MacKey Nonce Counter -> Bool
eq istep1(P1,P2,P3,K,K',N,C) =
        inv1(s,P1,P2,P3,K,K',N,C) implies inv1(s',P1,P2,P3,K,K',N,C).
```

where s' is the successor state. Next, we prove that after applying every different transition the claim is preserved. This step requires case split. For the fourth inductive case (transition rule `fkm22`) figure 3 presents the proof plan.

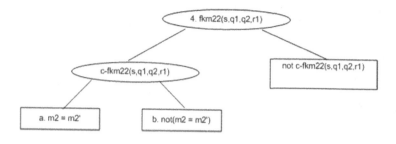

Fig. 3. The proof plan for the fourth transition

In the figure, edges mean case splits, ovals represent intermediate nodes, which mean case splitting in progress, and rectangles represent leaves, i.e. results of case splitting. For each rectangle, a fragment of a proof score is written. m2 and m2' correspond to (m2(p1,p2,p3,enc(k,c,p2),mac(k',n,c,enc(k,c,p2))) and m2(enemy,q1,q2,enc(k(q1),c(q1),q1),mac(kmac(q1),n(q1,q2,r1),c(q1), enc(k(q1),c(q1),q1)))) in the proof passage. The CafeOBJ system returned true for all subcases, which means that the proof is successful.

5.2   Proof Scores of Key Agreement Property

The property is expressed as an invariant as follows: At any reachable state s, base stations $b1,b2$, sensor nodes $n1,n2$, Bnkey bk, Nkey nk, random number r invariant(not(n(bk) = enemy)) and (not(b(bk) = ibase)) and (m3($b1,b2,n1$, enc(bk,nk),mac2(bk,n($n1,r$), $n2$,enc(bk,nk)))\in nw(S)) implies not(b2=ibase)
To prove the property, we follow the same methodology as above. An interesting proof passage that refers to the transition fkm31(s,bs1,q1,c1,mc2) is:

```
open ISTEP
  -- arbitrary objects
ops q1 q2 : -> Node .
op mc2 : -> Mac2 .
op c1 : -> Cipher .
op bs1 : -> Base .
  -- assumptions
  -- eq c-fkm31(s,bs1,q1,c1,mc2) = true .
eq (enc(bk,nk) \in ciphers(nw(s))) = true .
eq (mac2(bk,n(n1,r),n2,enc(bk,nk)) \in macs2(nw(s)) ) = true .
  -- subcase m3 = m3'
  -- eq (m3(b1,b2,n1,enc(bk,nk),mac2(bk,n(n1,r),n2,enc(bk,nk))) =
  -- (m3(ibase,bs1,q1,c1,mc2))) = true .
eq b1 = ibase .
eq b2 = bs1 .
eq n1 = q1 .
eq c1 = enc(bk,nk) .
```

```
eq mc2 = mac2(bk,n(n1,r),n2,enc(bk,nk)) .
-- sub-subcases
eq not(n(bk) = enemy) = true .
eq not(b(bk) = ibase) = true .
eq bs1 = ibase .
-- successor state
eq s' = fkm31(s,bs1,q1,c1,mc2) .
-- check
red istep1(b1,b2,n1,n2,bk,nk,r) .
close
```

The proof passage refers to the subcase where `c-fkm31(s,bs1,q1,c1,mc2) = true`
$\bigwedge$ `m3 = m3'` $\bigwedge$ `not(n(bk) = enemy)` $\bigwedge$ `not(b(bk) = ibase)` $\bigwedge$ `bs1 = ibase`.

6   Conclusion

We have applied the OTS/CafeOBJ technique for the modeling of security protocols for sensor networks. We have also verified that the protocols possess one important invariant property. Researchers from the OBJ community have applied
the method successfully in numerous systems specification and verification. Authentication protocols [8], real time systems[9], mutual exclusion algorithms[11],
railroad signaling systems [12], digital rights management systems[13] and mobile systems [14][15] are a subset of CafeOBJ applications.

References

1. Bjørner, D.: Logics of Formal Specification Languages - The Possible Worlds cum
   Domain Problem. In: Proceedings of 4th Panhellenic Symposium on Logic (2003)
2. Ogata, K., Futatsugi, K.: Proof Scores in the OTS/CafeOBJ Method. In: Najm, E.,
   Nestmann, U., Stevens, P. (eds.) FMOODS 2003. LNCS, vol. 2884, pp. 170–184.
   Springer, Heidelberg (2003)
3. Ogata, K., Futatsugi, K.: Some Tips on Writing Proof Scores in the OTS/CafeOBJ
   Method. In: Futatsugi, K., Jouannaud, J.-P., Meseguer, J. (eds.) Algebra, Meaning,
   and Computation. LNCS, vol. 4060, pp. 596–615. Springer, Heidelberg (2006)
4. Diaconescu, R., Futatsugi, K.: CafeOBJ Report. World Scientific, Singapore (1998)
5. Perrig, A., Szewczyk, R., Wen, V., Culler, D., Tygar, J.D.: SPINS: Security Protocols for Sensor Networks. In: Proceedings of MOBICOM 2001, pp. 189–199 (2001)
6. Goguen, J., Malcolm, G.: A hidden agenda. Technical Report CS97-538, University
   of California at San Diego (1997)
7. Dolev, D., Yao, A.C.: On the security of public key protocols. IEEE Trans. Inform.
   Theory IT-29, 198–208 (1983)
8. Ogata, K., Futatsugi, K.: Rewriting - based verification of authentication protocols.
   In: WRLA '02. ENTCS, vol. 71 (2002)
9. Ogata, K., Futatsugi, K.: Modeling and Verification of Real-Time Systems Based
   on Equations. Science of computer programming 66(2), 162–180 (2007)
10. Diaconescu, R., Futatsugi, K., Ogata, K.: CafeOBJ: Logical Foundations and
    Methodologies. Computing and Informatics 22, 1001–1025 (2003)

11. Ogata, K., Futatsugi, K.: Formal analysis of Suzuki and Kasami distributed mutual exclusion algorithm. In: FMOODS '02, pp. 181–195 (2002)
12. Seino, T., Ogata, K., Futatsugi, K.: Specification and Verification of a Single-Track Railroad Signaling in CafeOBJ. IEICE Trans. Fundamentals E84-A(6), 1471–1478 (2001)
13. Xiang, J., Kong, W., Futatsugi, K., Ogata, K.: Analysis of Positive Incentives for Protecting Secrets in Digital Rights Management. In: WEBIST '06 (2006)
14. Ouranos, I. Stefaneas, P., Frangos, P.: A Formal Specification Framework for ad hoc mobile communication networks. In: van Leeuwen, J., Italiano, G.F., van der Hoek, W., Meinel, C., Sack, H., Plášil, F., Bielikova, M. (eds.) SOFSEM 2007, vol. 2, pp. 91–102, Institute of Computer Science AS CR, Prague (2007)
15. Ouranos, I., Stefaneas, P., Frangos, P.: An Algebraic Framework for Modeling of Mobile Systems. IEICE Trans. Fundamentals E90-A(9) (to appear, 2007)

Nonassociativity à la Kleene

Jean-Marcel Pallo

Universit de Bourgogne, LE2I UMR 5158, BP 47870, 21078 DIJON-Cedex, France
pallo@u-bourgogne.fr

Abstract. First we recall the work of Suschkewitsch (1929) about the generalization of the associative law which is the starting point of the theory of quasigroups. Then we show that it is a particular case of the notion of relative associativity introduced by Roubaud in 1965. Thereafter we prove a coherence theorem over an infinite set of nonassociative operations. This result contains all the uppermentioned contributions. This allows to obtain a very general à-la-Kleene theorem on rational series which uses concatenations that can be associative or not.

1 Introduction

This paper is devoted to nonassociativity in formal language theory. More precisely, we establish a à-la-Kleene theorem. This result concerns rational series using concatenations that may be either associative or nonassociative.

There is a huge amount of literature about associative binary systems (monoids, semigroups...). On the contrary, very few contributions deal with nonassociative algebras except, obviously, Lie algebras. See for example the book's chapter "Nonassociative Structures" in [12] and the special issue on nonassociative algebras in [15]. However, the subject of several old papers is precisely the withdrawal of the associative axiom. There are at least two reasons for this droping out. First, the purpose of mathematical generalization (quasigroups, loops...[11,14,19]). It is the case in the paper of Suschkewitsch [33], the main idea of which is restated here in. Second, practical reasons specific to the domain of application (genetics, data analysis, quantum mechanics... [3,16]). In the area of linguistics, see for instance the paper of Roubaud [29] and the book of Harris [10, page 157]: in the word-sequence of a sentence, concatenation is semantically nonassociative.

The formal language theory has been developed mainly over free monoids generated by alphabets, namely over sets with an associative concatenation. Its natural achievement is the classical theory of formal power series on words [5,13,32]. A way to grasp nonassociativity lies in considering trees instead of words. This has sparked a large number of papers on tree grammars and tree languages [7]. Formal power series on trees appear as a generalization of the classical theory of formal power series on words [6]. Otherwise, linear languages in their full generality, defined with a nonassociative concatenation, have been introduced in [9], using notations of λ-calculus and functional programming.

Another approach has been introduced by Roubaud in 1965 [27,28,29]. If f is an associative operation, one can swap parentheses $f(x, f(y, z)) = f(f(x, y), z)$.

In a set of nonassociative operations, it is possible to swap parentheses provided that we also swap the operations $f(x, g(y, z)) = h(k(x, y), z)$. This equation has been studied in the framework of homogeneous groupods in [30] and as a functional equation in [1, page 311]. Starting from certain arbitrary word, the swap of parentheses needed to get them on the left side of the word can be achieved in several ways. It is therefore natural to require that the result be always the same. We obtain coherence conditions. In the context of category theory, the term "coherence conditions" refers to a set of diagrams whose commutativity implies the commutativity of a larger class of diagrams [17,18]. It is the case in this paper since we show that pentagone coherence implies coherence for wider diagrams. For this purpose we define an associative relative magma. Generalizing some results of [22,25], we prove that this magma verifies the above coherence conditions. Surprisingly, the generalization of Suschkewitsch is just a particular case of this magma! Rational series defined on this magma can be characterized à-la-Kleene in their full generality.

2   Suschkewitsch Generalization of the Associative Law

In [33], Anton Suschkewitsch observes that the proof of Lagrange theorem does not make any use of the associative law. This law can be replaced by his more general postulates, A and B namely. His nonassociative binary systems, such so-called "general groups", seem to be the predecessors of modern quasigroups. See [26] for an interesting historical point of view.

Postulate A of [33] can be writen as: in the equation $(X \star A) \star B = X \star C$ the element C depends upon the elements A and B only and not upon X.

Suschkewitsch also considers a special case of Postulate A which is however more general than the associative law. He states its Postulate B as: in the equation $X \star (A \star B) = (X \star A) \star \bar{B}$ the elements B and $\bar{B}$ depend only upon each other. Every B is completely defined by the corresponding $\bar{B}$ and conversely. The corresponding rewriting system has been studied in [8].

Let us recall a result of [33] that will be pointed out in the sequel. In Postulate A, let us denote C by $A \circ B$, i.e. $(X \star A) \star B = X \star (A \circ B)$. Let us prove that $\circ$ is an associative operation. Using this Postulate, we can write:

$((X \star A) \star B) \star C = (X \star (A \circ B)) \star C = X \star ((A \circ B) \circ C)$ and
$((X \star A) \star B) \star C = (X \star A) \star (B \circ C) = X \star (A \circ (B \circ C))$. Therefore we obtain:
$(A \circ B) \circ C = A \circ (B \circ C)$, i.e. $\circ$ is associative.

The law $(x * y) * z = x * (y \circ z)$ as a functional equation has been examined in [1, pages 253 and 315] and [2].

3   Roubaud Relative Associativity

Given a set $\Delta = \{\alpha, \beta, \gamma, \ldots\}$ of binary operations and $g, d : \Delta \times \Delta \to \Delta$, Jacques Roubaud introduces in [29] an axiom of relative associativity:

$\alpha(x, \beta(y, z)) = g(\alpha, \beta)(d(\alpha, \beta)(x, y), z)$ which verifies the following conditions for all $\alpha, \beta, \gamma \in \Delta$:

$$\begin{cases} g(\alpha, g(\beta, \gamma)) = g(g(\alpha, \beta), \gamma) \\ g(d(\alpha, g(\beta, \gamma)), d(\beta, \gamma)) = d(g(\alpha, \beta), \gamma) \\ d(d(\alpha, g(\beta, \gamma)), d(\beta, \gamma)) = d(\alpha, \beta) \end{cases}$$

Indeed, moving parentheses in $\alpha(x, \beta(y, \gamma(z, t)))$ at the left side can be obtained in the two following ways:

$\alpha(x, \beta(y, \gamma(z, t))) = g(\alpha, \beta)(d(\alpha, \beta)(x, y), \gamma(z, t)) =$
$g(g(\alpha, \beta), \gamma)(d(g(\alpha, \beta), \gamma)(d(\alpha, \beta)(x, y), z), t)$
and
$\alpha(x, \beta(y, \gamma(z, t))) = \alpha(x, g(\beta, \gamma)(d(\beta, \gamma)(y, z), t)) =$
$g(\alpha, g(\beta, \gamma))(d(\alpha, g(\beta, \gamma))(x, d(\beta, \gamma)(y, z)), t) =$
$g(\alpha, g(\beta, \gamma))(g(d(\alpha, g(\beta, \gamma)), d(\beta, \gamma))(d(d(\alpha, g(\beta, \gamma)), d(\beta, \gamma))(x, y), z), t)$

A convenient coherence condition should impose that these two results are equal. Thus g is associative.

For example, $g(\alpha, \beta) = \beta$ and $d(\alpha, \beta) = \alpha$ verify the above conditions. The law $\alpha(x, \beta(y, z)) = \beta(\alpha(x, y), z)$ as a relation on homogeneous groupods has been studied in [31].

If $(\Delta, +)$ is a monoid, $g(\mu, \nu) = \mu + \nu$ and $d(\mu, \nu) = \mu$ verify also these conditions. Note that exhausting all possible operations g and d is an open problem.

4   Notation and Definitions

Let $V = \{a, b, c, d \ldots\}$ be an alphabet, i.e. a finite nonempty set of letters and $V^\star$ the free monoid generated by V. Let $\mathcal{F} = \{f_p^\alpha, \alpha \in \Omega, p \in \mathbb{Z}\}$ a set of binary operations with index p in the set of relative integers and exponent α in a finite or infinite set of greek letters $\Omega = \{\alpha, \beta, \gamma, \mu, \nu, \ldots\}$.

We denote $\mathcal{M} = \mathcal{M}(\mathcal{F}, V^\star)$ the free magma over $V^\star$ equipped with the binary operations of $\mathcal{F}$, i.e. the smallest subset of the free monoid $(\mathcal{F} \bigcup V)^\star$ generated by the disjoint union of $\mathcal{F}$ and V, which contains $V^\star$ and satisfies the inductive rule: if $P, Q \in \mathcal{M}$ and $f_p^\alpha \in \mathcal{F}$ then $f_p^\alpha(P, Q) \in \mathcal{M}$. Elements of $\mathcal{M}$ are called generalized words (g-words in short).

Given $w \in \mathcal{M}$, we denote by $|w|$ the length of w, i.e. the number of letters of V in w, each letter is counted as many times it occurs. Let us denote $\mathcal{M}_n$ the set of g-words such that $|w| = n$. For example $f_2^\alpha(a, f_1^\beta(f_3^\beta(b, a), f_0^\alpha(c, b))) \in \mathcal{M}_5$.

Given $w \in \mathcal{M}$, we denote by $sk(w)$ the skeleton of w, i.e. the word obtained in deleting the operations of $\mathcal{F}$ as well as the variable separators (commas). For example, if $w = f_5^\mu(f_0^\nu(a, b), f_{-2}^\gamma(b, f_{-1}^\alpha(f_2^\eta(c, a), d)))$ then $sk(w) = ((ab)(b((ca)d)))$.

We call left g-word a g-word in which all operations of $\mathcal{F}$ occur at the left side of the g-word followed by the letters of V. For example
$f_2^\nu(f_{-1}^\mu(f_0^\nu(f_3^\alpha(f_{-4}^\varsigma(a, b), c), b), a), c)$ is a left g-word.

Definition 1. *The relative associative magma* $\mathcal{R} = \mathcal{R}(\mathcal{M})$ *is the magma* $\mathcal{M}$ *equipped with the relation (RAS): for all* $\alpha, \beta \in \Omega, p, q \in \mathbb{Z}, P, Q, R \in \mathcal{R}$ *we have*

$$f_p^\alpha(P, f_q^\beta(Q, R)) = f_{p+q}^\beta(f_p^\alpha(P, Q), R).$$

The relation above is the combination of both solutions of the coherence conditions appearing at the end of the previous Section. However, for the sake of simplicity in exposition, we prefer to consider as a model the standard case of integers with their addition rather than considering the general monoidal setting.

Let us notice that operations of $\mathcal{F}$ indexed by 0 are associative since from (RAS) we deduce $f_0^\alpha(P, f_0^\alpha(Q, R)) = f_0^\alpha(f_0^\alpha(P, Q), R)$.

If $p \neq 0$, the operations f_p^α are nonassociative since
$f_p^\alpha(P, f_p^\alpha(Q, R)) = f_{2p}^\alpha(f_p^\alpha(P, Q), R)$ and $f_p^\alpha(f_p^\alpha(P, Q), R) = f_p^\alpha(P, f_0^\alpha(Q, R))$.

The last equality is a generalization of Postulate A of Suschkewitsch [33], namely $(X \star A) \star B = X \star (A \circ B)$ where $\circ$ is associative!

Now let us consider Postulate B of Suschkewitsch [33] writen as: $f_1^\alpha(X, f_1^\alpha(A, B)) = f_1^\alpha(f_1^\alpha(X, A), \bar{B})$. From (RAS) we obtain $f_1^\alpha(f_1^\alpha(X, A), \bar{B}) = f_1^\alpha(X, f_0^\alpha(A, \bar{B}))$ and therefore $f_1^\alpha(A, B) = f_0^\alpha(A, \bar{B})$. We can easily prove that:
$$f_p^\alpha(A, B) = f_0^\alpha(A, \bar{B}^{-p}) \text{ and } f_0^\alpha(P, f_0^\alpha(Q^{-p}, R^{-p+q})) = f_0^\alpha(f_0^\alpha(P, Q^{-p}), R^{-p+q}).$$

Definition 2. *Let us denote* λ *the right unit in* $\mathcal{R}$, *i.e. for all* $P \in \mathcal{R}, \alpha \in \Omega, p \in \mathbb{Z}$ *we have* $f_p^\alpha(P, \lambda) = P$.

5   Coherence Results

Definition 3. *Let us define the relation* $\rightarrow$ *on* $\mathcal{M}$ *as the smallest preordering, left and right invariant with respect to* $\mathcal{F}$, *i.e. if* $P \rightarrow Q$ *then* $f_p^\alpha(P, R) \rightarrow f_p^\alpha(Q, R)$ *and* $f_p^\alpha(R, P) \rightarrow f_p^\alpha(R, Q)$ *for all* $R \in \mathcal{M}, \alpha \in \Omega, p \in \mathbb{Z}$, *and satisfying for all* $P, Q, R \in \mathcal{M}, \alpha, \beta \in \Omega, p, q \in \mathbb{Z}$:

$$f_p^\alpha(P, f_q^\beta(Q, R)) \rightarrow f_{p+q}^\beta(f_p^\alpha(P, Q), R).$$

Lemma 1. *We have* $f_p^\alpha(A, B) \rightarrow f_q^\beta(C, D)$ *iff either (1)*

$$\begin{cases} A \rightarrow C \\ B \rightarrow D \\ \alpha = \beta \\ p = q \end{cases}$$

or there exists $S \in \mathcal{M}$ *with* $|S| \geq 1$ *such that (2)*

$$\begin{cases} B \rightarrow f_{q-p}^\beta(S, D) \\ f_p^\alpha(A, S) \rightarrow C \end{cases}.$$

Proof. The conditions are obviously sufficient since
$f_p^\alpha(A, B) \rightarrow f_p^\alpha(A, f_{q-p}^\beta(S, D)) \rightarrow f_q^\beta(f_p^\alpha(A, S), D) \rightarrow f_q^\beta(C, D)$.

For proving the necessity, let us consider the relation $\prec$ on $\mathcal{M}$ defined by $f_p^\alpha(A, B) \prec f_q^\beta(C, D)$ iff conditions either (1) or (2) are verified.

$\prec$ is reflexive and invariant with respect to $\mathcal{F}$. Let us prove the transitivity of $\prec$, i.e. if $f_p^\alpha(A, B) \prec f_q^\beta(C, D)$ and $f_q^\beta(C, D) \prec f_r^\gamma(E, F)$ then we have $f_p^\alpha(A, B) \prec f_r^\gamma(E, F)$. Among the four cases to study, we only detail the following one. If there exist S and S' such that

$$\begin{cases} B \to f_{q-p}^\beta(S, D) \\ f_p^\alpha(A, S) \to C \end{cases}$$

and

$$\begin{cases} D \to f_{r-q}^\gamma(S', F) \\ f_q^\beta(C, S') \to E \end{cases}$$

then there exists S'' such that

$$\begin{cases} B \to f_{r-p}^\gamma(S'', F) \\ f_p^\alpha(A, S'') \to E \end{cases}.$$

Indeed $S'' = f_{q-p}^\beta(S, S')$ verifies:
$$B \to f_{q-p}^\beta(S, D) \to f_{q-p}^\beta(S, f_{r-q}^\gamma(S', F)) \to f_{r-p}^\gamma(f_{q-p}^\beta(S, S'), F) = f_{r-p}^\gamma(S'', F)$$
and $f_p^\alpha(A, S'') = f_p^\alpha(A, f_{q-p}^\beta(S, S')) \to f_q^\beta(f_p^\alpha(A, S), S') \to f_q^\beta(C, S') \to E$. $\quad\square$

Definition 4. *Given $w \in \mathcal{M}_n$, let us call associahedron $\mathcal{AS}_n(w)$ the diagram which is obtained from w by applying all possible $\to$ and $\overset{-1}{\to}$ relations.*

See for example the pentagone $\mathcal{AS}_4(f_p^\alpha(x, f_q^\beta(y, f_r^\gamma(z, t))))$ in Figure 1.

Theorem 1. *Given $w \in \mathcal{M}_n$, the associahedron $\mathcal{AS}_n(w)$ is coherent, i.e. there are no $w', w'' \in \mathcal{M}_n$ such that $w' \neq w''$ and $sk(w') = sk(w'')$.*

Proof. If $x, y, z, t, u, v \in V$, then $\mathcal{AS}_4(f_p^\alpha(x, f_q^\beta(y, f_r^\gamma(z, t))))$ is coherent: see Figure 1. See also the 14 elements of $\mathcal{AS}_5(f_p^\alpha(x, f_q^\beta(y, f_r^\gamma(z, f_s^\delta(u, v)))))$ below and on Figure 2.

$$f_p^\alpha(x, f_q^\beta(y, f_r^\gamma(z, f_s^\delta(u, v))))$$
$$f_{p+q}^\beta(f_p^\alpha(x, y), f_{r+s}^\delta(f_r^\gamma(z, u), v))$$
$$f_p^\alpha(x, f_{q+r}^\gamma(f_q^\beta(y, z), f_s^\delta(u, v)))$$
$$f_{p+q+r+s}^\delta(f_p^\alpha(x, f_q^\beta(y, f_r^\gamma(z, u))))$$
$$f_p^\alpha(x, f_{q+r+s}^\delta(f_{q+r}^\gamma(f_q^\beta(y, z), u), v))$$
$$f_{p+q+r}^\gamma(f_{p+q}^\beta(f_p^\alpha(x, y), z), f_s^\delta(u, v))$$
$$f_{p+q+r+s}^\delta(f_{p+q+r}^\gamma(f_{p+q}^\beta(f_p^\alpha(x, f_q^\beta(y, z)), u), v)$$

$$f_p^\alpha(x, f_q^\beta(y, f_{r+s}^\delta(f_r^\gamma(z, u), v)))$$
$$f_{p+q+r+s}^\delta(f_{p+q}^\beta(f_p^\alpha(x, y), f_r^\gamma(z, u)), v)$$
$$f_p^\alpha(x, f_{q+r+s}^\delta(f_q^\beta(y, f_r^\gamma(z, u)), v))$$
$$f_{p+q}^\beta(f_p^\alpha(x, y), f_r^\gamma(z, f_s^\delta(u, v)))$$
$$f_{p+q+r}^\gamma(f_p^\alpha(x, f_q^\beta(y, z)), f_s^\delta(u, v))$$
$$f_{p+q+r+s}^\delta(f_p^\alpha(x, f_{q+r}^\gamma(f_q^\beta(y, z), u)), v)$$
$$f_{p+q+r+s}^\delta(f_{p+q+r}^\gamma(f_{p+q}^\beta(f_p^\alpha(x, y), z), u), v)$$

By induction on n, suppose that for $n \geq 6$ and for some $v \in \mathcal{M}_n$ we have in $\mathcal{AS}_n(v)$: $w = f_p^\alpha(A, B) \to w' = f_{q'}^{\beta'}(C', D')$ and $w = f_p^\alpha(A, B) \to w'' = f_{q''}^{\beta''}(C'', D'')$, with $|w| = |w'| = |w''| = n$, $w' \neq w''$ and $sk(w') = sk(w'')$. Then we obtain $sk(C') = sk(C'')$ and $sk(D') = sk(D'')$. We apply Lemma 1.

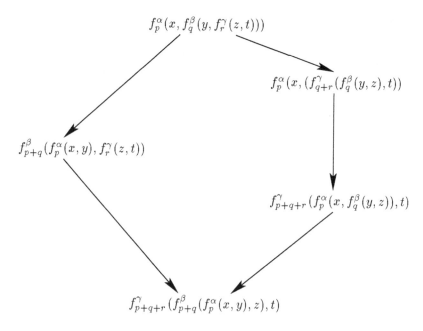

Fig. 1. The pentagon $\mathcal{AS}_4$

If $A \to C'$, $B \to D'$, $\alpha = \beta'$, $p = q'$ and if $A \to C''$, $B \to D''$, $\alpha = \beta''$, $p = q''$, then we have $|A| \leq n-1$, $|B| \leq n-1$ and by the inductive hypothesis $C' = C''$ and $D' = D''$ since $sk(C') = sk(C'')$ and $sk(D') = sk(D'')$. A contradiction follows.

If $A \to C'$, $B \to D'$, $\alpha = \beta'$, $p = q'$, then a g-word S'' verifying $B \to f^{\beta''}_{q''-p}(S'', D'')$ and $f^\alpha_p(A, C') \to C''$ cannot exist. Indeed $sk(D') = sk(D'')$ should imply $|S''| = |B| - |D''| = 0$.

Now let us assume that there exist S' and S'' such that $B \to f^{\beta'}_{q'-p}(S', D')$, $f^\alpha_p(A, S') \to C'$, $B \to f^{\beta''}_{q''-p}(S'', D'')$, $f^\alpha_p(A, S'') \to C''$. Let us denote $l(E)$ the unique g-word such that $E \to l(E)$ with $l(E)$ being a left g-word. $l(E)$ is uniquely defined by induction if $|E| \leq n-1$. The assumption $D' \neq D''$ implies $l(D') \neq l(D'')$ since $sk(D') = sk(D'')$. Therefore it implies also the existence in $\mathcal{AS}_{|B|}(B)$ of two g-words, namely $f^{\beta'}_{q'-p}(l(S'), l(D'))$ and $f^{\beta''}_{q''-p}(l(S''), l(D''))$, which have equal skeletons but are different since $l(D') \neq l(D'')$. This contradicts the inductive hypothesis. Therefore we obtain $D' = D''$. We have $B \to f^{\beta'}_{q'-p}(l(S'), D')$ and $B \to f^{\beta''}_{q''-p}(l(S''), D')$. These two g-words have the same skeletons in $\mathcal{AS}_{|B|}(B)$ with $|B| \leq n-1$. They are therefore equal: $\beta' = \beta''$, $q' = q''$ and $l(S') = l(S'')$. Now let us consider the following diagram:

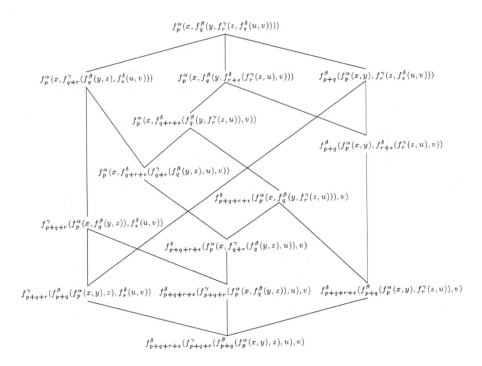

Fig. 2. The associahedron $\mathcal{AS}_5$

$$C' \leftarrow f_p^\alpha(A, S') \rightarrow f_p^\alpha(A, l(S')) = f_p^\alpha(A, l(S'')) \leftarrow f_p^\alpha(A, S'') \rightarrow C''$$

The induction hypothesis can be applied because $|C'| = |C''| \leq n - 1$. Then $sk(C') = sk(C'')$ implies $C' = C''$ and a contradiction holds.   □

Remark 1. The associahedrons $\mathcal{AS}_n(w)$ endowed with the ordering $\rightarrow$ are lattices for all n and $w \in \mathcal{M}_n$. This is an immediate consequence of the fact that the skeleton of $\mathcal{AS}_n(w)$ is the well-known n-th Tamari lattice. Tamari lattices have been extensively studied for algebraic and combinatorial purposes. A number of references on this subject are available in [24].

Remark 2. The rewrite ordering $\rightarrow$ is convergent. The unique normal form of w is the left g-word $l(w)$.

Theorem 2. *Given a set of nonassociative operations* $\Xi = \{f^\alpha, f^\beta, f^\gamma, \ldots\}$ *on* V^*, *the embedding* Ψ *of* $\mathcal{M}(\Xi, V^*)$ *in* $\mathcal{M}(\mathcal{F}, V^*)$ *with* $\mathcal{F} = \{f_p^\alpha, \alpha \in \Omega, p \in \mathbb{Z}\}$ *defined by* $\Psi(x) = x$ *for all* $x \in V^*$ *and* $\Psi(f^\alpha) = f_1^\alpha, \Psi(f^\beta) = f_1^\beta, \Psi(f^\gamma) = f_1^\gamma \ldots$ *is injective.*

Proof. The proof is by induction on the length of the g-words of $\mathcal{M}(\Xi, V^*)$. There are two types of g-words of length 3, namely $f^\alpha(x, f^\alpha(y, z))$ and $f^\alpha(f^\alpha(x, y), z)$ with $x, y, z \in V$. Their images are respectively $f_1^\alpha(x, f_1^\alpha(y, z)) =$

$f_2^\alpha(f_1^\alpha(x,y),z)$ and $f_1^\alpha(f_1^\alpha(x,y),z) = f_1^\alpha(x,f_0^\alpha(y,z))$. They are quite different. For g-words of length n, let $w = f^\alpha(A,B)$ and $w' = f^\alpha(A',B')$ such that $\Psi(w) = \Psi(w')$, i.e. $f_1^\alpha(\Psi(A),\Psi(B)) = f_1^\alpha(\Psi(A'),\Psi(B'))$. If we denote $l(C)$ the unique left g-word such that $C \rightarrow l(C)$, then $f_1^\alpha(\Psi(A),\Psi(B)) \rightarrow f_1^\alpha(l(\Psi(A')),l(\Psi(B')))$. If there exists S such that $\Psi(B) \rightarrow f_0^\alpha(S,l(\Psi(B')))$, then following Theorem 1 we obtain $l(\Psi(B)) = l(f_0^\alpha(S,l(\Psi(B'))))$. $l(\Psi(B))$ cannot contain operations indexed by 0, hence the contradiction. Therefore we have $\Psi(A) \rightarrow l(\Psi(A'))$, $\Psi(B) \rightarrow l(\Psi(B'))$ and by symmetry $\Psi(A) = \Psi(A')$ and $\Psi(B) = \Psi(B')$ proving that $w = w'$. □

Remark 3. The embedding of a nonassociative operation f^α into an infinite set $\{f_p^\alpha, p \in \mathbb{Z}\}$ may look strange. We had better embed into a finite set of operations. If one considers $\mathbb{Z}/n\mathbb{Z}$ in place of $\mathbb{Z}$, coherence Theorem 1 still holds. Unfortunately, in this case we obtain necessarily weak associativity equalities. For instance, if $n = 3$:

$$f_1^\alpha(x, f_1^\alpha(y, f_1^\alpha(z, f_1^\alpha(u,v)))) = f_1^\alpha(f_1^\alpha(x, f_1^\alpha(y, f_1^\alpha(z,u))), v)$$

since the normal forms of both members in this equality are:

$$f_1^\alpha(f_0^\alpha(f_2^\alpha(f_1^\alpha(x,y),z),u),v).$$

It is completely clear that the larger n, the less weak associativity relations. But by Theorem 2, we see that no weak associativity relation occurs as soon as we embed into a countable set.

6 Rational Formal Power Series

We use the classical notations on formal power series described in [5,13,32]. Given a semiring $\mathcal{A}$, we denote by $\mathcal{A}[[\mathcal{R}]]$ the set of formal series

$$s = \sum_{\sigma \in \mathcal{R}} <s,\sigma> \sigma$$

where $<s,\sigma> \in \mathcal{A}$.

The sum of two series is classicaly defined. The product $s = f_p^\alpha(s',s'')$ is defined by $<s,\sigma> = <s',\sigma'><s'',\sigma''>$ if $\sigma = f_p^\alpha(\sigma',\sigma'')$ and $<s,\sigma> = 0$ otherwise.

$s \in \mathcal{A}[[\mathcal{R}]]$ is proper if the coefficient of the right unit λ (i.e. the constant term of s) vanishes: $<s,\lambda> = 0$.

In this case, the series $s^{*\alpha} = \lambda + s + f_0^\alpha(s,s) + f_0^\alpha(s, f_0^\alpha(s,s)) + f_0^\alpha(s, f_0^\alpha(s, f_0^\alpha(s,s)) + \dots$ is defined. Since f_0^α is associative, we have also $s^{*\alpha} = \lambda + s + f_0^\alpha(s,s) + f_0^\alpha(f_0^\alpha(s,s),s) + f_0^\alpha(f_0^\alpha(f_0^\alpha(s,s),s),s) + \dots$.

Definition 5. *We call $s^{*\alpha}$ the Kleene star of the series $s \in \mathcal{A}[[\mathcal{R}]]$ with respect to $\alpha \in \Omega$.*

Remark 4. It is enough to observe that this Kleene star operation $s^{*\alpha}$ is defined from an associative operation f_0^α as in the classical case.

Lemma 2. *Let $r, s \in \mathcal{A}[[\mathcal{R}]]$ with s proper. Then the unique solution u of the left-linear equation $u = r + f_p^\alpha(u, s)$ is the series $u = f_p^\alpha(r, s^{*\alpha})$.*

Proof. One has $s^{*\alpha} = \lambda + f_0^\alpha(s^{*\alpha}, s)$ whence $f_p^\alpha(r, s^{*\alpha}) = f_p^\alpha(r, \lambda) + f_p^\alpha(r, f_0^\alpha(s^{*\alpha}, s))$. Since λ is a right unit and following (RAS): $f_p^\alpha(r, s^{*\alpha}) = r + f_p^\alpha(f_p^\alpha(r, s^{*\alpha}), s)$.

Conversely, from $u = r + f_p^\alpha(u, s)$ it follows that $u = r + f_p^\alpha(r + f_p^\alpha(u, s), s) = r + f_p^\alpha(r, s) + f_p^\alpha(u, f_0^\alpha(s, s)) = r + f_p^\alpha(r, s) + f_p^\alpha(u, f_0^\alpha(f_0^\alpha(s, s), s))) = \cdots$

Inductively and going to the limit, one gets $u = f_p^\alpha(r, s^{*\alpha})$ since s is proper. □

7   Kleene Theorem

Definition 6. *A formal series is rational if it is an element of the smallest subset $Rat[[\mathcal{M}]]$ of $\mathcal{A}[[\mathcal{R}]]$ containing V^* and closed for the sum, product and Kleene star operations $*_\alpha$ for all $\alpha \in \Omega$.*

Definition 7. *A left-linear system of order N with rational coefficients is a system of the form*

$$u_i = r_i + \sum_{1 \le j \le N} \sum_{1 \le k \le N_j} f_{p_{i,j,k}}^{\alpha_{i,j,k}}(u_j, s_{i,j,k})$$

with $1 \le i \le N$ where all $r_i, s_{i,j,k} \in Rat[[\mathcal{R}]]$, $\alpha_{i,j,k} \in \Omega$ and $p_{i,j,k} \in \mathbb{Z}$.

Theorem 3. *The components of the N-tuple solution of a left-linear system with proper rational coefficients are rational series. Conversely, a rational series can be obtained as a component of a N-tuple solution of such a system.*

Proof. The proof is by induction on N. According to Lemma 2, the solution of $u = r + f_p^\alpha(u, s)$ is $u = f_p^\alpha(r, s^{*\alpha})$ which is a rational series since $r, s \in Rat[[\mathcal{R}]]$. The solution u of the equation $u = r + f_p^\alpha(u, s) + f_q^\beta(u, t)$ is solution of the equation $u = f_q^\beta(r + f_p^\alpha(u, s), t^{*\beta})$ and we can write:

$u = f_q^\beta(r, t^{*\beta}) + f_q^\beta(f_p^\alpha(u, s), t^{*\beta})$. Thus $u = f_q^\beta(r, t^{*\beta}) + f_p^\alpha(u, f_{q-p}^\beta(s, t^{*\beta}))$ and the solution $u = f_p^\alpha(f_q^\beta(r, t^{*\beta}), (f_{q-p}^\beta(s, t^{*\beta}))^{*\alpha})$ is rational since $r, s, t \in Rat[[\mathcal{R}]]$. Similarly and by induction on M, the solution of

$$u = r + \sum_{1 \le k \le M} f_{p_k}^{\alpha_k}(u, s_k)$$

is rational if $r, s_k \in Rat[[\mathcal{R}]]$ for all $k \in [1, M]$. In a system $\mathcal{S}$ of order N, u_N is rationally computed from $u_1, u_2, \ldots, u_{N-1}$ and the induction hypothesis is applied.

Conversely, let us prove that the components which are solutions of left-linear systems with rational coefficients verify the conditions of Definition 6. Let us denote u_1 (respectively u_1') the first component of the N-tuple solution (respectively N'-tuple solution) of a system $\mathcal{S}$ (respectively $\mathcal{S}'$):

$$\mathcal{S} : u_i = r_i + \sum_{1 \leq j \leq N} \sum_{1 \leq k \leq N_j} f_{p_{i,j,k}}^{\alpha_{i,j,k}}(u_j, s_{i,j,k}), 1 \leq i \leq N$$

and

$$\mathcal{S}' : u_i' = r_i' + \sum_{1 \leq j \leq N'} \sum_{1 \leq k \leq N_j'} f_{p_{i,j,k}'}^{\alpha_{i,j,k}'}(u_j', s_{i,j,k}'), 1 \leq i \leq N'$$

where all $r_i, r_i', s_{i,j,k}, s_{i,j,k}' \in Rat[[\mathcal{R}]]$, $\alpha_{i,j,k}, \alpha_{i,j,k}' \in \Omega$ and $p_{i,j,k}, p_{i,j,k}' \in \mathbb{Z}$.

It is easy to exhibit a system which admits as solution $c_1 u_1 + c_1' u_1'$ with $c_1, c_1' \in \mathcal{A}$. Now, let $\tilde{u}_i = f_p^\alpha(u_1', u_i)$. Then

$$f_p^\alpha(u_1', u_i) = f_p^\alpha(u_1', r_i) + \sum_{1 \leq j \leq N} \sum_{1 \leq k \leq N_j} f_p^\alpha(u_1', f_{p_{i,j,k}}^{\alpha_{i,j,k}}(u_j, s_{i,j,k}))$$

and

$$f_p^\alpha(u_1', u_i) = f_p^\alpha(u_1', r_i) + \sum_{1 \leq j \leq N} \sum_{1 \leq k \leq N_j} f_{p+p_{i,j,k}}^{\alpha_{i,j,k}}(f_p^\alpha(u_1', u_j), s_{i,j,k}).$$

Thus $\tilde{u}_1 = f_p^\alpha(u_1', u_1)$ is the first component of the N-tuple solution of the system $\tilde{\mathcal{S}}$:

$$\tilde{u}_i = f_p^\alpha(u_1', r_i) + \sum_{1 \leq j \leq N} \sum_{1 \leq k \leq N_j} f_{p+p_{i,j,k}}^{\alpha_{i,j,k}}(\tilde{u}_j, s_{i,j,k}).$$

To conclude, let $\bar{u}_i = f_0^\alpha(u_1^{*\alpha}, u_i)$. Then

$$f_0^\alpha(u_1^{*\alpha}, u_i) = f_0^\alpha(u_1^{*\alpha}, r_i) + \sum_{1 \leq j \leq N} \sum_{1 \leq k \leq N_j} f_0^\alpha(u_1^{*\alpha}, f_{p_{i,j,k}}^{\alpha_{i,j,k}}(u_j, s_{i,j,k}))$$

and

$$f_0^\alpha(u_1^{*\alpha}, u_i) = f_0^\alpha(u_1^{*\alpha}, r_i) + \sum_{1 \leq j \leq N} \sum_{1 \leq k \leq N_j} f_{p_{i,j,k}}^{\alpha_{i,j,k}}(f_0^\alpha(u_1^{*\alpha}, u_j), s_{i,j,k}).$$

Thus $\bar{u}_1 = f_0^\alpha(u_1^{*\alpha}, u_1) = u_1^{*\alpha} - \lambda$ is the first component of the N-tuple solution of the system $\mathcal{S}^*$:

$$\bar{u}_i = f_0^\alpha(u_1^{*\alpha}, r_i) + \sum_{1 \leq j \leq N} \sum_{1 \leq k \leq N_j} f_{p_{i,j,k}}^{\alpha_{i,j,k}}(\bar{u}_j, s_{i,j,k}). \qquad \square$$

8   Generalization of the Relative Associativity

The coherence Theorem 1 about binary operations can be generalized to ternary operations by usual binarizations. The left (respectively right) binarization of a ternary operation t is obtained by replacing $t(P, Q, R)$ by $b(b(P, Q), R)$ (respectively $b(P, b(Q, R))$) where b is a binary operation. Since our binary operations have indices in $\mathbb{Z}$ and exponents in Ω, the indices and exponents of our forthcoming ternary operations will live in $\mathbb{Z} \times \mathbb{Z}$ and $\Omega \times \Omega$. More precisely, $g_{p,q}^{\alpha,\beta}(P, Q, R)$ (respectively $h_{p,q}^{\alpha,\beta}(P, Q, R)$) will come from the left binarization $f_p^\alpha(f_q^\beta(P, Q), R)$ (respectively right binarization $f_p^\alpha(P, f_q^\beta(Q, R))$).

By left binarization, the relation $\rightarrow$ of Definition 3 can be generalized as follows:

$$g_{p,q}^{\alpha,\beta}(P, Q, g_{r,s}^{\gamma,\delta}(R, S, T)) \xrightarrow{1} g_{p+r,q}^{\gamma,\beta}(P, g_{p+s-q,p-q}^{\delta,\alpha}(Q, R, S), T)$$
$$g_{p,q}^{\alpha,\beta}(P, g_{r,s}^{\gamma,\delta}(Q, R, S), T) \xrightarrow{2} g_{p,q+r}^{\alpha,\gamma}(g_{q+s,q}^{\delta,\beta}(P, Q, R), S, T)$$
$$g_{p,q}^{\alpha,\beta}(P, Q, g_{r,s}^{\gamma,\delta}(R, S, T)) \xrightarrow{3} g_{p+r,p+s}^{\gamma,\delta}(g_{p,q}^{\alpha,\beta}(P, Q, R), S, T)$$

with $\xrightarrow{3} = \xrightarrow{2} \circ \xrightarrow{1}$ and for all $P, Q, R, S, T \in \mathcal{M}$, $\alpha, \beta, \gamma, \delta \in \Omega$, $p, q, r, s \in \mathbb{Z}$.

See Figure 3 with $p, q, r, s, u, v \in \mathbb{Z}$, $\alpha, \beta, \gamma, \delta, \eta, \nu \in \Omega$ and $x_1, x_2, x_3, x_4, x_5, x_6, x_7 \in V$. By right binarization, the relation $\rightarrow$ of Definition 3 can be generalized as follows:

$$h_{p,q}^{\alpha,\beta}(P, Q, h_{r,s}^{\gamma,\delta}(R, S, T)) \xrightarrow{1} h_{p,q+r+s}^{\alpha,\delta}(P, h_{q,r}^{\beta,\gamma}(Q, R, S), T)$$
$$h_{p,q}^{\alpha,\beta}(P, h_{r,s}^{\gamma,\delta}(Q, R, S), T) \xrightarrow{2} h_{p+r+s,q-r-s}^{\delta,\beta}(h_{p,r}^{\alpha,\gamma}(P, Q, R), S, T)$$
$$h_{p,q}^{\alpha,\beta}(P, Q, h_{r,s}^{\gamma,\delta}(R, S, T)) \xrightarrow{3} h_{p+q+r,s}^{\gamma,\delta}(h_{p,q}^{\alpha,\beta}(P, Q, R), S, T)$$

with $\xrightarrow{3} = \xrightarrow{2} \circ \xrightarrow{1}$. See Figure 4.

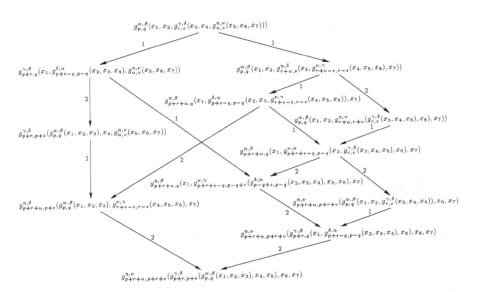

Fig. 3. The ternary associahedron obtained by left binarization

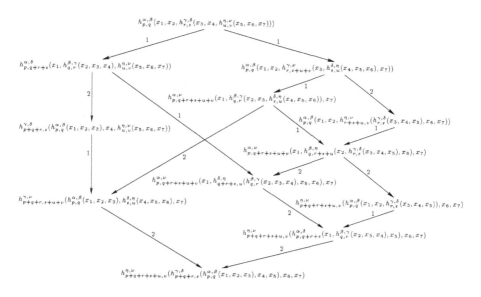

Fig. 4. The ternary associahedron obtained by right binarization

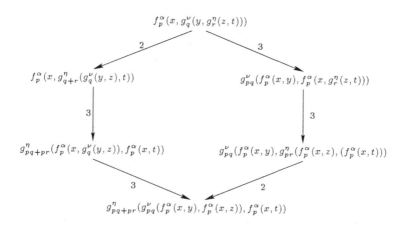

Fig. 5. f-g-g distributive diagram

Notice that the skeleton of Figures 3 and 4 is not a lattice but is a χ-lattice in the sense of Leutola-Nieminen [23].

Distributivity in the classical sense [20] can be generalized to "relative" distributibity as follows. The family of binary operations $\{f_p^\alpha, \alpha \in \Omega, p \in \mathbb{Z}\}$ is said relative distributive with respect to the family $\{g_q^\eta, \eta \in \Lambda, q \in \mathbb{Z}\}$ if the following conditions hold:

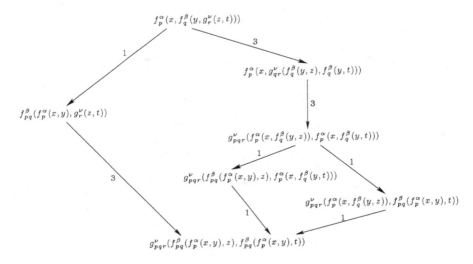

Fig. 6. f-f-g distributive diagram

$$f_p^\alpha(P, f_q^\beta(Q,R)) \xrightarrow{1} f_{pq}^\beta(f_p^\alpha(P,Q),R)$$
$$g_p^\eta(P, g_q^\nu(Q,R)) \xrightarrow{2} g_{p+q}^\nu(g_p^\eta(P,Q),R)$$
$$f_p^\alpha(P, g_q^\eta(Q,R)) \xrightarrow{3} g_{pq}^\eta(f_p^\alpha(P,Q), f_p^\alpha(P,R))$$
for all $P, Q, R \in \mathcal{M}$, $\alpha, \beta \in \Omega$, $\eta, \nu \in \Lambda$ and $p, q \in \mathbb{Z}$.
The corresponding diagrams are coherent: see Figures 5 and 6.

9   Conclusion

Theorem 3 characterizes -la-Kleene rational series defined in a very general way. This is done with concatenations that may be either associative or not. Moreover, our Kleene stars are defined uniquely on the basis of associative operations as in the classical case. This noteworthy fact follows from the formula $f_p^\alpha(f_p^\alpha(P,Q),R) = f_p^\alpha(P, f_0^\alpha(Q,R))$ which generalizes the Postulate A of Suschkewitsch [33]. The proof of Theorem 3 works since, pointing out the role of f, it follows from (RAS) that $\mathbf{f}(P, g(Q,R)) = h(\mathbf{f}(P,Q),R)$. Indeed the first part of the proof is based upon $f_q^\beta(f_p^\alpha(u,s),t) = f_p^\alpha(u, f_{q-p}^\beta(s,t))$. For the converse, we twice use the relation $f_p^\alpha(u, f_{p'}^{\alpha'}(s,t)) = f_{p+p'}^{\alpha'}(f_p^\alpha(u,s),t)$, first for p arbitrary, second for $p = 0$.

In a forthcoming paper, we shall use (RAS) in order to generalize a result on the Hadamard product of an algebraic language and a rational language [6,28].

Acknowledgments

I am very indebted to an anonymous referee for proofreading my paper and Jean-Paul Gauthier for linguistic help.

References

1. Ackél, J.: Lectures on Functional Equations and their Applications. Academic Press, New-York (1966)
2. Ackél, J., Hosszú, M.: On transformations with several parameters and operations in multidimensional spaces. Acta Math. Acad. Sci. Hungaricae 7, 327–338 (1956)
3. d'Adhémar, C.: Quelques classes de groupoídes non-associatifs. Math. Sci. Hum. 31, 17–31 (1970)
4. Alexandrakis, A., Bozapalidis, S.: Weighted grammars and Kleene's theorem. Inform. Process. Lett. 24, 1–4 (1987)
5. Berstel, J., Reutenauer, C.: Rational Series and their Langages. In: EATCS Monographs on Theoretical Computer Science, vol. 12, Springer, Heidelberg (1988)
6. Bozapalidis, S.: Context-free series on trees. Inform. Comput. 169, 186–229 (2001)
7. Gécseg, F., Steinby, M.: Tree Languages. In: Rosenberg, G., Salomaa, A. (eds.) Handbook of Formal Languages. Beyond Words, vol. 3, pp. 1–68. Springer, Heidelberg (1997)
8. Germain, C., Pallo, J.: Langages rationnels définis avec une concaténation non-associative. Theor. Comput. Sci. 233, 217–231 (2000)
9. Germain, C., Pallo, J.: Linear languages with a nonassociative concatenation. J. Autom. Lang. Comb. 7, 311–320 (2002)
10. Harris, Z.: A Theory of Language and Information: a Mathematical Approach. Oxford University Press, Oxford (1991)
11. Hausmann, B.A., Ore, O.: Theory of quasigroups. Amer. J. Math. 59, 983–1004 (1937)
12. Kostrikin, A.L., Shafarevich, I.R. (eds.): Algebra VI. Encyclopaedia of Mathematical Sciences, vol. 57. Springer, Heidelberg (1995)
13. Kuich, W.: Semirings and Formal Power Series: Their Relevance to Formal Languages and Automata. In: Rosenberg, G., Salomaa, A. (eds.) Handbook of Formal Languages. Word, Language, Grammar, vol. 1, pp. 609–677. Springer, Heidelberg (1997)
14. Kunen, K.: Quasigroups, loops and associative laws. J. Algebra 185, 194–204 (1996)
15. Lõhmus, J.: Preface to the Special Issue on Nonassociative Algebras, Quasigroups and Applications in Physics. Acta Applic. Math. 50, 1–2 (1998)
16. Lõhmus, J., Paal, E., Sorgsepp, L.: About nonassociativity in mathematics and physics. Acta Applic. Math. 50, 3–31 (1998)
17. Mac Lane, S.: Natural associativity and commutativity. Rice Univ. Stud. 49, 28–46 (1963)
18. Mac Lane, S.: Preface to "Coherence in Categories". Lecture Notes in Math., vol. 281. Springer, Heidelberg (1970)
19. Murdoch, D.C.: Quasigroups which satisfy certain generalized associative law. Amer. J. Math. 61, 509–522 (1939)
20. Pallo, J.M.: Word problem in distributive magmas. Fund. Inform. 4, 957–973 (1981)
21. Pallo, J.M.: Modéles associatif-relatif et commutatif cohérents appliqués aux langages réguliers. Calcolo 19, 289–300 (1982)
22. Pallo, J.M.: Coding binary trees by embedding into the Roubaud's magma. Rad. Mat. 2, 21–34 (1986)
23. Pallo, J.M.: The rotation χ-lattice of ternary trees. Computing 66, 297–308 (2001)
24. Pallo, J.M.: Generating binary trees by Glivenko classes on Tamari lattices. Inform. Process. Lett. 85, 235–238 (2003)

25. Pallo, J.M.: Permutoassociaédres d'arbres binaires étiquetés. Rad. Mat. 13, 5–14 (2004)
26. Pflugfelder, H.O.: Historical notes on loop theory. Comment. Math. Univ. Carolinae 41, 359–370 (2000)
27. Roubaud, J.: Types d'A-algèbres discrètes complètes: un théorème des fonctions implicites. C. R. Acad. Sc. Paris 261, 3005–3007 (1965)
28. Roubaud, J.: Sur un théorème de M. P. Schützenberger. C. R. Acad. Sc. Paris 261, 3265–3267 (1965)
29. Roubaud, J.: La notion d'associativité relative. Math. Sci. Hum. 34, 43–59 (1970)
30. Sade, A.: Théorie des systèmes demosiens de groupoïdes. Pacific J. Math. 10, 625–660 (1960)
31. Sade, A.: Groupoïdes en relation associative et semigroupes. Ann. Soc. Sc. Bruxelles 75, 52–57 (1961)
32. Salomaa, A.: Formal Languages and Power Series. In: Van Leeuwen, J. (ed.) Handbook of Theoretical Computer Science, ch. 3, vol. B, pp. 103–132. Elsevier, Amsterdam (1990)
33. Suschkewitsch, A.: On a generalization of the associative law. Trans. Amer. Math. Soc. 31, 204–214 (1929)

Restarting Tree Automata and Linear Context-Free Tree Languages

Heiko Stamer and Friedrich Otto

Fachbereich Elektrotechnik/Informatik, Universität Kassel
D-34109 Kassel, Germany
{stamer,otto}@theory.informatik.uni-kassel.de

Abstract. We derive a normal form for linear context-free tree grammars that involves only growing productions. Based on this normal form we then show that all linear context-free tree languages are recognized by restarting tree automata which utilize auxiliary symbols.

1 Introduction

Originally restarting automata were introduced to model the so-called *analysis by reduction*, which is a technique used in linguistics to analyze sentences of natural languages with free word order [9,13]. This technique consists in a stepwise simplification of a sentence in such a way that the syntactical correctness or incorrectness is not affected. After a finite number of steps either a correct simple sentence is obtained, or the core of an error is detected.

A restarting automaton has a finite-state control and a read/write window of a fixed size that works on a flexible tape delimited by a left and a right sentinel. It works in cycles, where in each cycle the current string is read and a single local rewrite is executed. After a finite number of cycles the automaton either halts and accepts, or it halts without accepting.

Actually, many variants of restarting automata have been introduced and studied. All these variants work on linear text, that is, on strings, although trees are often used in linguistics (as well as in formal language theory and its applications) to describe sentences of a language together with some structural information. Therefore, in [17] the notion of restarting automata has been extended from strings to trees, defining some basic types of *restarting tree automata*, and establishing some fundamental results on their expressive power and on the closure properties of the recognized families of tree languages.

Here we continue the study of restarting tree automata by showing that each linear context-free tree language is recognized by a restarting tree automaton that is allowed to use auxiliary symbols. Linear context-free tree grammars are of interest from a linguistic point of view, as linear, nondeleting, monadic context-free tree grammars generate the same class of string languages as tree adjoining grammars [7] and some other formalisms studied in linguistics (see, e.g., [18]). Our proof is based on the fact that each linear context-free tree grammar can be transformed into an equivalent linear context-free tree grammar that has only

S. Bozapalidis and G. Rahonis (Eds.): CAI 2007, LNCS 4728, pp. 275–289, 2007.
© Springer-Verlag Berlin Heidelberg 2007

growing productions. Fujiyoshi [5,6] has shown a similar result for linear context-free tree grammars that are monadic. Furthermore, Seki and Kato [16, Lemma 7] have shown a corresponding normalization result for macro grammars.

This paper is structured as follows. After giving the basic definitions and establishing the necessary notation in Section 2, we present the announced transformation of linear context-free tree grammars in Section 3. Then in Section 4 we establish our main result showing that each linear context-free tree language is recognized by a restarting tree automaton. The paper closes with some remarks and suggestions for future work.

2 Preliminaries

A *ranked alphabet* $\mathcal{F}$ is a finite nonempty set of symbols such that each $f \in \mathcal{F}$ has a unique nonnegative *arity* (or *rank*) denoted by $\mathrm{Rnk}(f)$. Symbols of arity zero are called *constants*. The subset of symbols of arity n is denoted by $\mathcal{F}_n$. Further, let $\mathcal{X} = \{ x_i \mid i \geq 1 \}$ be a countable set of *variables*, which are symbols of rank zero. Note that $\mathcal{X}$ is always assumed to be disjoint from any other ranked alphabet. Finally, let $\mathcal{X}_n = \{x_1, \ldots, x_n\}$ be the finite subset of $\mathcal{X}$ that consists of the first $n \geq 1$ variables (with respect to the natural ordering on the set of indices).

By $\mathcal{T}(\mathcal{F}, \mathcal{X})$ we denote the set of all *terms over $\mathcal{F}$ with variables in $\mathcal{X}$*. For $t \in \mathcal{T}(\mathcal{F}, \mathcal{X})$, $\mathrm{Var}(t)$ denotes the set of variables that occur in t. A term is *linear*, if no variable occurs more than once in it. Terms without variables are called *ground terms*. The set of ground terms over $\mathcal{F}$ is denoted by $\mathcal{T}(\mathcal{F})$.

A term can be considered as a finite ordered *ranked tree* whose leaves are labeled with variables or constants and whose internal nodes are labeled with symbols of positive arity. Positions in a term are represented by sequences of positive integers, where the empty sequence ε denotes the position at the root of the tree. The set of positions of a term $t \in \mathcal{T}(\mathcal{F}, \mathcal{X})$ is denoted by $\mathrm{Pos}(t)$. By $\mathrm{Top}(t)$ we denote the *outermost* symbol of t, which is the symbol at the root. For $p \in \mathrm{Pos}(t)$ we use the notation $t|_p$ to denote the *subterm* of t at position p. Further, by $t[u]_p$ we denote the term that is obtained from t by replacing the subterm $t|_p$ by the term u. Finally, a term t is called a *scattered subterm* of a term t', if t is homeomorphically embedded in t', that is, t can be obtained from t' by "striking out" some parts (see, e.g., [3]). The *size* $\|t\|$ and the *height* $\mathrm{Hgt}(t)$ of a term t are defined inductively as follows:

$$
\begin{aligned}
\|t\| &= 0, & \mathrm{Hgt}(t) &= 0, & &\text{if } t \in \mathcal{X}, \\
\|t\| &= 1, & \mathrm{Hgt}(t) &= 0, & &\text{if } t \in \mathcal{F}_0, \\
\|t\| &= 1 + \sum_{i=1}^{n} \|(t|_i)\|, & \mathrm{Hgt}(t) &= 1 + \max_{i=1,\ldots,n} \mathrm{Hgt}(t|_i), & &\text{if } n \geq 1 \text{ and} \\
& & & & &\mathrm{Top}(t) \in \mathcal{F}_n.
\end{aligned}
$$

A *substitution* is a mapping from $\mathcal{X}$ into $\mathcal{T}(\mathcal{F}, \mathcal{X})$ that is the identity on all but finitely many variables. Each substitution can uniquely be extended to the domain $\mathcal{T}(\mathcal{F}, \mathcal{X})$. We write substitutions always in prefix notation, that is, σt denotes the result of applying σ to the term $t \in \mathcal{T}(\mathcal{F}, \mathcal{X})$.

Let ϵ be a special constant from $\mathcal{F}_0$, and let $\Sigma_{\mathcal{F}} = \{\, a \mid a \in \mathcal{F}_0 \setminus \{\epsilon\}\,\}$ be the finite alphabet that contains all constants from $\mathcal{F}_0$ but ϵ. Then $\Sigma_{\mathcal{F}}^*$ denotes the set of all words over $\Sigma_{\mathcal{F}}$ including the empty word ε. The *yield* of a ground term $t \in T(\mathcal{F})$ is the string defined by the mapping Yld : $T(\mathcal{F}) \rightarrow \Sigma_{\mathcal{F}}^*$, where

$$\begin{aligned}
\text{Yld}(t) &= a, &&\text{if } t = a \text{ and } a \in \Sigma_{\mathcal{F}}, \\
\text{Yld}(t) &= \varepsilon, &&\text{if } t = \epsilon, \\
\text{Yld}(t) &= \text{Yld}(t|_1) \cdot \text{Yld}(t|_2) \cdots \text{Yld}(t|_n), &&\text{if } n \geq 1 \text{ and } \text{Top}(t) \in \mathcal{F}_n.
\end{aligned}$$

We abuse the notation Yld for the yield of a set of ground terms $E \subseteq T(\mathcal{F})$, that is, $\text{Yld}(E) = \{\,\text{Yld}(t) \mid t \in E\,\}$, which is a string language over $\Sigma_{\mathcal{F}}$.

A linear term $t \in T(\mathcal{F}, \mathcal{X}_n)$ satisfying the condition $\text{Var}(t) = \mathcal{X}_n$ is called an *n-context*, and by $t[t_1, \ldots, t_n]$ we denote the term that is obtained from t by replacing each variable $x_i \in \mathcal{X}_n$ by $t_i \in T(\mathcal{F}, \mathcal{X})$ ($1 \leq i \leq n$). The set of all *n*-contexts is denoted as $\text{Ctx}(\mathcal{F}, \mathcal{X}_n)$.

A *rewrite rule* is a pair of terms, denoted by $l \rightarrow r$, where $l, r \in T(\mathcal{F}, \mathcal{X})$, $l \notin \mathcal{X}$, and $\text{Var}(l) \supseteq \text{Var}(r)$. It is called *linear*, if both l and r are linear terms. A rule is called *nondeleting*, if $\text{Var}(l) = \text{Var}(r)$, that is, all variables of l occur in r. A *term rewriting system* (TRS) is a set $\mathcal{R}$ of rewrite rules. The induced *rewrite relation* $\rightarrow_{\mathcal{R}}$ over $T(\mathcal{F}, \mathcal{X})$ is the least relation containing $\mathcal{R}$ that is closed under subterm replacement and substitution. Thus, a term $t \in T(\mathcal{F}, \mathcal{X})$ rewrites to t', denoted as $t \rightarrow_{\mathcal{R}} t'$, if there exist a rewrite rule $(l \rightarrow r) \in \mathcal{R}$, a substitution $\sigma : \mathcal{X} \rightarrow T(\mathcal{F}, \mathcal{X})$, and a position $p \in \text{Pos}(t)$ such that $t|_p = \sigma l$ and $t' = t[\sigma r]_p$. By $\rightarrow_{\mathcal{R}}^*$ we denote the reflexive transitive closure of $\rightarrow_{\mathcal{R}}$. A TRS $\mathcal{R}$ is called *linear (nondeleting)*, if all its rewrite rules are linear (nondeleting).

A *context-free tree grammar* (CFTG) $G = (\mathcal{F}, \mathcal{N}, \mathcal{P}, S)$ consists of two disjoint ranked alphabets $\mathcal{F}$ and $\mathcal{N}$, a finite TRS $\mathcal{P}$, and an initial symbol $S \in \mathcal{N}_0$. The elements of $\mathcal{F}$ are terminal symbols, and those of $\mathcal{N}$ are nonterminal symbols. All rules (productions) from $\mathcal{P}$ are of the form $A(x_1, \ldots, x_n) \rightarrow t$, where $n \geq 0$, $A \in \mathcal{N}_n$, $\{x_1, \ldots, x_n\} = \mathcal{X}_n$, and $t \in T(\mathcal{F} \cup \mathcal{N}, \mathcal{X}_n)$ is a term. The *unrestricted derivation relation* $\Rightarrow_G$ and the reflexive transitive closure $\Rightarrow_G^*$ are induced by $\mathcal{P}$. The tree language generated by G is $L(G) = \{\, t \in T(\mathcal{F}) \mid S \Rightarrow_G^* t\,\}$.

For any class G of tree grammars, $\mathscr{L}(\mathsf{G})$ denotes the *class of tree languages* that are generated by grammars from G. A context-free tree grammar is called *linear* (lin-CFTG), if the corresponding TRS $\mathcal{P}$ is linear. It is called *monadic*, if $\mathcal{N} = \mathcal{N}_0 \cup \mathcal{N}_1$, that is, all nonterminal symbols are constants or unary functions, and it is called *regular* (RTG), if all its nonterminal symbols are constants, that is, $\mathcal{N} = \mathcal{N}_0$. A context-free tree grammar is called *nondeleting*, if the corresponding TRS $\mathcal{P}$ is nondeleting. It is called *growing*, if each rule of $\mathcal{P}$ is of the form $A(x_1, \ldots, x_n) \rightarrow t$, where $n \geq 0$, $A \in \mathcal{N}_n$, and $t \in T(\mathcal{F} \cup \mathcal{N}, \mathcal{X}_n)$ satisfying $\text{Var}(t) = \mathcal{X}_n$ and $||t|| \geq 2$, or of the form $S \rightarrow s$, where $s \in T(\mathcal{F} \cup \mathcal{N})$. Observe that a growing grammar is necessarily nondeleting. A set of ground terms $E \subseteq T(\mathcal{F})$ is a *(linear) context-free tree language*, if there exists a (linear) context-free tree grammar G such that $L(G) = E$.

We recall some results on context-free tree languages. Guessarian has defined an equivalent type of automaton – the so-called *pushdown tree automaton* [8]. This automaton is equipped with an auxiliary pushdown storage and works in top-down manner. Later Schimpf and Gallier introduced a *tree pushdown automaton* that processes its input tree from the leaves to the root (bottom-up). It was shown that this automaton yields another equivalent representation for context-free tree languages [15]. Moreover, for every context-free tree language $E \subseteq T(\mathcal{F})$ the corresponding yield language $\text{Yld}(E) \subseteq \Sigma_{\mathcal{F}}^*$ is an indexed language [2,4]. The converse statement is also true, since our definition of the Yld-mapping incorporates the special constant $\epsilon \in \mathcal{F}_0$. Moreover, it was shown that the derivation mode used (inside-out or outside-in, respectively) is irrelevant for linear context-free tree grammars [10]. We will use this fact extensively in Section 3.

For context-free tree grammars several normal forms have been proposed (see, e.g., [1,11,12,14]). Here we only need the following variant which was originally introduced by Maibaum [11] and later specified by Schimpf and Gallier [15] in more detail. A context-free tree grammar is in *Chomsky normal form* (CNF), if each production is of one of the following types

$$F(x_1, \ldots, x_n) \to G(H_1(x_1, \ldots, x_n), \ldots, H_m(x_1, \ldots, x_n)),$$
$$F(x_1, \ldots, x_n) \to f(x_{j_1}, \ldots, x_{j_m}),$$
$$F(x_1, \ldots, x_n) \to x_k,$$

where $m \geq 0$, $j_1, \ldots, j_m, k \in \{1, \ldots, n\}$, $F \in \mathcal{N}_n$, $G \in \mathcal{N}_m$, $H_i \in \mathcal{N}_n$ $(1 \leq i \leq m)$, $f \in \mathcal{F}_m$, and $x_i \in \mathcal{X}_n$ $(1 \leq i \leq n)$. Maibaum and Schimpf have shown that every context-free tree grammar G can be rewritten as a grammar G' in Chomsky normal form such that $L(G) = L(G')$ holds.

Of course, to convert an arbitrary linear context-free tree grammar into a similar normal form, the first of the above types of productions must be modified slightly to

$$F(x_1, \ldots, x_n) \to G(H_1(x_{j_{1,1}}, \ldots, x_{j_{1,h_1}}), \ldots, H_m(x_{j_{m,1}}, \ldots, x_{j_{m,h_m}})),$$

where $x_{j_{1,1}}, \ldots, x_{j_{1,h_1}}, \ldots, x_{j_{m,1}}, \ldots, x_{j_{m,h_m}}$ are distinct variables from $\mathcal{X}_n$, $F \in \mathcal{N}_n$, $G \in \mathcal{N}_m$, and $H_i \in \mathcal{N}_{h_i}$ for some integer $h_i \in \{0, \ldots, n\}$ $(1 \leq i \leq m)$ such that $\sum_{i=1}^m h_i \leq n$ holds. However, such a restricted form can be obtained easily by adjusting Maibaum's construction accordingly.

3   Growing Linear Tree Grammars

In this section we show that each linear context-free tree grammar can be transformed into an equivalent linear context-free tree grammar that is growing. This transformation is achieved in two steps: first we show how to transform a linear context-free tree grammar into an equivalent nondeleting linear context-free tree grammar, and then we present a transformation into an equivalent growing linear context-free tree grammar.

Proposition 1. *From a given linear context-free tree grammar G a nondeleting linear context-free tree grammar G' can be constructed such that $L(G) = L(G')$ holds.*

Proof. Let $G = (\mathcal{F}, \mathcal{N}, \mathcal{P}, S)$ be a linear context-free tree grammar in Chomsky normal form. We will construct a sequence of linear context-free tree grammars $G = G^0, G^1, \ldots, G^\ell = G'$, where $G^k = (\mathcal{F}, \mathcal{N}^k, \mathcal{P}^k, S)$ for $0 \le k \le \ell$, such that all these grammars generate the same tree language, and $G' = G^\ell$ is nondeleting, that is, all rules $(l \to r) \in \mathcal{P}^\ell$ are linear context-free productions satisfying the condition $\mathrm{Var}(l) = \mathrm{Var}(r)$. Throughout this construction we will maintain a set $\mathcal{P}'$ that will contain all those rules which have already been processed.

Our construction consists of three transformation rules T_1, T_2, and T_3. First T_1 is applied as long as it is applicable, then T_2 is applied iteratively as long as possible, and then the same is done with T_3. Once this process terminates, the linear context-free tree grammar obtained has the desired properties. We start with the tree grammar $G^0 = G$, that is, $\mathcal{N}^0 := \mathcal{N}$, $\mathcal{P}^0 := \mathcal{P}$, and $\mathcal{P}' = \emptyset$. Below we describe the various transformation rules in detail. If G^k is the current tree grammar, then the next transformation step will generate the tree grammar G^{k+1} from G^k. It starts by taking $\mathcal{N}^{k+1} := \mathcal{N}^k$ and $\mathcal{P}^{k+1} := \mathcal{P}^k$.

– *Transformation rule T_1:*
Choose a projection rule $F(x_1, \ldots, x_n) \to x_j$ from $\mathcal{P}^{k+1}$, delete it from $\mathcal{P}^{k+1}$, and add it to the set $\mathcal{P}'$. Now consider all rules $(l \to r) \in \mathcal{P}^{k+1}$ that contain an occurrence of the symbol F in their right-hand side r. Let R consist of all terms that are obtained by replacing one or more subterms of r with outermost symbol F by their j-th subterm, respectively, that is, if $r|_p = F(s_1, \ldots, s_n)$, then $r[s_j]_p$ is contained in R. For all $r' \in R$, if $l \ne r'$ and $(l \to r') \notin \mathcal{P}'$, then add the rule $l \to r'$ to the set $\mathcal{P}^{k+1}$. Here the test $(l \to r') \notin \mathcal{P}'$ is used to ensure that no rule is introduced into $\mathcal{P}^{k+1}$ that has already been processed previously. This completes the description of transformation rule T_1.

Unfortunately, this transformation will in general destroy the Chomsky normal form, as may obtain rules like $G(x_1, \ldots, x_m) \to H(x_1, \ldots, x_m)$ or even new projection rules like $G(x_1, \ldots, x_m) \to x_j$. However, the process of iterating T_1 will terminate eventually, since for each rule $l \to r'$ introduced, we have $||r'|| < ||r||$. Thus, there exists a point from where on no new projection rules can occur, that is, all projection rules that are generated from that point on are already contained in $\mathcal{P}'$.

– *Transformation rule T_2:*
Choose a rule $F(x_1, \ldots, x_n) \to f(x_{j_1}, \ldots, x_{j_{\bar{n}}})$ from $\mathcal{P}^{k+1}$ such that $0 \le \bar{n} < n$ and $1 \le j_i \le n$ for all $i = 1, \ldots, \bar{n}$. If $\bar{F}_\alpha \notin \mathcal{N}^{k+1}$, add a fresh nonterminal $\bar{F}_\alpha$ of arity $\bar{n}$ to $\mathcal{N}^{k+1}$, delete the above rule, and create the new rule $\bar{F}_\alpha(x_{\mu_1}, \ldots, x_{\mu_{\bar{n}}}) \to f(x_{j_1}, \ldots, x_{j_{\bar{n}}})$, where $\alpha = \{1, \ldots, n\} \setminus \{j_1, \ldots, j_{\bar{n}}\}$ is a label that indicates the arguments removed, $1 \le \mu_1 < \mu_2 < \cdots < \mu_{\bar{n}} \le n$, and $\mu_i = \pi(j_i)$ for all $1 \le i \le \bar{n}$ and an appropriate permutation π of the index set $\{1, \ldots, n\}$. By replacing the variable x_{μ_i} by x_i for all $i = 1, \ldots, \bar{n}$, we obtain

the *normalized* variant of this rule, which is added to $\mathcal{P}^{k+1}$. Further, the rule $F(x_1, \ldots, x_n) \to f(x_{j_1}, \ldots, x_{j_{\hat{n}}})$ is put into $\mathcal{P}'$.

Next, we consider all productions $(l \to r) \in \mathcal{P}^{k+1}$ with at least one occurrence of F in the right-hand side. Let $p_1, \ldots, p_m \in \mathrm{Pos}(r)$ be those positions for which $\mathrm{Top}(r|_{p_i}) = F$ holds. Build all possible variants of the rule $l \to r$, where the subterm $F(t_1, \ldots, t_n)$ of r is replaced by $\bar{F}_\alpha(t_{\mu_1}, \ldots, t_{\mu_{\hat{n}}})$ at some of the positions $p_1, \ldots, p_m$. All these new rules are added to $\mathcal{P}^{k+1}$. Note that the original production $l \to r$ must remain in $\mathcal{P}^{k+1}$, as there may exist other rules with left-hand side $F(x_1, \ldots, x_n)$. However, if no such rule exists, then $l \to r$ has become a useless production, and we can remove it from $\mathcal{P}^{k+1}$.

– *Transformation rule T_3:*
Choose a rule $F(x_1, \ldots, x_n) \to H(t_1, \ldots, t_m)$ from $\mathcal{P}^{k+1}$ such that $t_1, \ldots, t_m \in \mathcal{T}(\mathcal{N}, \{x_1, \ldots, x_n\})$ and $\hat{n} := |\bigcup_{i=1}^{m} \mathrm{Var}(t_i)| < n$. We delete this rule from $\mathcal{P}^{k+1}$ and add it to $\mathcal{P}'$. If $\hat{F}_\alpha \notin \mathcal{N}^{k+1}$, where $\alpha = \{1, \ldots, n\} \smallsetminus \{ j \mid x_j \in \bigcup_{i=1}^{m} \mathrm{Var}(t_i) \}$ is a label that indicates the arguments removed, then we add a fresh nonterminal $\hat{F}_\alpha$ of arity $\hat{n}$ to $\mathcal{N}^{k+1}$. Further, if $\hat{F}_\alpha(x_{\mu_1}, \ldots, x_{\mu_{\hat{n}}}) \neq H(t_1, \ldots, t_m)$, then we create the rule $\hat{F}_\alpha(x_{\mu_1}, \ldots, x_{\mu_{\hat{n}}}) \to H(t_1, \ldots, t_m)$, where $1 \leq \mu_1 < \mu_2 < \cdots < \mu_{\hat{n}} \leq n$ and μ_i is appropriately chosen ($1 \leq i \leq \hat{n}$). As in transformation T_2 we normalize this rule by replacing the variable x_{μ_i} by x_i for all $i = 1, \ldots, \hat{n}$. If the resulting rule is not contained in $\mathcal{P}'$, then we add it to $\mathcal{P}^{k+1}$. Again, for all productions $(l \to r) \in \mathcal{P}^{k+1}$ with at least one occurrence of F in the right-hand side r, we enlarge $\mathcal{P}^{k+1}$ by all possible combinations that are obtained by replacing a subterm $F(s_1, \ldots, s_n)$ of r by the term $\hat{F}_\alpha(s_{\mu_1}, \ldots, s_{\mu_{\hat{n}}})$. These new rules $l \to r'$ are only inserted into $\mathcal{P}^{k+1}$, if $l \neq r'$, $(l \to r') \notin \mathcal{P}^{k+1}$, and $(l \to r') \notin \mathcal{P}'$, that is, if they are nontrivial, if they are not already contained in $\mathcal{P}^{k+1}$, and if they have not already been processed before.

Termination: Note that each transformation removes a rule, or it replaces a rule $l \to r$ by some rules $l' \to r'$ such that $\mathrm{Rnk}(\mathrm{Top}(l')) < \mathrm{Rnk}(\mathrm{Top}(l))$ holds, or a subterm of r is replaced in r' by a term with an outermost symbol of smaller arity. Moreover, the number of newly introduced nonterminals is bounded by the initial grammar G^0. Finally, loops are avoided by using the set $\mathcal{P}'$. It follows that each sequence of transformations T_1, T_2, and then T_3 terminates after finitely many steps.

Correctness: All transformations preserve linearity and context-freeness. The projection rules of G are removed by T_1, and no new projection rules are introduced by T_2 or T_3, as these transformations always replace nonterminals by other nonterminals. Analogously, rules of the form $F(x_1, \ldots, x_n) \to f(x_{j_1}, \ldots, x_{j_{\hat{n}}})$, which are replaced by T_2, cannot be reintroduced by T_3. Thus, when the transformation process terminates after ℓ steps, then $\mathcal{P}^\ell$ contains no deleting rule. Hence, G^ℓ is a nondeleting linear context-free tree grammar. It remains to show that $L(G^\ell) = L(G)$. For that it suffices to prove $L(G^k) = L(G^{k+1})$.

Claim 1. $L(G^k) \subseteq L(G^{k+1})$.

Proof. Let $S \Rightarrow^*_{G^k} u_1 \Rightarrow_{G^k} v_1 \Rightarrow^*_{G^k} t$ be a derivation of minimal length of the ground term $t \in \mathcal{T}(\mathcal{F})$, where $u_1 \Rightarrow_{G^k} v_1$ is the first step that uses a rewrite rule $(l \to r) \in G^k \setminus G^{k+1}$. Hence, there exist a position $p \in \mathrm{Pos}(u_1)$ and a substitution σ such that $u_1|_p = \sigma(l)$ and $v_1 = u_1[\sigma(r)]_p$. Note that for linear context-free tree grammars the derivation strategy used is of no concern [10]. Thus, we may assume that some steps of a derivation occur adjacent to each other. Now consider each transformation rule separately:

- T_1: Then $(l \to r)$ is of the form $F(x_1, \ldots, x_n) \to x_j$, that is, $u_1|_p = F(t_1, \ldots, t_n)$ and $v_1 = u_1[t_j]_p$. Let $u'_1 \Rightarrow_{G^k} v'_1$ be the derivation step at which the occurrence of the nonterminal F at position $p \in \mathrm{Pos}(u_1)$ is generated, that is, at this step a rule $(l_1 \to r_1) \in \mathcal{P}^k$ is used such that the corresponding subterm $r_1|_q$ of r_1 is of the form $F(s_1, \ldots, s_n)$. Now $\mathcal{P}^{k+1}$ contains the rule $l_1 \to r_1[s_j]_q$. By applying this rule to u'_1 we obtain a G^{k+1}-derivation of v_1 from u'_1.

- T_2: Then $(l \to r)$ is of the form $F(x_1, \ldots, x_n) \to f(x_{j_1}, \ldots, x_{j_{\bar n}})$, where $\bar n < n$. Thus, $u_1|_p = F(t_1, \ldots, t_n)$ and $v_1|_p = f(t_{j_1}, \ldots, t_{j_{\bar n}})$. Again, let $u'_1 \Rightarrow_{G^k} v'_1$ be the derivation step at which the occurrence of the nonterminal F at position $p \in \mathrm{Pos}(u_1)$ is generated, that is, at this step a rule $(l_1 \to r_1) \in \mathcal{P}^k$ is used such that the corresponding subterm $r_1|_q$ of r_1 is of the form $F(s_1, \ldots, s_n)$. Now $\mathcal{P}^{k+1}$ contains the added rule $l_1 \to r_1[\bar F_\alpha(s_{\mu_1}, \ldots, s_{\mu_{\bar n}})]_q$. By using this rule instead of the original rule $l_1 \to r_1$, we obtain a derivation $S \Rightarrow^*_{G^{k+1}} u_1[\bar F_\alpha(t_{\mu_1}, \ldots, t_{\mu_{\bar n}})]_p$. As G^{k+1} also contains the normalized form of the rule $\bar F_\alpha(x_{\mu_1}, \ldots, x_{\mu_{\bar n}}) \to f(x_{j_1}, \ldots, x_{j_{\bar n}})$, we see that $u_1[\bar F_\alpha(t_{\mu_1}, \ldots, t_{\mu_{\bar n}})]_p \Rightarrow_{G^{k+1}} v_1$ holds.

- T_3: Then $(l \to r)$ is of the form $F(x_1, \ldots, x_n) \to H(t_1, \ldots, t_m)$ such that $\hat n := |\bigcup_{i=1}^{m} \mathrm{Var}(t_i)| < n$, that is, $\mathrm{Top}(u_1|_p) = F$ and $\mathrm{Top}(v_1|_p) = H$. As before, let $u'_1 \Rightarrow_{G^k} v'_1$ be the derivation step at which the occurrence of the nonterminal F at position $p \in \mathrm{Pos}(u_1)$ is generated, that is, at this step a rule $(l_1 \to r_1) \in \mathcal{P}^k$ is used such that the corresponding subterm $r_1|_q$ of r_1 is of the form $F(s_1, \ldots, s_n)$. Now $\mathcal{P}^{k+1}$ contains the added rule $l_1 \to r_1[\hat F_\alpha(s_{\mu_1}, \ldots, s_{\mu_{\bar n}})]_q$. By using this rule instead of the original rule $l_1 \to r_1$, we obtain a derivation $S \Rightarrow^*_{G^{k+1}} u_1[\hat F_\alpha(t_{\mu_1}, \ldots, t_{\mu_{\bar n}})]_p$. As G^{k+1} also contains the normalized form of the rule $\hat F_\alpha(x_{\mu_1}, \ldots, x_{\mu_{\bar n}}) \to H(t_1, \ldots, t_m)$, we see that $u_1[\hat F_\alpha(t_{\mu_1}, \ldots, t_{\mu_{\bar n}})]_p \Rightarrow_{G^{k+1}} v_1$ holds.

Proceeding by induction we obtain a derivation $S \Rightarrow^*_{G^{k+1}} t$.   □

Claim 2. $L(G^k) \supseteq L(G^{k+1})$.

Proof. Note that the new nonterminal symbols of the form $\bar F_\alpha$ or $\hat F_\alpha$ are only put at places where the symbol F occurred previously. Thus, the additional rules of the form $(l \to r) \in \mathcal{P}^{k+1}$, where $\mathrm{Top}(l) = \bar F_\alpha$ or $\mathrm{Top}(l) = \hat F_\alpha$, will not lead to more ground terms. Also the effect of the rules introduced by transformation rule T_1 can be simulated by the original rules of G^k.   □

This completes the proof of Proposition 1.   □

We illustrate the transformation described in Proposition 1 by an example.

Example 1. Consider the linear context-free tree grammar $G = (\mathcal{F}, \mathcal{N}, \mathcal{P}, S)$, where $\mathcal{F} = \{f(\cdot, \cdot), s(\cdot), a\}$, $\mathcal{N} = \{F(\cdot, \cdot), B(\cdot), A, S\}$, and $\mathcal{P}$ contains only productions in Chomsky normal form:

$$F(x_1, x_2) \to F(B(x_1), B(x_2)), \qquad B(x_1) \to s(x_1),$$
$$F(x_1, x_2) \to f(x_1, x_2), \qquad\qquad\quad A \to a,$$
$$F(x_1, x_2) \to x_2, \qquad\qquad\qquad\; S \to F(A, A).$$

Obviously, the generated tree language is

$$L(G) = \{f(s^n(a), s^n(a)) \mid n \geq 0\} \cup \{s^n(a) \mid n \geq 0\}.$$

Table 1 shows the transformation steps performed until the nondeleting grammar $G' = (\mathcal{F}, \mathcal{N}^3, \mathcal{P}^3, S)$ is obtained, where the set of nonterminals is $\mathcal{N}^3 = \{F(\cdot, \cdot), \hat{F}_{\{2\}}(\cdot), B(\cdot), A, S\}$, and $\mathcal{P}^3$ contains the following nondeleting productions:

$$F(x_1, x_2) \to F(B(x_1), B(x_2)), \qquad B(x_1) \to s(x_1),$$
$$F(x_1, x_2) \to f(x_1, x_2), \qquad\qquad\quad A \to a,$$
$$S \to F(A, A), \qquad\qquad\qquad S \to A,$$
$$S \to \hat{F}_{\{1\}}(A),$$
$$\hat{F}_{\{1\}}(x_1) \to B(x_1), \qquad\qquad \hat{F}_{\{1\}}(x_1) \to \hat{F}_{\{1\}}(B(x_1)).$$

Table 1. Transformations performed in Example 1

k	added nonterminals	rules added to $\mathcal{P}^k$	rules added to $\mathcal{P}'$
1 T_1		$S \to A$ $F(x_1, x_2) \to B(x_2)$	$F(x_1, x_2) \to x_2$
2 T_3	$\hat{F}_{\{1\}}(\cdot)$	$\hat{F}_{\{1\}}(x_1) \to B(x_1)$ $S \to \hat{F}_{\{1\}}(A)$ $F(x_1, x_2) \to \hat{F}_{\{1\}}(B(x_2))$	$F(x_1, x_2) \to B(x_2)$
3 T_3		$\hat{F}_{\{1\}}(x_1) \to \hat{F}_{\{1\}}(B(x_1))$	$F(x_1, x_2) \to \hat{F}_{\{1\}}(B(x_2))$

Proposition 2. *From a given linear context-free tree grammar G a growing linear context-free tree grammar G' can be constructed such that $L(G) = L(G')$ holds.*

Proof. Let $G = (\mathcal{F}, \mathcal{N}, \mathcal{P}, S)$ be the given linear context-free tree grammar. Because of Proposition 1 we can assume that G is nondeleting. We now apply the following two transformations iteratively to derive G' from G. If neither of them is applicable anymore the process terminates. The system G' obtained at that point has the intended properties. Of course, during the construction we can remove useless productions, too.

– *Transformation rule T_4:*
A rule of the form $F(x_1, \ldots, x_n) \to H(x_{j_1}, \ldots, x_{j_n})$, where $F, H \in \mathcal{N}_n$, is called a *unit production* of G. To remove these unit productions from $\mathcal{P}$ we apply the well-known technique from the Chomsky normal form construction for context-free (string) grammars. However, for tree grammars the algorithm is slightly more sophisticated, as we must keep track of permuted arguments when we determine equivalent nonterminals. For example, if G contains the rules

$$E(x_1, x_2) \to F(x_2, x_1), \quad F(x_1, x_2) \to H(x_1, x_2), \quad H(x_1, x_2) \to E(x_2, x_1),$$

then $F(x_1, x_2) \Rightarrow_G H(x_1, x_2) \Rightarrow_G E(x_2, x_1) \Rightarrow_G F(x_1, x_2)$ implying that F and H are equivalent, while E is equivalent to F and to H only modulo commutation of the arguments. To take care of this issue we can associate an appropriate label with each nonterminal which represents the corresponding permutation of its arguments. In the example above we would need two representatives, e.g., $E_{(1,2)}$ and $E_{(2,1)}$. Then equivalent nonterminals are replaced by a unique representative, the unit productions are deleted, and for each of the remaining rules all possible rules are created in which each nonterminal is replaced by one of its representatives. In our example each E would have to be replaced by $E_{(1,2)}$ as well as by $E_{(2,1)}$. Of course, we remove a unit production only, if $F \neq S$.

– *Transformation rule T_5:*
If $F(x_1, \ldots, x_n) \to f(x_{j_1}, \ldots, x_{j_n})$ is a rule, where $F \in \mathcal{N}_n \setminus \{S\}$ and $f \in \mathcal{F}_n$ $(n \geq 0)$, then we remove this production from $\mathcal{P}$. For each of the remaining rules with occurrences of the symbol F in the right-hand side, we enlarge $\mathcal{P}$ by adding all combinations of that rule in which some occurrences of $F(t_1, \ldots, t_n)$ in the right-hand side are replaced by the term $f(t_{j_1}, \ldots, t_{j_n})$.

Note that the above transformations preserve the property of being nondeleting. It is now rather obvious that this process terminates, and that the resulting context-free tree grammar G' has all the intended properties. □

Example 2. Starting with the grammar G' from Example 1 we have to apply the transformation rule T_4 for the unit productions $\hat{F}_{\{1\}}(x_1) \to B(x_1)$ and $S \to A$. First, we remove $\hat{F}_{\{1\}}(x_1) \to B(x_1)$ and obtain the additional rules

$$\hat{F}_{\{1\}}(x_1) \to B(B(x_1)) \quad \text{and} \quad S \to B(A).$$

For the unit production $S \to A$ we obtain no additional rules, because S does not occur in the right-hand side of any rule. Note that the unit production $S \to A$ is necessary, and that it is therefore not removed by transformation T_4.

Next the rules

$$F(x_1, x_2) \to f(x_1, x_2), \quad B(x_1) \to s(x_1), \quad \text{and} \quad A \to a$$

have to be processed according to transformation rule T_5. By applying T_5 to the first of these productions we remove the rule $F(x_1, x_2) \to f(x_1, x_2)$ and insert the rules

$$F(x_1, x_2) \to f(B(x_1), B(x_2)), \qquad\qquad S \to f(A, A)$$

into $\mathcal{P}$. When we remove the production $B(x_1) \to s(x_1)$, we obtain many new rules:

$$F(x_1, x_2) \to F(B(x_1), s(x_2)), \qquad F(x_1, x_2) \to F(s(x_1), B(x_2)),$$
$$F(x_1, x_2) \to F(s(x_1), s(x_2)), \qquad F(x_1, x_2) \to f(B(x_1), s(x_2)),$$
$$F(x_1, x_2) \to f(s(x_1), B(x_2)), \qquad F(x_1, x_2) \to f(s(x_1), s(x_2)),$$
$$\hat{F}_{\{1\}}(x_1) \to \hat{F}_{\{1\}}(s(x_1)),$$
$$\hat{F}_{\{1\}}(x_1) \to B(s(x_1)), \qquad\qquad \hat{F}_{\{1\}}(x_1) \to s(B(x_1))$$
$$\hat{F}_{\{1\}}(x_1) \to s(s(x_1)), \qquad\qquad S \to s(A).$$

Finally, the removal of the last rule $A \to a$ leads to the following additional productions:

$$S \to F(A, a), \qquad S \to F(a, A), \qquad S \to F(a, a),$$
$$S \to f(A, a), \qquad S \to f(a, A), \qquad S \to f(a, a),$$
$$S \to \hat{F}_{\{1\}}(a), \qquad S \to s(a), \qquad S \to a.$$

At this point no productions with left-hand side A or B are left. Therefore, all productions containing these nonterminals have become useless, that is, we can now safely remove these productions. In this way we obtain the growing grammar with the following productions:

$$F(x_1, x_2) \to F(s(x_1), s(x_2)), \qquad F(x_1, x_2) \to f(s(x_1), s(x_2)),$$
$$\hat{F}_{\{1\}}(x_1) \to \hat{F}_{\{1\}}(s(x_1)), \qquad \hat{F}_{\{1\}}(x_1) \to s(s(x_1)),$$
$$S \to F(a, a), \qquad\qquad S \to f(a, a),$$
$$S \to \hat{F}_{\{1\}}(a), \qquad\qquad S \to s(a),$$
$$S \to a.$$

4   Restarting Tree Automata

Recently, restarting automata [9,13] have been generalized from strings to trees [17]. Formally, a *top-down restarting tree automaton* (RRWWT-*automaton*, for short) is given by a six-tuple $\mathcal{A} = (\mathcal{F}, \mathcal{G}, \mathcal{Q}, q_0, k, \Delta)$, where $\mathcal{F}$ is a ranked input alphabet, $\mathcal{G} \supseteq \mathcal{F}$ is a ranked working alphabet, $\mathcal{Q} = \mathcal{Q}_1 \cup \mathcal{Q}_2$ is a finite set of states such that $\mathcal{Q}_1 \cap \mathcal{Q}_2 = \emptyset$, $q_0 \in \mathcal{Q}_1$ is the initial state and simultaneously the restart state, $k \geq 1$ is the height of the read/write-window, and $\Delta = \Delta_1 \cup \Delta_2$ is a finite term rewriting system on $\mathcal{G} \cup \mathcal{Q}$. The rule set Δ_1 only contains bounded top-down transitions of the form $q(t) \rightarrow t[q_1(x_1), \ldots, q_n(x_n)]$, where $n \geq 1$, $t \in \mathrm{Ctx}(\mathcal{G}, \mathcal{X}_n)$, $1 \leq \mathrm{Hgt}(t) \leq k$, and $q, q_1, \ldots, q_n \in \mathcal{Q}_1$, and k-height bounded final transitions of the form $q(t) \rightarrow t$, where $t \in T(\mathcal{G})$, $0 \leq \mathrm{Hgt}(t) \leq k$, and $q \in \mathcal{Q}_1$. The rule set Δ_2 contains transitions of the following types:

1. Size-reducing *rewrite transitions*, that is, linear rewrite rules of the form

$$q(t) \rightarrow t'[q_1(x_1), \ldots, q_n(x_n)],$$

    where $n \geq 1$, $t \in T(\mathcal{G}, \mathcal{X}_n)$, $t' \in \mathrm{Ctx}(\mathcal{G}, \mathcal{X}_n)$, $q \in \mathcal{Q}_1$, and $q_1, \ldots, q_n \in \mathcal{Q}_2$, and size-reducing *final rewrite transitions* of the form $q(t) \rightarrow t'$, where $q \in \mathcal{Q}_1$ and $t, t' \in T(\mathcal{G})$. For both these types of transitions it is required that $||t|| > ||t'||$ and $\mathrm{Hgt}(t) \leq k$.
2. k-height bounded top-down transitions of the form

$$q(t) \rightarrow t[q_1(x_1), \ldots, q_n(x_n)],$$

    where $n \geq 1$, $t \in \mathrm{Ctx}(\mathcal{G}, \mathcal{X}_n)$, $1 \leq \mathrm{Hgt}(t) \leq k$, and $q, q_1, \ldots, q_n \in \mathcal{Q}_2$, and k-height bounded final transitions of the form $q(t) \rightarrow t$, where $t \in T(\mathcal{G})$, $0 \leq \mathrm{Hgt}(t) \leq k$, and $q \in \mathcal{Q}_2$.

The *partial move relation* $\rightarrow_\Delta$ and its reflexive transitive closure $\rightarrow_\Delta^*$ are induced by the TRS Δ, while the *final move relation* $\rightarrow_{\Delta_1}$ and its reflexive transitive closure $\rightarrow_{\Delta_1}^*$ are induced by Δ_1. We use the notation $u \hookrightarrow_\mathcal{A} v$ $(u, v \in T(\mathcal{G}))$ to express the fact that there exists a cycle that starts with the configuration $q_0(u)$ and finishes with the stateless configuration v, that is, $q_0(u)(\rightarrow_\Delta^* \setminus \rightarrow_{\Delta_1}^+)v$. The relation $\hookrightarrow_\mathcal{A}^*$ is the reflexive transitive closure of $\hookrightarrow_\mathcal{A}$. The *tree language* recognized by the RRWWT-automaton $\mathcal{A}$ is

$$L(\mathcal{A}) = \left\{ t \in T(\mathcal{F}) \mid \exists t' \in T(\mathcal{G}) \text{ such that } t \hookrightarrow_\mathcal{A}^* t' \text{ and } q_0(t') \rightarrow_{\Delta_1}^* t' \right\}.$$

The *(auxiliary) simple tree language* recognized by $\mathcal{A}$ is $S_\mathcal{F}(\mathcal{A}) = \{ t \in T(\mathcal{F}) \mid q_0(t) \rightarrow_{\Delta_1}^* t \}$ and $S_\mathcal{G}(\mathcal{A}) = \{ t \in T(\mathcal{G}) \mid q_0(t) \rightarrow_{\Delta_1}^* t \}$, respectively.

A restarting tree automaton is called an RWWT-*automaton*, if all its rewrite transitions are of the special form $q(t) \rightarrow t'[x_1, \ldots, x_n]$, where $n \geq 1$, $q \in \mathcal{Q}_1$, $t \in T(\mathcal{G}, \mathcal{X}_n)$, and $t' \in \mathrm{Ctx}(\mathcal{G}, \mathcal{X}_n)$ such that $||t'|| < ||t||$ and $\mathrm{Hgt}(t) \leq k$. In this case the subset $\mathcal{Q}_2$ of the set of states $\mathcal{Q}$ and the top-down and final transitions from the rule set Δ_2 are superfluous, as the automaton does not propagate state information in affected branches after a rewrite has been performed. A restarting tree automaton is an RRWT-*automaton*, if its working alphabet $\mathcal{G}$ coincides with its input alphabet $\mathcal{F}$, that is, no auxiliary symbols are available. It is an RRT-*automaton*, if it is an RRWT-automaton for which the right-hand side of every rewrite transition is a scattered subterm of the corresponding left-hand side. Analogously, we obtain the RWT- and the RT-*automaton* from the RWWT-automaton.

Example 3. Let $\mathcal{A} = (\mathcal{F}, \mathcal{G}, \mathcal{Q}, q_0, k, \Delta)$, where $\mathcal{F} = \mathcal{G} = \{f(\cdot, \cdot), g(\cdot), h(\cdot), a\}$, $\mathcal{Q} = \{q_0, q_1\}$, $k = 3$, and the TRS Δ is given by the following rules:

$$q_0(f(g(h(a)), g(h(a)))) \to f(g(h(a)), g(h(a))),$$
$$q_0(f(g(x_1), g(x_2))) \to f(g(q_1(x_1)), g(q_1(x_2))),$$
$$q_1(g(x_1)) \to g(q_1(x_1)),$$
$$q_1(g(h(h(x_1)))) \to h(x_1).$$

Then $\mathcal{A}$ is an RT-automaton, and it is not hard to see that

$$L(\mathcal{A}) = \{\, f(g^n(h^n(a)), g^n(h^n(a))) \mid n \geq 1 \,\},$$

as $\mathcal{A}$ reduces both branches simultaneously until the term $f(g(h(a)), g(h(a))) \in T(\mathcal{F})$ is obtained, which belongs to the simple tree language $S_{\mathcal{F}}(\mathcal{A})$.

Based on the normal form for linear context-free tree grammars presented in Proposition 2 we now derive our main result.

Theorem 1. *Given a linear context-free tree grammar G an RWWT-automaton $\mathcal{A}$ can be constructed such that $L(G) = L(\mathcal{A})$ holds.*

Proof. Let $G = (\mathcal{F}, \mathcal{N}, \mathcal{P}, S)$ be a linear context-free tree grammar. By Proposition 2 we can assume that G is growing. Let C_S be the set of all constants $c \in (\mathcal{F}_0 \cup \mathcal{N}_0)$ such that $\mathcal{P}$ contains a rule $S \to c$. For each rule $(l \to r) \in \mathcal{P}$, where the initial symbol S occurs at least once in the right-hand side r, we enlarge $\mathcal{P}$ by all combinations of that rule in which some occurrences of S in the right-hand side are replaced by symbols from C_S.

We construct an RWWT-automaton $\mathcal{A} = (\mathcal{F}, \mathcal{G}, \mathcal{Q}, q_0, k, \Delta)$ by taking $\mathcal{G} := \mathcal{F} \cup \mathcal{N}$, $\mathcal{Q} := \{q_0, q_1\}$, and by defining Δ as follows. Recall that $\Delta = \Delta_1 \cup \Delta_2$. For each production from $\mathcal{P}$ of type $F(x_1, \ldots, x_n) \to t$, where $n \geq 0$, $F \in \mathcal{N}_n$, $\|t\| > 1$, and $t \in T(\mathcal{F} \cup \mathcal{N}, \mathcal{X}_n)$, we add the linear rewrite transitions

$$q_0(t) \to F(x_1, \ldots, x_n) \text{ and } q_1(t) \to F(x_1, \ldots, x_n)$$

to Δ_2. Note that all these rewrite transitions are size-reducing. For each constant $t \in C_S$, we add the final top-down transition $q_0(t) \to t$ to Δ_1. Additionally, we put the transition $q_0(S) \to S$ into Δ_1. Further, for each symbol $F \in (\mathcal{F}_n \cup \mathcal{N}_n)$, where $n > 0$, Δ_1 contains the rules $q_0(F(x_1, \ldots, x_n)) \to F(q_1(x_1), \ldots, q_1(x_n))$ and $q_1(F(x_1, \ldots, x_n)) \to F(q_1(x_1), \ldots, q_1(x_n))$. Here state q_1 is used to guarantee that the final transitions can only be applied at the root of a tree.

The automaton $\mathcal{A}$ simulates all derivations of G nondeterministically and in reverse order. Let $t \in L(G)$ be a ground term generated by G, and let

$$S \Rightarrow_G t_1 \Rightarrow_G^* \cdots \Rightarrow_G^* t_i \Rightarrow_G t_{i+1} \Rightarrow_G^* \cdots \Rightarrow_G^* t_\ell = t$$

be a derivation in G. The automaton guesses in each cycle the correct production from $\mathcal{P}$. Then $\mathcal{A}$ applies the corresponding reverse transition on t_{i+1} to obtain

t_i and restarts. Finally, the automaton reaches either a constant $t_1 \in C_S$ and accepts, since $t_1 \in S_{\mathcal{G}}(\mathcal{A})$, or it reaches S and accepts by the transition $q_0(S) \to S$ from Δ_1. On the other hand, for each accepting computation

$$t \hookrightarrow_{\mathcal{A}} t_{\ell-1}, \ \ldots, \ t_2 \hookrightarrow_{\mathcal{A}} t_1, \ q_0(t_1) \to^*_{\Delta_1} t_1, \text{ where } t_1 \in C_S \cup \{S\},$$

there is a corresponding derivation starting with $S \Rightarrow^\varepsilon_G t_1 \Rightarrow_G t_2 \Rightarrow^*_G \cdots$. Hence, we have $t \in L(G)$ if and only if $t \in L(\mathcal{A})$.   □

Observe that the constructed RWWT-automaton needs at most a read/write-window of height $k = 2$, because we start with a grammar in Chomsky normal form and the applied transformations from Section 3 always preserve or even decrease the height of the right-hand sides.

Example 4. Using the above construction we obtain the RWWT-automaton

$$\mathcal{A} = (\mathcal{F}, \mathcal{F} \cup \{F(\cdot, \cdot), \hat{F}_{\{1\}}(\cdot)\}, \{q_0, q_1\}, q_0, 2, \Delta),$$

from the growing linear context-free tree grammar of Example 2, where Δ_1 contains the following top-down and final transitions for all $i \in \{0, 1\}$:

$$\begin{aligned}
q_i(F(x_1, x_2)) &\to F(q_1(x_1), q_1(x_2)), & q_i(\hat{F}_{\{1\}}(x_1)) &\to \hat{F}_{\{1\}}(q_1(x_1)), \\
q_i(f(x_1, x_2)) &\to f(q_1(x_1), q_1(x_2)), & q_i(s(x_1)) &\to s(q_1(x_1)), \\
q_0(a) &\to a, & q_0(S) &\to S.
\end{aligned}$$

Additionally, Δ_2 contains the following rewrite transitions for all $i \in \{0, 1\}$:

$$\begin{aligned}
q_i(F(s(x_1), s(x_2))) &\to F(x_1, x_2), & q_i(\hat{F}_{\{1\}}(s(x_1))) &\to \hat{F}_{\{1\}}(x_1), \\
q_i(f(s(x_1), s(x_2))) &\to F(x_1, x_2), & q_i(s(s(x_1))) &\to \hat{F}_{\{1\}}(x_1),
\end{aligned}$$

and $q_i(t) \to S$ for all $t \in \{s(a), \hat{F}_{\{1\}}(a), F(a, a), f(a, a)\}$. It is easily seen that $L(\mathcal{A}) = L(G) = \{f(s^n(a), s^n(a)) \mid n \geq 0\} \cup \{s^n(a) \mid n \geq 0\}$.

The tree language $L = \{f(g^n(h^n(a)), g^n(h^n(a))) \mid n \geq 1\}$ considered in Example 3 is not context-free according to the duplication theorem of Arnold and Dauchet [1]. Thus, we obtain the following corollary.

Corollary 1. $\mathscr{L}(\text{lin-CFTG}) \subseteq \mathscr{L}(\text{RWWT})$, *and this inclusion is proper.*

5   Conclusion

In Figure 1 the known inclusion results between the classes of tree languages specified by the various types of restarting tree automata and some classical families of tree languages are depicted. We have seen that restarting tree automata with auxiliary symbols accept all linear context-free tree languages. Thus, they are quite expressive. However, it is currently not known whether in fact all context-free tree languages can be accepted by restarting tree automata, or whether there

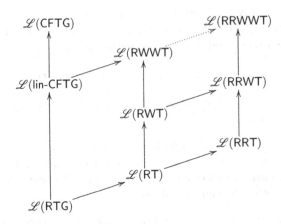

Fig. 1. Inclusions between language classes defined by restarting tree automata and language classes generated by various tree grammars. An arrow denotes a proper inclusion, while a dotted arrow denotes an inclusion that is not known to be proper.

exists a context-free tree language that is not accepted by any restarting tree automaton. Also it is not known whether there exists a restricted class of restarting tree automata that accepts exactly the linear context-free tree languages.

For use in applications like verification we would need variants of restarting tree automata for which the emptiness problem is efficiently decidable. However, the corresponding classes of tree languages should still be closed under intersection with regular tree languages. Currently it is not known how to achieve both these properties simultaneously.

Another possible restriction of the restarting tree automaton that the authors currently explore is the so-called *single-path restarting tree automaton*, which walks down a single path only. Of course, this model is much more limited, because *only one* size-reducing rewrite is possible in each cycle, and also the regular control is restricted to a single path only. It would be interesting to know about the expressive power of these automata and how they relate to the types studied in [17] and in the present paper. These are open questions that will be addressed in our future work.

References

1. Arnold, A., Dauchet, M.: Un théorème de duplication pour les forêts algébriques. Journal of Computer and System Sciences 13, 223–244 (1976)
2. Aho, A.V.: Indexed grammars—An extension of context-free grammars. Journal of the ACM 15, 647–671 (1968)
3. Baader, F., Nipkow, T.: Term Rewriting and All That. Cambridge University Press, Cambridge (1998)
4. Fischer, M.J.: Grammars with macro-like productions. In: IEEE Conf. Record, 9th Ann. Symp. on Switching and Automata Theory, pp. 131–142. IEEE Computer Society Press, Los Alamitos (1968)

5. Fujiyoshi, A.: Restrictions on monadic context-free tree grammars. In: COLING 2004, Proc., pp. 78–84 (2004)
6. Fujiyoshi, A.: Linearity and nondeletion on monadic context-free tree grammars. Information Processing Letters 93, 103–107 (2005)
7. Fujiyoshi, A., Kasai, T.: Spinal-formed context-free tree grammars. Theory of Computing Systems 33, 59–83 (2000)
8. Guessarian, I.: Pushdown tree automata. Mathematical Systems Theory 16, 237–263 (1983)
9. Jančar, P., Mráz, F., Plátek, M., Vogel, J.: On restarting automata with rewriting. In: Păun, G., Salomaa, A. (eds.) New Trends in Formal Languages. LNCS, vol. 1218, pp. 119–136. Springer, Heidelberg (1997)
10. Kepser, S., Mönnich, U.: Closure properties of linear context-free tree languages with an application to optimality theory. Theoretical Computer Science 354, 82–97 (2006)
11. Maibaum, T.S.E.: A generalized approach to formal languages. Journal of Computer and System Sciences 8, 409–439 (1974)
12. Maibaum, T.S.E.: Pumping lemmas for term languages. Journal of Computer and System Sciences 17, 319–330 (1978)
13. Otto, F.: Restarting automata and their relations to the Chomsky hierarchy. In: Ésik, Z., Fülöp, Z. (eds.) DLT 2003. LNCS, vol. 2710, pp. 55–74. Springer, Heidelberg (2003)
14. Rounds, W.C.: Tree-oriented proofs of some theorems on context-free and indexed languages. In: Second Annual ACM Symp. on Theory of Computing, Proc., pp. 109–116. ACM Press, New York (1970)
15. Schimpf, K.M., Gallier, J.H.: Tree pushdown automata. Journal of Computer and System Sciences 30, 25–40 (1985)
16. Seki, H., Kato, Y.: On the generative power of multiple context-free grammars and macro grammars. Information Science Technical Report, NAIST-IS-TR2006007, Nara Institute of Science and Technology (2006)
17. Stamer, H., Otto, F.: Restarting tree automata. In: van Leeuwen, J., Italiano, G.F., van der Hoek, W., Meinel, C., Sack, H., Plášil, F. (eds.) SOFSEM 2007. LNCS, vol. 4362, pp. 510–521. Springer, Heidelberg (2007)
18. Vijay-Shankar, K., Weir, D.J.: The equivalence of four extensions of context-free grammars. Mathematical Systems Theory 27, 511–546 (1994)

Author Index

Lecture Notes in Computer Science

Sublibrary 1: Theoretical Computer Science and General Issues

For information about Vols. 1– 4528
please contact your bookseller or Springer

Vol. 4697: L. Choi, Y. Paek, S. Cho (Eds.), Advances in Computer Systems Architecture. XIII, 400 pages. 2007.

Vol. 4688: K. Li, M. Fei, G.W. Irwin, S. Ma (Eds.), Bio-Inspired Computational Intelligence and Applications. XIX, 805 pages. 2007.

Vol. 4684: L. Kang, Y. Liu, S. Zeng (Eds.), Evolvable Systems: From Biology to Hardware. XIV, 446 pages. 2007.

Vol. 4683: L. Kang, Y. Liu, S. Zeng (Eds.), Advances in Computation and Intelligence. XVII, 663 pages. 2007.

Vol. 4681: D.-S. Huang, L. Heutte, M. Loog (Eds.), Advanced Intelligent Computing Theories and Applications. XXVI, 1379 pages. 2007.

Vol. 4672: K. Li, C. Jesshope, H. Jin, J.-L. Gaudiot (Eds.), Network and Parallel Computing. XVIII, 558 pages. 2007.

Vol. 4671: V.E. Malyshkin (Ed.), Parallel Computing Technologies. XIV, 635 pages. 2007.

Vol. 4669: J.M. de Sá, L.A. Alexandre, W. Duch, D. Mandic (Eds.), Artificial Neural Networks – ICANN 2007, Part II. XXXI, 990 pages. 2007.

Vol. 4668: J.M. de Sá, L.A. Alexandre, W. Duch, D. Mandic (Eds.), Artificial Neural Networks – ICANN 2007, Part I. XXXI, 978 pages. 2007.

Vol. 4666: M.E. Davies, C.J. James, S.A. Abdallah, M.D. Plumbley (Eds.), Independent Component Analysis and Blind Signal Separation. XIX, 847 pages. 2007.

Vol. 4665: J. Hromkovič, R. Královič, M. Nunkesser, P. Widmayer (Eds.), Stochastic Algorithms: Foundations and Applications. X, 167 pages. 2007.

Vol. 4664: J. Durand-Lose, M. Margenstern (Eds.), Machines, Computations, and Universality. X, 325 pages. 2007.

Vol. 4661: U. Montanari, D. Sannella, R. Bruni (Eds.), Trustworthy Global Computing. X, 339 pages. 2007.

Vol. 4649: V. Diekert, M.V. Volkov, A. Voronkov (Eds.), Computer Science – Theory and Applications. XIII, 420 pages. 2007.

Vol. 4647: R. Martin, M.A. Sabin, J.R. Winkler (Eds.), Mathematics of Surfaces XII. IX, 509 pages. 2007.

Vol. 4646: J. Duparc, T.A. Henzinger (Eds.), Computer Science Logic. XIV, 600 pages. 2007.

Vol. 4644: N. Azémard, L. Svensson (Eds.), Integrated Circuit and System Design. XIV, 583 pages. 2007.

Vol. 4641: A.-M. Kermarrec, L. Bougé, T. Priol (Eds.), Euro-Par 2007 Parallel Processing. XXVII, 974 pages. 2007.

Vol. 4639: E. Csuhaj-Varjú, Z. Ésik (Eds.), Fundamentals of Computation Theory. XIV, 508 pages. 2007.

Vol. 4638: T. Stützle, M. Birattari, H. H. Hoos (Eds.), Engineering Stochastic Local Search Algorithms. X, 223 pages. 2007.

Vol. 4630: H.J. van den Herik, P. Ciancarini, H.H.L.M.(J.) Donkers (Eds.), Computers and Games. XII, 283 pages. 2007.

Vol. 4628: L.N. de Castro, F.J. Von Zuben, H. Knidel (Eds.), Artificial Immune Systems. XII, 438 pages. 2007.

Vol. 4627: M. Charikar, K. Jansen, O. Reingold, J.D.P. Rolim (Eds.), Approximation, Randomization, and Combinatorial Optimization. XII, 626 pages. 2007.

Vol. 4624: T. Mossakowski, U. Montanari, M. Haveraaen (Eds.), Algebra and Coalgebra in Computer Science. XI, 463 pages. 2007.

Vol. 4623: M. Collard (Ed.), Ontologies-Based Databases and Information Systems. X, 153 pages. 2007.

Vol. 4621: D. Wagner, R. Wattenhofer (Eds.), Algorithms for Sensor and Ad Hoc Networks. XIII, 415 pages. 2007.

Vol. 4619: F. Dehne, J.-R. Sack, N. Zeh (Eds.), Algorithms and Data Structures. XVI, 662 pages. 2007.

Vol. 4618: S.G. Akl, C.S. Calude, M.J. Dinneen, G. Rozenberg, H.T. Wareham (Eds.), Unconventional Computation. X, 243 pages. 2007.

Vol. 4616: A.W.M. Dress, Y. Xu, B. Zhu (Eds.), Combinatorial Optimization and Applications. XI, 390 pages. 2007.

Vol. 4614: B. Chen, M. Paterson, G. Zhang (Eds.), Combinatorics, Algorithms, Probabilistic and Experimental Methodologies. XII, 530 pages. 2007.

Vol. 4613: F.P. Preparata, Q. Fang (Eds.), Frontiers in Algorithmics. XI, 348 pages. 2007.

Vol. 4600: H. Comon-Lundh, C. Kirchner, H. Kirchner (Eds.), Rewriting, Computation and Proof. XVI, 273 pages. 2007.

Vol. 4599: S. Vassiliadis, M. Bereković, T.D. Hämäläinen (Eds.), Embedded Computer Systems: Architectures, Modeling, and Simulation. XVIII, 466 pages. 2007.

Vol. 4598: G. Lin (Ed.), Computing and Combinatorics. XII, 570 pages. 2007.

Vol. 4596: L. Arge, C. Cachin, T. Jurdziński, A. Tarlecki (Eds.), Automata, Languages and Programming. XVII, 953 pages. 2007.

Vol. 4595: D. Bošnački, S. Edelkamp (Eds.), Model Checking Software. X, 285 pages. 2007.

Vol. 4590: W. Damm, H. Hermanns (Eds.), Computer Aided Verification. XV, 562 pages. 2007.

Vol. 4588: T. Harju, J. Karhumäki, A. Lepistö (Eds.), Developments in Language Theory. XI, 423 pages. 2007.

Vol. 4583: S.R. Della Rocca (Ed.), Typed Lambda Calculi and Applications. X, 397 pages. 2007.

Vol. 4580: B. Ma, K. Zhang (Eds.), Combinatorial Pattern Matching. XII, 366 pages. 2007.

Vol. 4576: D. Leivant, R. de Queiroz (Eds.), Logic, Language, Information and Computation. X, 363 pages. 2007.

Vol. 4547: C. Carlet, B. Sunar (Eds.), Arithmetic of Finite Fields. XI, 355 pages. 2007.

Vol. 4546: J. Kleijn, A. Yakovlev (Eds.), Petri Nets and Other Models of Concurrency – ICATPN 2007. XI, 515 pages. 2007.

Vol. 4545: H. Anai, K. Horimoto, T. Kutsia (Eds.), Algebraic Biology. XIII, 379 pages. 2007.

Vol. 4533: F. Baader (Ed.), Term Rewriting and Applications. XII, 419 pages. 2007.